Radical by Nature

The Green Assault on Liberty, Property, and Prosperity

Thomas J. McCaffrey

Radical by Nature: The Green Assault on Liberty, Property, and Prosperity

STAIRWAY PRESS—SEATTLE

Cover Design by Guy Corp
www.GrafixCorp.com

STAIRWAY≡PRESS

www.StairwayPress.com
1500A East College Way #554
Mount Vernon, WA 98273 USA

Dedication

To my wife, Bobbie McCaffrey,
whose love and support made this book possible,
and to my parents,
Thomas and Marilyn McCaffrey

Acknowledgements

I wish to thank Ed Cline and Hanania Dover for their critical readings of the manuscript.

Contents

Preface

A THOUSAND YEARS from now, the explosion of wealth which occurred on the North American continent from the founding of the United States until the present will shine like a supernova. It will shine not just because of the quantity of wealth produced, a quantity unimagined by earlier peoples, but also because of the quality of that wealth. It is not static wealth, like the gold amassed by the potentates throughout history, but wealth intended for human *use*, like locomotive engines and refrigerators and ball-point pens. In the case of industrial plants and machinery, it is wealth intended to produce more wealth. This explosion of wealth will shine, moreover, for its wide distribution. In the United States today, anyone willing to hold a job can eat better, live longer, enjoy better health, and possess more material comforts than even the wealthy of most earlier eras.

The engine of America's wealth has been technology. America's has been an industrial economy, and industrialism is driven by technological innovation. If wealth-production has been one of America's most distinctive characteristics, then technology has been her most distinctive cultural artifact. What sculpture and architecture were to ancient Greece, what public works were to ancient Rome, what painting and sculpture were to Renaissance Italy, technology has been to the United States. From the cotton gin to the assembly line, from the light bulb to nuclear energy, from the polio vaccine to genetic engineering, America's is a culture that has expressed itself most distinctively in technological innovation. It is an essential element of the American identity.

Francis Bacon, in 1620, first articulated the idea of engaging in a program of scientific research in order to develop new forms of useful technology. Bacon is a spiritual father of the Industrial Revolution and, by extension, of the United States. His ideas about man's "commanding nature" through technological innovation are a part of the America's DNA. Those ideas are important to an understanding of environmentalism because environmentalism is, above all else, an attack on the premises underlying Bacon's famous dictum, "Nature, to be commanded, must be obeyed." Chapter 1 begins with a discussion of Bacon's ideas.

The other characteristic of America that historians will study a thousand years from now will be her government, both means and ends. The means are historically distinctive, the written constitution, the separation of powers, and the rest. But the ends of American government, the purpose it is intended to serve, "to secure these rights," is an idea that has changed the world.

John Locke introduced the idea of natural rights, rights inherent in a person by virtue of his humanity, in his *Two Treatises of Government* in 1689. His idea echoes in our "Life, Liberty, and the Pursuit of Happiness." It manifests itself in our understanding that slavery is morally wrong in a fundamental way that transcends the enactments of legislatures. The United States was the first country in history to secure the inherent rights of the individual to speak his mind and to practice his religion free from compulsion by other men. The idea of inherent rights is also part of the DNA of the United States.

The most fundamental of all rights are property rights. Without secure property rights, no other rights are possible. A primary purpose of the Framers of the U.S. Constitution was to establish a polity in which the rights of private property would be secure. America's unparalleled record of technological innovation, her corresponding development into an industrial powerhouse, and the resulting explosion of wealth that she has produced, all attest to the Framers' success at making property rights secure, since none of these things would have been possible had they failed. Secure property rights are as much a part of the American identity as are technological innovation and industrialism.

In his *Second Treatise*, John Locke identified the central role of property rights in any polity committed to individual liberty. I discuss Locke's ideas of natural rights and property rights in the latter part of Chapter 1.

Chapter 2 consists of a detailed account of the place of property rights in the constitutional scheme of the United States, as well as a discussion of how their place in that scheme developed and changed over the first century of the Constitution's existence. I also identify a certain divergence of opinion among the Framers and among the jurists who interpreted the Constitution as to just how extensive should be the security afforded to property rights. This divergence would help make possible the extinguishing of property rights that is taking place in the United States today at the hands of environmentalists.

Property rights would remain secure in the United States for about a hundred years. But toward the end of the nineteenth century there arose a movement, Progressivism, whose specific purpose was to dismantle the constitutional protections surrounding property rights. Progressives believed that government management of economic matters was preferable to private management of them, and they recognized property rights as an obstacle to their plans. Chapter 3 consists of a discussion of Progressivism.

The conservation movement was an important wing of the Progressive movement. Conservationists shared the Progressives' antipathy toward private property, but conservation was concerned specifically with natural resources. Conservation held that natural resources should be the collective property of the people as a whole. Conservationists thus opposed, in principle, private ownership of land, though they never pressed this idea to its logical conclusion, the abolition of private land ownership altogether. That would be left to the environmental movement, which would grow out of the conservation movement in the 1960s. Chapter 4 discusses the conservation movement.

Within the conservation movement there existed a minority composed of nature preservationists, a minority epitomized by John Muir, founder of the Sierra Club. Conservation may have advocated government management of natural resources, but it advocated nevertheless that they be *used* for human benefit. The preservationists, in contrast, opposed the use of natural resources for human benefit in important instances. Nature preservationism stands opposed, in principle, to the Baconian "command of nature" that is epitomized by America's heritage of technological innovation and industrialism. Chapter 5 discusses nature preservationists from Thoreau to Aldo Leopold.

In the 1960s, nature preservationists would come to dominate the conservation movement, and environmentalism would be born out of the process. Environmentalism would set out, in turn, to eradicate the two most distinctive elements of the American identity, its cultivating of technological innovation, industrialism, and wealth production, and its grounding in private property rights. Environmentalism thus represents the disassembling of the American identity. This is the topic of Chapters 6 through 11. Chapter 12 concludes the book.

Some readers might prefer to skip the philosophical and historical context provided in Chapters 1 through 3 and begin reading at Chapter 4.

Part I: Historical Context

1: Bacon and Locke

WHEN FRANCIS BACON wrote "Nature, to be commanded, must be obeyed," he had in mind the inventions of the magnetic compass, the telescope, and gunpowder. Inspired by these, he proposed a program of what we today would call "scientific research," aimed at developing new technological wonders. Technology was not new to the world in 1620, of course. The use of tools goes back to the dawn of man; tool making is a function of man's nature as a thinking being. What was new in Bacon's proposal was the idea of making a conscious, systematic effort to discover new knowledge of nature, on the one hand, for the practical purpose of developing new forms of technology, on the other.

The word "science" did not exist in 1620. To be sure, Galileo was already practicing science, as had many others before him, but such genuine science had not yet forged an identity of its own apart from the field of study out of which it would grow, which was called "natural philosophy." Natural philosophy looked back to Aristotle. It studied things in nature in order to discover their "essences." It relied on syllogistic reasoning as much as on direct observation. Its purpose was philosophical rather than utilitarian. Aristotelian natural philosophy, which was still regnant in Bacon's time, did not, and indeed could not, extend man's "command" of nature in a practical way. This Bacon set out to change, advocating among other things the use of experiments and quantitative measurement to help man understand nature in ways that would prove *useful* to him.

In stating that "nature, to be commanded, must be obeyed," Bacon assumed that it would be *possible* for man to command nature on a greater scale than he was doing in 1620. In the event, the revolution in thought to which he helped give rise, now known as the Enlightenment, would eventually issue in an explosion of invention and technology, which we call the Industrial Revolution. This would extend man's command of nature far beyond what anyone could have imagined possible in 1620.

But in addition to assuming that it would be possible for man to command nature on a much greater scale, Bacon also implied that it would be *right* for him to do so. He could not have foreseen the Industrial Revolution, and we cannot know if he would have approved of it. But the assumptions underlying his dictum, that it is possible for man to command nature on a grand scale and that it would be right for him to do so, would become bedrock premises upon which the edifice of industrialism would rise.

In 1688-89 the English deposed their king, Charles II, and replaced him with William of Orange from the Netherlands. In forcibly removing the Stuarts, the English rejected the European model of divine right monarchy and paved the way for eventual Parliamentary supremacy. But there was nothing explicit in English constitutional history to justify such an act as the *coup d'état* of 1688. John Locke's *Two Treatises of Government*, which he had written some years earlier but did not publish until 1690, provided a moral justification for the Glorious Revolution.

Locke's *Second Treatise* changed the course of political philosophy. In it he developed his revolutionary conception of natural rights. The idea of rights was not new with Locke. The idea of specific limitations on the power of the English Crown, such as the trial by jury requirement, was at least as old as Magna Carta. Initially, the rights of Magna Carta were grounded in nothing more than the physical force which the English barons brought to bear on King John to extort his "agreement" to the document. Over time, these rights acquired the legitimacy afforded by tradition. What Locke achieved was to take this idea of traditional rights and provide it with a moral base, one grounded in the nature of man. The idea of *natural* rights was new in the history of mankind.

Locke started from the idea of a "state of nature." If there were no government, if men lived in a state of nature relative to each other, there would still be right and wrong, things men ought and ought not to do. Locke referred to these moral principles collectively as the "law of nature." The means that men have to discover this natural law is their faculty of reason. Locke said that the law of nature *is* reason. Or, as Jerome Huyler has suggested, the fundamental moral tenet implicit in Locke's conception of natural law is "follow reason." [1]

Locke held that, because men have an innate drive to preserve their own lives, it must be a law of nature that they so act. And if nature dictates that they so act, then they must have a *right* so to act. [2] Locke's political philosophy derives from this most fundamental of rights, the right of self-preservation. And since all men are equally human, all have an equal right of self-preservation. From this follows the obligation to refrain from harming other men.

> *And Reason, which is that Law [of nature], teaches all*
> *Mankind, who will but consult it, that being all equal*

McCaffrey

*and independent, no one ought to harm another in his
Life, Health, Liberty, or Possessions.*[3]

All men are equal, for Locke, precisely in the sense that they possess equal rights in the state of nature, "*Equality ... being that equal Right that every Man hath, to his Natural Freedom*, without being subjected to the Will or Authority of any other Man."[4]

What kind of "harm" would constitute a violation of a man's rights? The use of physical force against him. "*Force without Right, upon a Man's Person, makes a State of War*."[5] Specifically, it is the aggressor, he who first resorts to the use of force, who violates rights. "And therefore it is lawful for me to treat him, as one who has put *himself into a state of war* with me, i.e., kill him if I can; for to that hazard does he justly expose himself, whoever introduces a State of War, and is *aggressor* in it."[6] If, in the realm of human affairs, reason is the law of nature, then the introduction of the use of physical force is, for Locke, its antithesis. The initiation of the use of force against a man constitutes a violation of his natural right of self-preservation.

In the state of nature, in the complete absence of government, it is the responsibility of each individual to enforce the law of nature for himself. In practice, this means that it falls to each individual to exact punishment or reparations from anyone who initiates the use of force against him. But because someone who did so violate a man's rights would evince himself to be a potential threat to *all* men, *any* man could justifiably take it upon himself to punish such a violator, much as one could take it upon oneself to hunt down a wild animal that had attacked a neighbor or his livestock.

There are obvious problems with enforcing the law of nature in this way. Men do not agree on the "statutes" of the law of nature, says Locke, nor do they make reliable judges or executors in their own cases. It is to overcome these problems that men

r">
12

institute government. The purpose of government is to give effect to the law of nature, and, thus, to men's natural right of self-preservation. The written laws of a polity, therefore, should be consistent with the requirements of the law of nature. They should be made known to all, and, since all men possess the natural right of self-preservation to the same extent, the laws should apply to all men equally and should be enforced impartially.

In any society based on Locke's principles, one which abjures the initiation of the use of physical force in human relations, the basis for social interaction will be voluntary cooperation. This is the society of contract. In such a society, the only positive obligations that exist among men are those they contract voluntarily. Locke extends this principle to men's relationship to government. Under his conception, government is a kind of agent hired by the people to perform a specific function, to secure their natural rights. The government's responsibility to the people is a fiduciary one. It follows, said Locke, that the only legitimate government in a society of contract is one to which the people consent voluntarily.

In the thought of Locke there are two bases of political legitimacy: government must originate in the consent of the governed, and it must secure their natural rights. A government which failed on either of these points would not be a legitimate one. No government, even one which enjoyed the consent of an overwhelming majority of its citizens, could be legitimate if it failed to secure their "unalienable" rights:

> For a Man, not having the Power of his own Life, cannot, by Compact, or his own consent, enslave himself to any one, nor put himself under the Absolute, Arbitrary Power of another, to take away his Life, when he pleases.[7]

Consent of the governed, in and of itself, can never constitute the sole basis of political legitimacy.

Elsewhere Locke explained that the authority of government to intervene by force in the lives of individuals is only the authority delegated to it by the individuals that constitute the citizenry.[8] (Therefore, that authority extends no further than to what the law of nature would authorize those private individuals to do in the state of nature. In the state of nature, the law of nature authorizes only the *defensive* use of force, or its use in retaliation against those who initiate its use. This would then be the moral limit of its use by government, although Locke himself never explicitly drew this conclusion.)

According to Locke, when a number of free individuals living in the state of nature agree to form a government, every man within the territory affected is entitled to vote as to which form of government to adopt, and everyone is morally obligated to abide by the majority's decision. A government chosen in this manner, said Locke, can fairly claim to enjoy the consent of the governed. Locke's ideas that political legitimacy requires the consent of every man and that this requirement is satisfied by including every man in a process of decision-making by majority rule seem to sanction democracy as the proper form of government. Indeed, Locke tended to favor the rule of democratic majorities, since he was writing to counter abuses by monarchical rule. He therefore tended to portray democratic majorities as a welcome counterpoise to monarchy.

On the other hand, he was not unaware of the possibility of abuses of the rights of individuals by democratic majorities. He said that democratic majorities are not morally free to enact any measures they choose, any more than monarchies are. Locke considered even democratic governments to be a potentially dangerous threat to the rights of individuals.

As a countermeasure to the potential for majority tyranny, Locke advocated certain *institutional* safeguards. Among these were the rule of law, equality before the law—specifically for lawmakers themselves, who were to be given short terms of office and returned to civilian life forthwith—and the separation of the legislative from the executive, with the former enjoying superiority. Locke did not believe that paper declarations of the rights of individuals would ever be sufficient to secure those rights. He preferred instead such institutional safeguards as would equalize the forces within a polity, creating a kind of political balance of power. Locke's attention to the *engineering* of a proper balance of political energy, his emphasis upon the *architecture* necessary for a political edifice to withstand internal and external stresses and strains, we will see repeated by the Americans a century later at the Constitutional Convention in Philadelphia.

Locke was clear that when institutional safeguards fail, majorities are not morally permitted to act arbitrarily. Acts of government that are contrary to the law of nature, that violate, rather than secure, the natural rights of individuals, are not morally binding upon the governed. A government that violated the natural rights of a citizen would initiate a state of war between itself and that individual. If a government engaged in a long-standing pattern of such abuses, if it failed in its fiduciary responsibility to its clients, the people, then it would be incumbent on those people to "fire" the government in favor of a new one. This we call the right of revolution.

For Locke, the fundamental purpose of government is to secure the right of the individual to preserve his own life. In practice, this means securing the individual's autonomy, his freedom to act according to the dictates of his own mind. The autonomy which Locke accorded to the individual distinguishes his thought from that of all earlier political philosophers. Locke had borrowed the idea of a right of self-preservation originating in the state of nature from Thomas Hobbes. But, whereas Hobbes had advocated an all-powerful sovereign, effectively wiping out the fundamental liberty he had attributed to the individual in the state of nature, Locke more logically sought to carry over the individual's natural autonomy into civil society. The difference between the two on this point stems from their opposing views of the efficacy of reason.[9]

Hobbes believed that human reason does not provide man with valid knowledge of reality. Thus, men have no means of reaching a common understanding on matters of mutual concern, and their only means of settling differences is violence. They tend to pursue their self-interest with no rational regard for other men. When left to themselves, in a state of nature, men tend to carry on a continuous war of all against all. Hobbes believed that the only way to avoid this war was for men to surrender their autonomy to an all-powerful sovereign who would, in effect, force them to live in peace.

Locke, in contrast, believed in the efficacy of reason. He believed that most men are capable of understanding the need to abide by certain fundamental moral principles, if those principles are given effect in a proper political context. This led Locke to fundamentally different political conclusions than Hobbes had reached. Where Hobbes would institute an all-powerful government, effectively depriving the individual of all freedom, Locke would erect a government strictly limited, mainly to keeping the peace among private persons, thus reserving to the individual a broad sphere in which he would enjoy the liberty to think for himself, to act accordingly, and thereby to secure his own self-preservation.

The basis that Locke provided for the right of self-preservation, however, man's natural propensity to pursue self-preservation, is inadequate. Animals have a similar propensity, but only man has rights. Man's rights therefore must be grounded in something unique to man—which is, of course, his faculty of reason. Locke himself came close to making this argument. He recognized that reason is man's distinctive means of knowing reality, and that it is, therefore, his guide to securing his self-preservation. He recognized that man is a creature of free will, and that the seat of the will lies in man's faculty of rationality. He even goes so far as to say:

> *The Freedom then of Man and Liberty of acting according to his own Will, is grounded on his having reason, which is able to instruct him in that Law he is to govern himself by, and make him know how far he is left to the freedom of his own will.*[10]

Man's liberty, in other words, is grounded in his having reason. But he meant that man's reason *permits* that a man be left free, because it is by means of his reason that he is able to control his actions so as not to pose a threat to other men. This is indeed true, but what Locke never did say is that man's dependence on reason *requires* that he be free.[11]

An animal is guided toward self-preservation by instinct. Its strategy for keeping itself alive is predetermined. All animals of a given species in a given environment tend to keep themselves alive in about the same way. There are no individualistic beavers building their shelters in revolutionary new ways like Frank Lloyd Wrights of the riverbank. But man's strategy of self-preservation is not predetermined. Because he can reason, man has the ability to discover new and better ways of keeping himself alive. There simply is no single right way for a man to live.

This is not to suggest that there are no objectively valid principles by which a man ought to live. Indeed, there are such principles, and one of the most fundamental is that a man ought always act in accordance with reality. Since man's means of knowing reality is his faculty of reason, this principle can be stated alternatively as "Always act rationally." It is precisely because a man's life depends on his always acting rationally that he requires freedom, because to act rationally means to act in accordance with the dictates of one's own judgment, and one can only act in accordance with one's own judgment if one is free to do so, i.e., if one is not physically constrained from doing so by other men. Man's faculty of reason both *permits* him to be free, as Locke pointed out, and it *requires* that he be free.

The right of self-preservation is the right to think for oneself and to act accordingly. It is the right to do everything necessary to preserve one's life according to one's own best judgment. To say that one *has* a right to preserve one's own life is to say:

a.   that it *is* right that one do so;
b.   that one is morally entitled to expect other men not forcibly to prevent one's doing so;
c.   that one is entitled to require government to enforce that expectation; and
d.   that one is entitled to expect government itself to refrain from forcibly preventing one's doing so.

The right of self-preservation is the most fundamental of rights; it is the right from which all other rights derive. One of those other rights, though derivative, is itself indispensable to self-preservation; it is the right of private property.

> *The State of Nature has a Law of Nature to govern it,*
> *which obliges every one: and Reason, which is that*
> *Law, teaches all Mankind, who will but consult it, that*
> *being all equal and independent, no one ought to harm*
> *another in his Life, Health, Liberty, or Possessions.*[12]

In places, Locke uses the term "property" to refer collectively to the sum total of men's natural rights, to their rights, considered collectively, to "Life, Liberty, and Estates." In other places, he uses it to refer exclusively to their right to their "Possessions" or their "Estates." Under the former usage, the securing of their "property" constitutes the sole reason for men to institute government. Under the latter, it constitutes an important part of their reason for doing so. Either way, it is clear that Locke considered the securing of one's material possessions to be indispensable to securing one's self-preservation.

As Locke had it, just as the right of self-preservation originates in the state of nature, so also does the right of private property. Just as a man has a right to act to preserve his own life, so, and for the same reason, has he a right to appropriate to himself those things, including land, on which he depends for his survival. According to Locke, what gives a man warrant to appropriate any specific item, or any specific plot of land, is his labor. A man owns his labor, so anything with which he "mixes his labor" thereby becomes his property. Through his labor, a man has a right, in the state of nature, to amass as much property as he can make use of, but no more than this.

Locke held that a man does not have a right to gather so much fruit, for example, that some of it spoils before he can use it. But of money, say, gold and silver, a man may accumulate as much as he is able to, because money does not spoil. Money, said Locke, comes into being by virtue of the mutual consent of men, but *outside the bounds of civil society*. It is money that gives rise to inequality of wealth among men, and Locke saw such inequality of wealth as existing *prior to civil society*. So property and inequality of wealth are things to which man has a natural right, a right that precedes the instituting of civil society.

For Locke, a natural right is a right that pertains in the state of nature. But earlier I defined "right" as incorporating four distinct propositions:

a. that it is right for one to do such-and-such;
b. that one is morally entitled to expect other men not to forcibly prevent one's so doing;
c. that one is entitled to require government to enforce the latter expectation; and
d. that one is entitled to expect government itself not to forcibly prevent one's so doing.[13]

The first two of these propositions are purely moral, and they apply, therefore, in the state of nature (as well as in civil society). But the second pair of propositions are necessarily *political*; they apply, that is, only under government. By this definition, therefore, the concept "natural right," comprising as it does all four propositions, does not apply in the state of nature. There are no rights in the state of nature. Natural rights are "natural" not because they apply in the state of nature, as Locke held, but, rather, because they derive from the nature of man.

The natural right of private property, for example, derives from man's right of self-preservation. If it is right that a man do everything he can to preserve his own life, then it is right that he labor to produce the material goods on which his life depends. And if it is right for him so to labor, then:

a. it is right that he should possess, use, and dispose of the product of his labor as he chooses;
b. he is morally entitled to expect other men not to forcibly prevent his so doing;
c. he is entitled to require government to enforce this expectation, and
d. he is entitled to expect government itself not to forcibly prevent his so doing.

Note that these four propositions apply uniquely to rational beings. Rights derive, in other words, from man's nature *as a rational being*. (This is the reason that animals do not have rights.)

On the status of the right of property in civil society, Locke said:

> For in Governments the Laws regulate the right of property, and the possession of land is determined by positive constitutions. [14]

Did he mean that in civil society the right to own land is merely conventional rather than natural, meaning that governments are morally free to alter or abridge property rights in land as they choose? No. Locke had obviously been at pains to show that both property itself, and inequality of property, are not products of civil society but enjoyed an existence prior to it.

Locke was unequivocal that the primary purpose of men's entering into civil society is the preservation of their property. Clearly, and about this he was explicit, it would be absurd for civil society to be permitted to nullify the very purpose for which it was voluntarily called into being.

> The Supream Power cannot take from any Man any part of his Property without his own consent. For the preservation of Property being the end of Government, and that for which Men enter into Society, it necessarily supposes and requires, that the People should have Property, without which they must be suppos'd to lose that by entering Society, which was the end for which they entered into it, too gross an absurdity for any Man to own. [15]

What he meant when he said that in civil society "possession of land is determined by positive constitutions" is that under government a man may own more land than he can make use of, but that he would not hold the surplus by natural right. The land he did make use of he would indeed own by natural right. This seems to be how Thomas Jefferson understood Locke, as I shall discuss in the next chapter.

In his disquisition on property, Locke acknowledged the connection between property and self-preservation. And, while he fully appreciated the importance of individual autonomy, nowhere did he elaborate on the connection between individual liberty and the right of property. This is a connection worth exploring. In broad outline: in order to preserve his life, a man must be able to act freely; and in order to be able to act freely, he must be free to acquire, possess, use, and dispose of property as he chooses.

The whole range of "natural" rights, including, for example, freedom of religion, freedom of speech, and the right to bear arms, are possible only in a context of private property. Try to imagine, for example, freedom of the press in the total absence of private property. The only alternative to privately owned property is collectively owned property, which must mean, in practice, government-owned property. What could possibly be the meaning of "freedom of the press" if the government owned everything used by the press, i.e., the land, the buildings, the machinery, the paper, even the ink, as well as the trucks used to transport the newspapers and the retail outlets which distribute them? Under such circumstances, the government would *be* the press, so "freedom of the press" would be a nullity.

The status of the right to bear arms would be similar. If the government owned all the raw materials out of which arms are manufactured, all the factories that made them, and all the facilities for transporting and distributing them, then the government would be the arms supplier for the nation. As such, it would have total control over the people's "right to bear arms," which it could extinguish at will.

Or consider freedom of religion. Separation of church and state would be impossible if the government owned all the land and all the buildings in which religious worship could take place. Any attempt to keep the state fully and consistently separated from religion would necessarily mean that religious worship could occur in no place whatever. (Consider the way in which religious expression today is being banished from the public square under precisely this logic.)

In the absence of property rights virtually no other natural rights are possible. The extinguishing of freedom of religion, freedom of speech, and the whole range of individual rights in the Union of Soviet Socialist Republics was not an aberration but a fully logical development that followed from the original abolition of private property. Property rights are the *sine qua non* of all other natural rights. Without property rights, there can be no such thing as individual liberty.

Consider that in the United States today, in those instances in which the right of private property is relatively secure, other rights are appreciably more secure than they are in those instances in which property rights have been diminished. For example, privately owned property used for commercial purposes ought to be every bit as private as property used for a person's home. But private property used for commerce has long been classified as "public." In consequence, it is subject to a degree of government regulation that would be unthinkable in the case of a private home.

As of this writing, for example, the government may not require the owner of a private home to install facilities at his own expense to accommodate the wheelchair-bound. Yet the government may require this of the owner of private property used for a "public," meaning commercial, use. Today, the government may not prohibit an individual from smoking a cigarette in his own home, but it may prohibit him from doing so in a privately owned restaurant open to the "public," even if the smoker has the permission of the restaurant's owner. The right of free speech, when employed in advertising a product or service for commercial purposes, may also be regulated, and thus violated, by government in ways that non-commercial speech may not (yet) be regulated.

As another example, consider the regulation of the airwaves. The airwaves themselves have long been classified by the government as public property. In consequence, the U.S. Government is able to regulate the content of what passes over the airwaves in ways that it may not regulate speech that emanates from privately owned printing presses and is distributed on privately owned paper, in private trucks, to private retail outlets. The Federal Communications Commission, for example, is able to require, as a condition of issuing a broadcast license, that private companies include specified amounts of "children's programming." Yet imagine the uproar that would ensue (one would hope) if the government instituted a Federal Newspaper Commission charged with licensing newspaper publishers on condition that they agree to include certain kinds of content in their papers.

Then there are the public roadways. As of this writing, the government is not yet permitted to enter a private home without probable cause to discover whether illegal drugs are being used. But on the *public* roadways armed agents of the government are now permitted to stop motorists without probable cause in order to ensure that they are not intoxicated.

Two developments in the spring of 2014 illustrate the dependence of other rights upon property rights. The right to bear arms, which is a corollary of the right of self-defense, has been under furious assault by the liberal establishment in the

United States. Unable to defeat head-on the Second Amendment's protection of that right, the Left might have hit upon a way indirectly to render the Amendment moot. In October of 2013, the Doe Run Company announced that they would shut down their lead smelting plant in Herculaneum, Missouri rather than install expensive new air pollution control equipment mandated by the EPA.[16] The smelter is the last in the United States that produces lead bullion, used to make ammunition, from raw lead ore.[17] The closing will make it more expensive to supply ammunition to American gun owners, and it presages the eventual extinction of the practical ability to exercise the right to bear arms. When property rights are so eroded that the government is able to regulate businesses out of existence, in this case in the name of preventing pollution, then no rights are secure.

Just as the right to bear arms has been under assault recently, so has freedom of religion. A statute in Arizona that would have guaranteed the right of any business owner, such as a Christian baker of wedding cakes, the right to refuse service for religious reasons to any customer, such as a homosexual couple planning to get married, was defended by its proponents as a defense of religious liberty, which it was. But in a free society, any business owner should be free to deny service to any customer for any reason whatsoever, because his commercial enterprise is his private property. It should be no different than denying someone entry into one's own home. If secure property rights had been in place in Arizona, there would have been no need for a statute to protect certain rights of religious property owners (thus making the exercise of those rights dependent upon one's religious affiliation, which would have been unjust to atheists).

We are currently witnessing the banishing of religious expression from the public square in the United States. Since commercial enterprises have long been considered "public," it was only a matter of time before the Left went after business owners for practicing religious-based discrimination. Property rights afford one the indispensable, physical ground on which to stand when exercising one's rights. Without that ground, as the Arizona case illustrates, individuals can literally be forced by the government to violate their most sacred beliefs.

These are a few examples out of many that suggest that, in the absence of full property rights, other rights are less secure than they would otherwise be. Individual autonomy without private property is a chimera. The economy of the United States today is only partially based on private property, and that proportion is diminishing by the year. Yet the extent to which we still enjoy a degree of property rights—and, therefore, freedom—allows us the illusion that we can continually chip away at the right of private property without endangering such rights as freedom of religion, speech, and press.

If property rights are necessary to the autonomy of the individual, they are also largely sufficient to the purpose.[18] Imagine a polity in which the right of private property were fully and consistently sacrosanct, in which all property were private except that required for essential government functions, and in which any individual could use his property in any way he chose so long as he refrained from physically harming the person or property of another individual. Under such circumstances, an owner of a newspaper would enjoy complete freedom of the press even in the absence of a specific constitutional protection of that right. His freedom to use his land, buildings, equipment, paper, ink, delivery trucks, and retail outlets as he chose, so long as he refrained from physically assaulting anyone, would amount to complete and total freedom of the press.

Likewise, a religious congregation that was free to use its property, its land and the church upon it, entirely as it chose would, in effect, enjoy complete freedom of religious worship. And the right of its members to possess, use, and dispose of their money as they chose would prevent the government from requiring them to contribute to the support of their own religion or any other one; it would prevent the government, in other words, from effecting an establishment of religion.

The need to enumerate such specific rights as freedom of religion and freedom of speech arises only in the absence of a full, consistent, and properly implemented system of property rights. By "properly implemented" I mean a system based on the understanding that the only justification for government interference with a man's freedom to possess, use, and dispose of his property must be that he has initiated some sort of physical harm against another individual or his property. The right of private property *is* freedom. It is not merely a part of what constitutes freedom, much less is it a dispensable appendage to freedom. The right to possess, use, and dispose of the product of one's labor and anything else one owns is both *necessary* and *sufficient* to render the individual autonomous. It is therefore indispensable to his self-preservation.

John Locke and Francis Bacon were both part of the same intellectual movement. Locke wrote, almost echoing Bacon:

> He that first invented printing, discovered the use of the compass, or made public the virtue and right use of quinine, did more for the propagation of knowledge, for the supplying and increase of useful commodities, and saved more from the grave, than those who built colleges, workhouses, and hospitals.[19]

A direct outgrowth of Bacon's call to arms in the campaign to command nature was the Royal Society of London for Improving Natural Knowledge, founded at Gresham College, Oxford in 1660. With such members as Isaac Newton and Robert Boyle, the Society explicitly dedicated itself to discovering new knowledge of the world and utilizing it to extend men's technological command of nature. Boyle, for example, discovered that the volume of a confined gas decreases proportionally as the pressure increases. John Locke was one of the Society's first Fellows. He was also a friend and long-time correspondent of Newton. And he was an executor of Boyle's estate and fulfilled a promise to his "chief scientific mentor" by publishing in 1692 Boyle's "History of the Air," which Locke received in fragmentary form and largely rewrote himself.[20]

Locke's own intellectual output, though not technological in nature, represented very much the sort of effort to extend men's mastery of the world that Bacon had advocated. As Neal Wood has noted, Francis Bacon had contemplated a wide-ranging, unified science comprehending all fields of intellectual endeavor. He compiled a "Catalogue of Particular Histories" in which he listed each separate field of knowledge to be included. The seventy-eighth item in Bacon's Catalogue was the science of knowledge itself, what we call epistemology, and which Bacon recognized needed to be explored and placed on a solid logical footing. As Wood points out, Locke answered this call from Bacon with his *Essay Concerning Human Understanding*.[21]

Another necessary requirement of the Baconian enterprise was the removal of a certain artificial obstacle to men's ability to make use of their faculty of reason, that obstacle being established religion. Established religion aims at enforcing established ways of thinking. The example *nonpareil* of man's commanding nature is the Industrial Revolution. It has taught us, as nothing else possibly could have, that the advancement of man's command of nature thrives in the presence of free minds, countless numbers of them, extending the boundaries of man's knowledge on a thousand fronts into realms of endeavor previously unimagined. Where Baconism gives rise to a Galileo, discovering hard astronomical evidence that the earth revolves around the sun, established religion answers with the Inquisition, forcing Galileo to deny the conclusions of his own mind. What the Baconian project required was the antithesis of what established religion aspires to. Locke fulfilled that requirement with his *Letters on Toleration*, in which he made the case for disestablishing religion.

Having validated man's fundamental means of understanding reality, and having made the case for freeing that means from artificial control, Locke then addressed a third requirement of the Baconian project, the need for freedom of *action*. It avails men little to be able to think their own thoughts and to communicate them freely if they are not free to act upon them as well. Locke's exploration of *The True Original, Extent, and End of Civil-Government* (the subtitle of his *Two Treatises*) represents an attempt to discover the natural laws governing human association. In it he identified the political principles, including the centrality of property rights, on which individual freedom of action depends.

Just as Boyle had sought to understand the laws governing the behavior of gasses so as to extend men's mastery of them, Locke sought to discover the laws of human nature in order to afford men the ability to alter and fashion their political institutions to accord better with those laws. Wrote Locke, "God … hath also given [men] reason to make use of [the world] to the best advantage of Life, and convenience." [22] With these words, Locke implies that man *can* command nature, to use Bacon's term, and that it is *right* for him to do so. Property rights would prove indispensable to man's making use of the world. They would enable him to command nature on an industrial scale.

A century after the appearance of *Two Treatises*, a nation would come into being uniquely poised to pursue the Baconian command of nature and fundamentally committed to securing the property rights necessary to achieve it.

Notes

1. Jerome Huyler, Locke in America (Lawrence, KS: University Press of Kansas, 1995), 95.
2. John Locke, *Two Treatises of Government*, ed. Peter Laslett (London: Cambridge University Press, 1960), I, 86.
3. Ibid., II, 6, 8-11.
4. Ibid., II, 54, 10-15. Emphasis in original.
5. Ibid., II, 19, 26-27. Emphasis in original.
6. Ibid., II, 18. Emphasis in original.
7. Ibid., II, 23, 4-8. Emphasis in original.
8. Ibid., II, 135, 6-23.
9. W.T. Jones, *A History of Western Philosophy* (New York, Chicago, San Francisco, Atlanta: Harcourt Brace Jovanovich, Inc., 1969), 3, 136.
10. Locke, Two Treatises, II, 63, 1-5. Emphasis in original.
11. See Ayn Rand, "Man's Rights" in *The Virtue of Selfishness*, (New York, NY: New American Library, 1961).
12. Ibid., II, 6, 6-11.
13. One can have a right to do specific things that are not morally right. It is morally wrong to lie, but in order for a person to be free to speak the truth, he must be free to tell lies.
14. Locke, *Two Treatises*, II, 50, 16-18.
15. Ibid., II, 138, 1-8. Emphasis in original.
16. *North America's Largest Lead Producer to Spend $65 Million for Environmental Violations at Missouri Facilities*, United States Environmental Protection Agency, press release dated October 8, 2010, accessed December 17, 2014,
http://www.epa.gov/region07/cleanup/doe_run/pdf/hq_news_release_R336-100810.pdf.
17. There are still several smelters that produce recycled lead.
18. There would remain the question of the exercise of individual rights, such as the right of free speech, in places that are legitimately public, such as government buildings.

19. John Locke, *Essay Concerning Human Understanding*, ed. Alexander Campbell Fraser (New York, NY: Dover Publications, 1959) IV, 12, 12.

20. Huyler, Locke, 69.

21. Neal Wood, *The Politics of Locke's Philosophy* (Berkeley, CA: University of California Press, 1983), 77.

22. Locke, Two Treatises, II, 26, 1-3.

2: The United States

[A]ny polis which is truly so called, and is not merely one in name, must devote itself to the end of encouraging goodness. Otherwise, a political association sinks into a mere alliance.... Otherwise, too, law becomes a mere covenant ... "a guarantor of men's rights against one another"—instead of being, as it should be, a rule of life such as will make the members of a polis good and just.[1]
—Aristotle

JOHN LOCKE, AS we have seen, departed fundamentally from this view of government. He held that the purpose of government is indeed merely to guarantee "men's rights against one another." A government committed to securing individuals' rights of life, liberty, and property is antithetical to one committed to ensuring that individuals lead good and just lives. To embrace one is necessarily to reject the other. If government exists to ensure that the governed lead "virtuous" lives, then there can be no such thing as individual rights. But if government exists to secure those rights, then it may not, beyond enforcing prohibitions against physically injuring others or their property, force people to lead lives of "virtue."

Adams

When the Americans dedicated their new nation to the "unalienable Rights" of "Life, Liberty and the pursuit of Happiness," they implicitly rejected the older, Aristotelian conception of the purpose of government. John Adams was explicit in his rejection of that conception. He saw man as a fundamentally rational being possessed of free will. To deny such a being his liberty for the sake of forcing him to live virtuously would be to violate his nature. "[A]n enemy to liberty," he wrote, is "an enemy to human nature." [2]

Among the important books of the time, Montesquieu's *Spirit of the Laws* figured prominently. In it the Frenchman elaborated a conception of virtue that emphasized frugal living and the subordination of one's own interests to the public good. Adams, in his *Defence of the Constitutions of the United States*, criticized Montesquieu for attempting to design a constitution for an unselfish kind of man who never did and never would exist. [3]

Adams leveled similar criticism against the ancient Spartan conception of virtue, which was enjoying a vogue among some European and American reformers. He criticized the Spartan lawgiver, Lycurgus, for stifling human nature in his effort to shape a new, public-spirited man. "[I]t was necessary," said Adams, "to extinguish every other appetite, passion, and affection in human nature." He criticized…

> …the equal division of property; the banishment of gold and silver; the prohibition of travel and intercourse with strangers; the prohibition of arts, trades, and agriculture; the discouragement of literature; the public meals; the incessant warlike exercises; the doctrine that every citizen was the property of the state, and that parents should not educate their own children. [4]

Adams objected, in other words, to the violence done to human nature, and specifically to the natural human requirement for liberty, in the name of cultivating some "celestial" notion of virtue. Spartans lived "as if fighting and intriguing, and not life and happiness, were the end of man and society." [5]

Private property, fully and consistently institutionalized, is the distinguishing characteristic of the society of individual liberty. Men who have the freedom to possess, use, and dispose of the product of their own labor are free to place their own interests ahead of any notion of the public good and to pursue such earthly values as material well-being, wealth, and luxury. Adams advocated a political philosophy that deals with man as he is, not as he allegedly ought to be. If man is a rational, volitional being, then his government ought to secure his freedom to act according to his own judgment. And if a man judges it to be in his own interest to pursue material comfort, wealth, and luxury, then government ought to secure his freedom to act accordingly. Adams described his *Defence* as "an attempt to place Government upon the only Philosophy which can ever support it, the real constitution of human nature, not upon some wild Visions of its perfectibility." [6]

Property was central to Adams' ideas about government. Throughout history, he discovered, republics had been characterized by a continual tension between the propertied and the unpropertied, or between the more and the less propertied. The former would use their wealth to purchase political power at the expense of the latter, and the latter would convert their superior numbers into political power with which to plunder the propertied. Adams, the architect of constitutions, saw the solution in a certain *structure* of government, a counterbalancing of forces between a lower house of the legislature representing the people and an upper house representing the propertied, with a strong executive to mediate their differences.

Adams was clear that liberty and property are inseparable. He wrote:

> *Res Populi, and the original meaning of the word republic … had more relation to property than liberty. It signified a government, in which the property of the public, or people, and every one of them, was secured and protected by law. This idea, indeed, implies liberty; because property cannot be secure unless the man be at liberty to acquire, use, or part with it, at his discretion, and unless he have his personal liberty of life and limb, motion and rest, for the purpose. It implies, moreover, that the property and liberty of all men, not merely of a majority, should be safe; for the people, or public, comprehends more than a majority, it comprehends all and every individual.*[7]

Recognizing that it must be one or the other, liberty or classical "virtue," Adams chose liberty as the proper end of government. Liberty, in turn, is inseparable from property; Adams was as aware of this as John Locke had been. The American Founders' commitment to individual liberty is nowhere more clearly expressed than in the Declaration of Independence.

Jefferson

> *We hold these truths to be self-evident, that all men are created equal, that they are endowed by their Creator with certain unalienable Rights; that among these are Life, Liberty and the pursuit of Happiness. That to secure these rights, Governments are instituted among Men, deriving their just powers from the consent of the governed, that whenever any Form of Government becomes destructive of these ends it is the Right of the*

> *People to alter or abolish it, and to institute new*
> *Government, laying its foundation on such principles*
> *and organizing its powers in such form, as to them shall*
> *seem most likely to effect their Safety and Happiness.*[8]

It would be difficult to compose a more concise and eloquent summary of Locke's political philosophy than this. All the more curious, then, is Jefferson's substitution of "pursuit of Happiness" for "property" in the familiar trio of rights. If liberty implies, and requires, property, then Jefferson's commitment to liberty would be problematic if he was not equally committed to securing the rights of property.

Jefferson's use of "pursuit of Happiness" instead of "property" suggests that he might have meant implicitly to deny to private property the status of an "unalienable" right. Elsewhere he is more explicit. In a letter he wrote in 1813, in which he addressed the question whether inventors have a natural right to exclusive ownership of their inventions, Jefferson discussed the status of property as a natural right. He began by saying that it is debatable ("moot") "Whether any kind of property is derived from nature at all." He continued:

> *It is agreed by those who have seriously considered the*
> *subject, that no individual has, of natural right, a*
> *separate property in an acre of land, for instance. By an*
> *universal law, indeed, whatever, whether fixed or*
> *movable, belongs to all men equally and in common, is*
> *the property for the moment of him who occupies it; but*
> *when he relinquishes the occupation, the property goes*
> *with it. Stable ownership is the gift of social law, and it*
> *is given late in the progress of society.*[9]

The last line of this quotation reminds us of Locke's statement that "For in Governments the Laws regulate the right of property, and the possession of land is determined by positive constitutions." In a letter he wrote in 1816, Jefferson wrote:

> *[A] right to property is founded in our natural wants, in the means with which we are endowed to satisfy these wants, and the right to what we acquire by those means without violating the similar rights of other sensible beings.*[10]

Here Jefferson is clear that private property is grounded upon certain natural needs of man, on the (natural) means he possesses of fulfilling those needs, i.e., his labor, and on the right he thereby acquires to the product of that labor. But the key to understanding his ideas on property is the thought with which he ends the sentence just quoted, that the natural right to the product of one's labor is somehow *conditioned upon* one's not "violating the similar rights of other sensible beings."

Jefferson is suggesting here that the natural right to property is grounded in *and maintained by* the productive use *and only* the productive use of one's property, just as Locke believed. As Jean Yarbrough has pointed out, Jefferson held the right to labor to be more fundamental than the right of property.[11] It is the right to labor that gives rise to the right to the product of one's labor. The right to labor thus gives rise to the right to occupy a patch of ground somewhere. But the right to that patch of ground exists, as a *natural* right, only so long as one makes productive use of it. As soon as one vacates the land or otherwise ceases to make productive use of it, any claim one had to the land that was based in natural right ceases as well.

For a man to defend by force land he is not using productively can, for Jefferson, violate the law of nature; it can violate other men's rights to labor—upon that land. "Whenever there is in any country uncultivated lands and unemployed poor, it is clear that the laws of property have been so far extended as to violate natural right." [12] (Jefferson had in mind here the vast, largely unused private estates of the French nobility.) One's title to lands left "uncultivated," especially when there are unemployed poor, may be protected under the laws of "stable ownership" within civil society, but not by natural right.

This does not mean that Jefferson denied altogether the natural right status of private property. In a "Prospectus" he wrote later in life recommending the economist DeStrutt de Tracy, Jefferson explicitly referred to a *natural* right of private property:

> If the overgrown wealth of an individual be deemed dangerous to the state, the best corrective is the law of equal inheritance to all in equal degree; and the better as this enforces a law of nature, while extra-taxation violates it. [13]

He continues:

> To take from one, because it is thought that his own industry and that of his father's has acquired too much, in order to spare to others, who, or whose fathers have not exercised equal industry and skill, is to violate arbitrarily the first principle of association—the guarantee to every one of his industry and the fruits acquired by it. [14]

"Extra-taxation" violates the "*guarantee*" to each man of the freedom to control his own labor and to own the product of that labor, which is the "first principle of association." We are reminded of Locke's argument that it would be "too gross an absurdity" for the "Supream Power" to take "from any Man any part of his *Property* without his own consent," given that it was to preserve their property that men entered civil society to begin with.[15] Jefferson did indeed subscribe to the idea of a natural right of private property, but it encompassed only a part of what we usually consider to be the right of property. It did not include land that one did not occupy or use productively.[16]

Jefferson considered private property to be the foundation of the whole republican enterprise. The yeoman farmer, largely self-sufficient and beholden to no one, Jefferson saw as the backbone of liberty. The greater the proportion of the population that farmed their own land, the better would be the prospects for long-term liberty in the United States. It was his own desire to expand the supply of land available to future generations of yeoman farmers that lead Jefferson to purchase the Louisiana Territory. It was his belief in the fundamental importance of the yeoman farmer that led him to oppose Hamilton's plan to accelerate the development of manufacturing in America (through just the kind of property-state alliance that Jefferson had always opposed). Jefferson believed that the employees of large-scale manufacturing concerns do not make virtuous, independent-minded republicans, so he opposed any attempt to stimulate the growth of manufacturing through government intervention.

Jefferson believed that a man has a natural right to occupy land that he uses productively, and a natural right to the product of his labor upon that land. This conception may not accord natural right status to the full bundle of the rights usually associated with "stable ownership," but it is nonetheless an important part of the foundation on which to erect a polity committed to individual liberty. If all the property devoted to productive use in the United States were rendered completely inviolable to government interference, Americans would enjoy a degree of freedom unknown in the world today. On the other hand, Jefferson's exclusion of "property" from its proper place with life and liberty in the Declaration of Independence has helped to obscure the fact that the right of private property enjoyed, in the estimation of Jefferson and virtually all of his fellow Founders, a place among the most fundamental rights of man.

The State Bills of Rights

Jefferson once said that there was nothing original in the Declaration of Independence, that he had merely intended to express the "common sense" of the matter. But was Jefferson's somewhat complicated conception of the natural right of private property indeed the "common sense" of the matter? What was the prevailing view during the Revolutionary period as to the natural status of that right? The bills of rights of many of the state constitutions adopted in the years following the Declaration suggest an answer. Six of these included a reference to the Lockean trio of life, liberty, and property. The 1780 Declaration of the Rights of the Inhabitants of Massachusetts was typical. Article I states:

> All men are born free and equal, and have certain natural, essential, and unalienable rights; among which may be reckoned the right of enjoying and defending their lives and liberties; that of acquiring, possessing, and protecting property; in fine, that of seeking and obtaining their safety and happiness.[17]

Judging by this and other similar passages in the state bills of rights, many of America's Revolutionary leaders understood the right of property to be, without qualification, a fundamental, natural right. Jefferson's more complicated understanding of the "natural" status of the property right was probably not the "common sense" of the matter.

The Constitution

We The People of the United States, in Order to form more perfect Union, establish Justice, insure domestic Tranquility, provide for the common defence, promote the general Welfare, and secure the Blessings of Liberty to ourselves and our Posterity, do ordain and establish this Constitution for the United States of America.[18]

The U.S. Constitution is explicit in its commitment to liberty, but it is noteworthy for the absence within it of any mention of natural rights, on the one hand, or of property rights, on the other. The same is true of the U.S. Bill of Rights. We must ask, then, whether the Framers of the Constitution and the authors of the Bill of Rights meant to deny a "higher law" status to the rights of individuals. And we must ask, especially, whether they intended not to render property rights inviolable. Was it their intention that the U.S. Government, subject only to the exceptions spelled out in the Fifth Amendment, and that the state governments, subject only to their own constitutions and bills of rights, should have the authority to abridge or abolish property rights at will? If indeed they did so intend, then it is possible that they did not intend to establish a nation of free individuals in a manner consistent with the principles of John Locke.

The Framers and Natural Rights

That all men are by nature equally free and independent, and have certain inherent rights, of which, when they enter into a state of society, they cannot, by any compact, deprive or divest their posterity; namely, the enjoyment of life and liberty, with the means of acquiring and possessing property, and persuing and obtaining happiness and safety.[19]

These words from the Virginia Declaration of Rights would have "prefixed" the U.S. Constitution if James Madison had had his way. They would have tied the Constitution unequivocally to the Lockean principles of the Declaration of Independence, and they would have formally endorsed the idea of private property as a fundamental, natural right. But the Framers rejected Madison's opening, and they chose not explicitly to endorse the idea of natural rights in general nor the idea of a natural right of private property. We must now ask what place, if any, they intended property rights to have in their constitutional scheme.

The Framers and Property Rights

The men who framed the Constitution of the United States were probably unanimous in their belief that the protection of property rights was a fundamental reason for the existence of government. Even though they intentionally chose not to endorse the idea of natural rights, nor to name as the purpose of their frame of government the familiar trio of life, liberty, and property that would have been suggestive of Lockean natural rights, there probably was not one among them who would have disagreed that the securing of property rights was a fundamental purpose of the government they were erecting.

Said Gouverneur Morris:

> Men don't unite [to form governments] for life or liberty, they possess both in the savage state in the highest perfection they unite for the protection of property. [sic] [20]

John Rutledge and Pierce Butler of South Carolina agreed with Morris that the protection of property is the primary purpose of government.[21] So did Rufus King of Massachusetts.[22] James

Madison was, as we have seen, a thorough-going Lockean and an advocate of a natural right of private property. Likewise on both counts James Wilson, another leader of the Convention. Likewise George Mason, John Dickenson, and Elbridge Gerry.[23] And William Paterson of New Jersey, on circuit for the U.S. Supreme Court in 1795, would write:

> [T]he right of acquiring and possessing property, and having it protected, is one of the natural, inherent, and unalienable rights of man.[24]

Alexander Hamilton, though apparently skeptical about the idea of natural rights, was unequivocal that the protection of property is a fundamental purpose of government. Hugh Williamson encouraged his fellow North Carolinians to ratify the new constitution by noting that its Framers "imagined that we had been securing both liberty and property on a more stable foundation."[25] And William R. Davie of North Carolina, Charles Pinckney of South Carolina, and Abraham Baldwin of Georgia all believed the protection of property rights to be important enough to merit the instituting of an upper house of the federal legislature specifically to represent the interests of men of property.

It has never been a serious question among historians whether the men who framed the Constitution considered the protection of property rights to be a fundamental purpose of government. What has been disputed, due to the Framers' failure explicitly to endorse property rights in some form or other, is whether they intended to render property rights inviolable in the way, for example, that freedom of speech and press have been rendered (relatively) inviolable in our own day.

The Constitutional Convention

The power of changing the relative situation of debtor and creditor, of interfering with contracts, a power which comes home to every man, touches the interest of all, and controls the conduct of every individual in those things which he supposes to be proper for his own exclusive management, had been used to such excess by the State legislatures [under the Articles of Confederation], as to break in upon the ordinary intercourse of society, and destroy all confidence between man and man. This mischief had become so great, so alarming, as not only to impair commercial intercourse, and threaten the existence of credit, but to sap the morals of the people, and destroy the sanctity of private faith. To guard against the continuance of the evil ... was one of the important benefits expected from a reform of the government.[26]

The revolution announced by the Declaration of Independence did not end with the signing of the treaty of peace with Britain in 1783. The men who authored the state constitutions and bills of rights of the era showed a correct understanding of the proper *ends* of government, the securing of men's rights of life, liberty, and property. But their early attempts at designing the governmental *means* of achieving those ends were not equal to the task. Among their failings, the governments they designed proved inadequate to secure men's rights of property. It was, in part, the recognition of this failing that led to the effort to devise a new political architecture for the nation, an effort that eventually would issue in the framing and ratification of the U.S. Constitution.

But if the inadequacy of the state constitutions at securing that right was so important a reason for the framing of the U.S.

Constitution, why then is there no blanket protection accorded to property rights in the Preamble or the body of the Constitution, or in the U.S. Bill of Rights? A full understanding of the answer to this question requires that we examine the conditions that gave rise to the effort to replace America's first frame of national government, the Articles of Confederation, with a new constitution.

When the thirteen colonies declared their independence from Britain, they did so as "one people," to use the Declaration's phrase.[27] But in 1776 this people did not have a common government. Each of the new states had its own structure of government left over from the colonial days, and most states soon adopted new, republican frames of government. But the only semblance of a common, national government that existed in 1776 was the Continental Congress, to which each state voluntarily sent representatives. The Congress was not a formal government at all, and the states would not have a formal central government until the last of them ratified the Articles of Confederation in 1781.

The government of the Articles was a confederation government, rather than a true national government, with ultimate power residing in each of the separate states, rather than in the central government. Before a decade would pass, Americans' experience with this form of government would convince enough of them of its inadequacies that they would scrap the Articles in favor of a unique type of hybrid constitution, something more consolidated than a confederation of independent states, but less so than a fully consolidated national government like that of England.

The War for Independence and its aftermath had persuaded many Americans that the loose, confederation form of government was inadequate to the task of directing the national defense against a foreign enemy. In addition, the Americans' frustration with British efforts after the war to restrict Americans' foreign trade had led many of them to believe that more centralized and energetic direction than the Articles could afford was needed to secure Americans' rights to engage in foreign commerce.

A third weakness of the confederation form of government also related to freedom of commerce, in this case, the interstate variety. Certain of the states lacked the port facilities to carry on trade with Europe. North Carolina, for example, though a coastal state, was effectively cut off from the sea by her barrier islands. She had to conduct her foreign trade through Virginia and South Carolina. The latter two states took advantage of North Carolina by levying "imposts" on all goods passing through their states into or out of North Carolina.

Delaware and western New Jersey likewise suffered at the hands of Philadelphia; eastern New Jersey and western Connecticut at the hands of New York; and eastern Connecticut at the hands of Rhode Island and Massachusetts. New York was the worst offender, though she directed most of her attention to foreign trade, especially imports, as opposed to interstate trade. Virginia, on the other hand, was notorious for taxing interstate trade, whether bound for Virginians or just passing through.[28] In the minds of many Americans, these "trade wars" pointed to a need for a central superintending authority to establish freer movement of goods among the states, an authority distinctly lacking under the Confederation government.

These three shortcomings of the Confederation government, its ineffectuality at defending against foreign military threats and its inability to address foreign and interstate commercial restrictions, formed, in the minds of men like Alexander Hamilton and others of a national perspective during the 1780s, a compelling reason either to overhaul the Articles of Confederation or to replace them altogether with a more "vigorous" central government armed with the authority to address such problems. But these inadequacies were not, in and of themselves, enough to give rise to a movement to amend or replace the Articles sufficient to carry the effort through to reality. By late 1786, however, a new incentive appeared which tipped the scale and led, eventually, to the Constitutional Convention of 1787.

Britain's restricting of America's trade with her helped give rise in the mid-1780s to economic hard times and a wave of bankruptcies throughout the States. This wave gave rise, in turn, to a rash of legislation in the states intended to mitigate the effects of the economic downturn, legislation intended specifically to ease the condition of debtors at the expense of creditors. In 1786, for example, seven states succumbed to debtor pressure and inflated their money supplies with issues of paper.[29]

Rhode Island, the worst offender, legislated that creditors had to accept its depreciated paper money at face value, and it provided that violations of this statute could be litigated by trial without jury.[30] New Jersey tried to force acceptance of its paper by stipulating that if any creditor refused to accept it, payment of the debt in question would be suspended for twelve years.[31] Such "stay" or "moratory" laws, suspending or rescheduling the payment of debts, were common.

In 1782 the State of Maryland stayed the payment of all debts until January of 1784. It also stipulated that during this period debtors could use as legal tender land, slaves, or almost any kind of produce.[32] Statutes thus authorizing the payment of debts with property instead of money were common. South Carolina became famous in this regard for its "pine barren law," which required creditors to accept payment in land or other property, though the land proffered was often worthless.[33]

Massachusetts largely resisted the pressure from debtors to enact such legislation, and the result was Shay's Rebellion. In the western part of the state, farmers who had been beset by creditors armed themselves and marched on the courts in an effort to halt legal proceedings against themselves. The rash of legislation blatantly infringing private contracts and undermining the security of property rights had already disturbed advocates of liberty and property throughout the country, but the specter of armed rebellion in an effort to bring about further such legislation sent shock waves up and down the land.

The legislative depredations of the Confederation period had their roots, in the estimation of some leaders, in the political *architecture* of the states. The state governments under the Confederation had weak executive and judicial branches, with the preponderance of power concentrated in the legislatures. And the balance of power within the legislatures themselves tended to reside in the lower houses. Because these lower houses were, by design, the branch most directly answerable to the people at large, they were the department most responsive to the momentary passions of the masses. If the mass of the people in the respective states wanted to use the machinery of government to override the property rights of a minority of the citizenry, there was little in the way of structural checks in the state governments to stop them. The moratory and paper money laws and other legalized looting of the Confederation period were due, said Elbridge Gerry, to "an excess of Democracy." [34]

Liberty, Property, and Democracy

In his famous Federalist No. 10, James Madison addressed the problem of "factions," or what we today call "special interests," under popular government. The Founders saw the propensity of "factions" to seek special treatment from government as a dangerous threat to good government, an essential characteristic of which they believed to be equal protection under the law. In terms reminiscent of John Adams, Madison noted that "the most common and durable source of factions has been the various and unequal distribution of property." [35] He meant that wherever there are a wealthy few and a less wealthy many, each of these "factions" will tend to urge government policies that favor their own short-range interests at the expense of the other group's. Wealthy industrialists, for example, might favor protective tariffs that restrict competition from foreign manufactured goods, thus causing non-wealthy consumers to have to pay higher prices for domestically produced goods.

Madison and Adams and others among the Founders saw differences in wealth as an unavoidable consequence of individual liberty.

> The diversity in the faculties of men from which the rights of property originate, is not less an insuperable obstacle to a uniformity of interests. The protection of these faculties is the first object of government. From the protection of different and unequal faculties of acquiring property, the possession of different degrees and kinds of property immediately results; and from the influence of these on the sentiments and views of the respective proprietors ensues a division of the society into different interests and parties. [36]

Inevitably, men who are equally free to exercise unequal faculties will accumulate "different degrees and kinds of property;" some men will become wealthy and others will not.

A government committed to preserving men's equal rights of self-preservation will, inevitably, face and have to defuse the problem of economic factions. At the time he authored his Federalist entrees, Madison was most concerned, because of recent events, about the threat to the minority faction of creditors posed by the majority faction of debtors, a threat exemplified by "[a] rage for paper money, for an abolition of debts, for an equal division of property, or for any other improper or wicked project." [37]

All men have an equal right to pursue their self-preservation, but they are distinctly *unequal* in their native and their cultivated abilities. Says Madison, "the first object of government" is to protect men's equal rights to pursue their self-preservation by means of their unequal "faculties." Madison is not saying here that men of different degrees of wealth *in fact* have different long-range interests. He makes the point explicitly:

> There is no maxim in my opinion which is more liable to be misapplied, and which therefore more needs elucidation than the current one that the interest of the majority is the political standard of right and wrong. Taking the word "interest" as synonymous with "Ultimate happiness," in which sense it is qualified with every necessary moral ingredient, the proposition is no doubt true. But taking it in the popular sense, as referring to immediate augmentation of property and wealth, nothing can be more false. In the latter sense, it would be the interest of the majority in every community to despoil & enslave the minority of individuals; and in a federal community to make a similar sacrifice of the minority of the component states. In fact, it is only reestablishing under another name and a more spe[c]ious form, force as the measure of right. [38]

Madison recognized that a great many men tend to consult only their apparent short-range advantage, and that, therefore, men of modest wealth will often believe it in their interest to take advantage of their superior numbers to expropriate by political means the wealth of a more propertied minority. Madison is clear, first, that such men mistake the true nature of their long-range interest, and, second, that regardless of the size of their political majority, such men are not morally justified in their actions. He argues, in Lockean fashion, that force thus employed is anathema to right. In order to be just, a government must find a way to secure all individuals in the enjoyment of their property, however little or much it might be. An important failing of the Articles of Confederation was that, due to "an excess of Democracy" at the state level, the state governments were unable to secure the property rights of minorities against the depredations of democratic majorities.

The Strategy of the Framers

The four principal shortcomings of the Articles of Confederation that the Framers of the U.S. Constitution set out to remedy were the impotence of the federal government in the face of foreign threats, both military and commercial, the proliferation of state-sponsored obstacles to interstate trade, and the threat to the security of private property posed by the actions of democratically controlled state legislatures. Their solution involved a unique splitting of "sovereignty," or power, between the state governments and the new central government. Their strategy, in general terms, was to erect a strong central government that would be fiscally independent of and legally superior to the state governments. They then set about to make this new central government sufficiently independent of the people of the states as to render it largely immune to the kinds of democratic pressures that had so compromised the state legislatures under the Articles.

Having thus erected a central government substantially independent of the state governments and insulated from the democratic pressures emanating from the general electorate, the Framers then removed from the state governments and vested in the new central government exclusive authority over certain areas that had been problematic under the Articles. These included national defense, foreign and interstate commerce, and certain matters relating to the security of private property, such as the issuing of money. Finally, the Framers explicitly prohibited to the states—and to the new federal government—certain types of legislation that violated the rights of property and which had proliferated under the Articles.

Article I, Section 8 of the U.S. Constitution authorizes Congress to "lay and collect Taxes, Duties, Imposts and Excises." This clause makes the federal government fiscally independent of the state governments. Article VI states that "This Constitution, and the Laws of the United States which shall be made in pursuance thereof, under the Authority of the United States, shall be the supreme Law of the Land; and the judges in every State shall be bound thereby, any Thing in the constitution or Laws of any State to the Contrary notwithstanding." This clause renders enactments of the federal government legally superior to those of the states. Together, these two clauses make the federal government substantially independent of the state governments.

Next the Framers set about to make the new federal government less susceptible to democratic pressures emanating from the people at large than the state legislatures had been under the Confederation. Their first tactic was to balance the power of the department of government most exposed to democratic pressures, the lower house of the federal legislature, with a sufficiently powerful upper house, of different composition, which would be insulated from the democratic pressures affecting the lower house. This last the Framers accomplished by placing the election of members of the upper house at one remove from the people at large—into the hands of the state legislatures, and by electing those members for terms of six years, rather than the two years served by members of the lower house.

As a further check on the power of the federal legislature, and especially its lower house, the Framers devised an office of chief executive, to be elected indirectly like members of the Senate, but by electors chosen as the respective state legislatures should direct. This chief executive would exercise a qualified power to veto the enactments of the federal legislature.

A final check on legislative power would come from the new, independent federal judiciary. To doubly insulate this judiciary from democratic pressures, they would be appointed by the chief executive, subject to approval by the upper house of the federal legislature, and they would serve for life. Whether the Framers intended the federal judiciary to act as a check upon the *federal* legislature is a matter about which experts disagree. That they intended it to be a check upon the democracy-prone legislatures of the states is beyond dispute.

Having made their new federal government independent of and superior to the governments of the states, and having insulated it from democratic pressures emanating from the people at large, the Framers then transferred from the states to the federal government exclusive control over certain areas of political endeavor. The first of the powers over which the Framers gave the federal government exclusive control was the power to carry on national defense and to conduct America's foreign relations. This was a logical power to invest in the central government, and it was the Framers' answer to the first of the problems under the Articles that the Framers had set out to solve.

The next two powers which the Framers transferred from the states to the federal government were the powers to regulate foreign and interstate commerce. To an important extent, the transferring of both of these powers to the federal government was done in the interest of securing individuals' freedom to acquire and dispose of property. Many of the Framers believed that the United States would never achieve free trade with other nations until it had the power to retaliate in kind against other nations' trade barriers. Unifying and making exclusive in the federal government the power to regulate the nation's foreign trade was thus, in the minds of these men, a means of breaking down foreign, especially British, trade barriers.

In the event, this would prove to be a double-edged sword, as effective at erecting American trade barriers as it was at breaking down foreign ones. Americans' freedom of commerce would have been better served had the Framers simply prohibited any regulation of foreign trade by either the states or the federal government.

On the other hand, the transferring from the states to the federal government of the exclusive power to regulate trade between the states has had incalculably beneficial results. Nearly a hundred years would pass before the federal government would exercise this power to any great extent. And with the states thus prohibited from intervening in this area, Americans would enjoy almost a century during which there was virtually no government intervention in interstate trade.

The transferring of these two powers to the federal government was the Framers' answer to the second and third of the problems blamed on the Articles of Confederation, the problems of artificial obstacles to foreign trade and of state-sponsored obstacles to interstate trade.

The last of the major problems the Framers intended to address, the problem which most immediately gave rise to the Constitutional Convention, was the tendency of the state legislatures to pass enactments violative of the rights of property, enactments such as moratory and legal tender laws. One means that the Framers employed to solve this problem was, as they did in the cases of foreign and domestic trade, to transfer certain powers from the states to the federal government. In this case, the powers in question amounted to control over the issuing of money. The Constitution prohibits the states from coining money, from issuing paper money ("Bills of Credit"), and from legislating that anything other than gold or silver should be accepted as legal tender.

Sound money is necessary for the carrying on of trade on any level above primitive barter. It is indispensable in a system with the highly developed division of labor that characterizes capitalism. In the interest of sound money, the Framers deprived the democracy-prone states of all control over money and vested it exclusively in the now less-democratic federal government. The Framers' concern to provide for sound money is an unmistakable indication of their intention to establish a frame of government capable of maintaining an economy based on private property.

The Three Powers of Government

As we have seen, the Framers' primary strategy for securing individual liberty, and especially the rights of property, against government infringement consisted of the structure of the government they devised. Part of that structure, the U.S. Congress, clearly resembles the English Houses of Parliament. The ruling principle of the English system is the supremacy of the legislative branch; Parliament is virtually unchecked in its power, save for its ultimate need to answer to the voters at the ballot box. The American Framers intended for the U.S. Congress to be the ultimate power in their system of government as well, but they were determined to place limits on the power of Congress that Parliament was not subjected to. The qualified veto power of the American president is one such check.

The English invented limited government. The "Rights of Englishmen," such as the *habeas corpus* and trial by jury, placed limits on what the government could do in enforcing the law against transgressors. But these rights were not judicially enforceable against acts of Parliament; had Parliament decided to abolish, say, the right to a jury trial, no British court could have declared Parliament's action unconstitutional and therefore void. The Americans, in this respect, gave new meaning to the term "limited government."

The first step toward improving on the English form of limited government was, of course, the idea of a written constitution as the fundamental law of the land. The next step was to render the written constitution enforceable by courts as against the acts of legislatures. Whether the Famers of the U.S. Constitution intended for its clauses to be court-enforceable as against Congress itself is problematic. But there is no doubt that they intended those clauses to carry, as against Congress, at least the weight that the traditional "Rights of Englishmen" carried under the unwritten English Constitution. There also is no doubt that the Framers intended the provisions of the U.S. Constitution to be enforceable by federal courts against acts of the *state* legislatures. In this scheme, certain clauses of the Constitution are, therefore, crucial to the securing of the rights of property, especially against the kind of the legalized looting by the state legislatures that had characterized the Confederation period.

"Congress shall make no law respecting an establishment of religion, or prohibiting the free exercise thereof." This one sentence bars the federal government entirely from regulating the sphere of religion. It ensures that that sphere will be one entirely of private, voluntary relations. It does not bar government from performing its proper function within that sphere; if a sect were to forcibly abduct new recruits, government would be free to intervene, liberate the abductees, and retaliate against the abductors. This provision of the First Amendment bars government from *initiating* the use of force against individuals in the religious sphere. It is a crucially important limitation upon the power of government, and it is a perfect example of how to limit that power within a specific sphere of human endeavor.

But the governments of the United States, federal, state, and local, have never been fully and properly limited. The First Amendment effectively (more or less) prohibits those governments from interfering with men's rights to think as they choose and to communicate their thoughts freely. But the Constitution does not now, and never has, fully and effectively prohibited governments from interfering with men's rights to *act* as they see fit by securing men's rights of property. The simplest and most direct way to achieve this would be a simple prohibition: "Neither Congress nor the states shall make any law respecting private property."

But there are two factors that make the securing of property rights against government a much more complicated matter than the securing of religious freedom is. First, government has a role in adjudicating disputes in the realm of property and contract rights that it does not have in the realm of religion. The second factor is the existence of three "powers" of government that touch upon property rights. These are the powers of taxation, eminent domain, and police. All three are a legacy of pre-Lockean conceptions of government.

The power to tax is the power to expropriate the wealth of citizens for the purpose of paying the expenses of the government. The premise underlying the idea of *forcible* taxation is that, because every individual enjoys the benefits of government, each may be compelled to contribute to paying the cost of those benefits. The power of eminent domain is the power of government to expropriate the property, usually land, of specific individuals. Governments expropriate land for public buildings, military installations, dams, bridges, and roadways, among other things. The premise behind *forcible* expropriation from specific individuals is that the need of the public takes precedence over the needs or wants of individuals.

Thomas McCaffrey

The third "power" of government that touches property is the police power. Whereas the taxing and eminent domain powers are relatively well-defined, the police power is ill-defined, amorphous, and, today, nearly all-encompassing. One way to define it, as it touches property rights, is as a power to do almost anything except what the taxing and eminent domain powers do. Traditionally, the police power existed to enact statutes for the public health, safety, morals, and welfare. It derives, in part, from the English common law of nuisance, under which the government could prohibit certain activities, such as operating a tannery in a residential district, which had the *potential* to violate the rights of individuals by such means as sending invasive odors or fumes or noise onto their property.

In addition to guarding health, safety, and morals, English law also regulated the production and distribution of wealth. Today in the United States both fields of regulation, health, safety, and morals, on the one hand, and the production and distribution of wealth, on the other, are subsumed under the police power. But as recently as 1904, according to Ernst Freund, the applicability of the police power to the latter sphere was still "debatable." Wrote Freund:

> *Broadly speaking, there are therefore three spheres of activities, conditions and interests which are to be considered with reference to the police power; a conceded sphere affecting safety, order and morals, covered by an ever-increasing amount of restrictive legislation; a debatable sphere, that of the proper production and distribution of wealth, in which legislation is still in an experimental stage, and an exempt sphere, that of moral, intellectual and political movements, in which our institutions proclaim the principle of individual liberty.*[39]

58

Today the first and second of these spheres are very much within the purview of the police power. It is difficult to grasp the present range of this power without enumerating at least a few of the kinds of regulation included in it: zoning ordinances, building codes, licensing and permit laws, public health regulations, vice laws, traffic laws, local and state labor laws, local and state minimum wage laws, anti-noise ordinances, anti-pollution laws and regulations of all sorts, a wide range of laws and regulations pertaining to wildlife, local and state statutes regulating the sale of alcohol and tobacco, laws regulating the sale and use of firearms, fire regulations, seatbelt and bicycle helmet laws, laws regulating food processing, and local and state laws requiring that commercial enterprises make physical accommodations for the disabled.

Whereas taxation and eminent domain are powers to expropriate property entirely, the police power does not do this. What the police power takes from individuals is not property, strictly speaking, but liberty. It is the restricting of individuals' liberty to *use* their property (and sometimes their liberty to dispose of it) that the police power affects.

One interpretation of the meaning of the term "police power" (there are many) excludes from its purview any concern with actions that unequivocally constitute violations of individuals' rights. Under this interpretation, for example, murder, rape, and robbery are not police power matters. Wrote Freund in *The Police Power*:

> *The peculiar province of the criminal law is the punishment of acts intrinsically vicious, evil, and condemned by social sentiment; the province of the police power is the enforcement of merely conventional standards, so that in the absence of legislative action, there would be no possible offense.*[40]

59

Freund's point is that police power regulations, such as an ordinance prohibiting the sale of alcoholic beverages on Sundays, prohibit actions that do not necessarily constitute violations of any specific individual's rights. Even at their best, police power regulations grounded in the common law of nuisance prohibit actions that would constitute violations of the rights of individuals only *if* those individuals objected to the invasion of their property. Foul odors from a tannery, for example, constitute trespass-like violations only if the owner of property invaded by the odors objects to the invasion of his property.[41]

All three of these powers, taxation, eminent domain, and police, are in essence powers for government to *commence* the use of force against individuals who have not necessarily violated anyone's rights. (Recall Locke's—correct—argument that the powers possessed by government are only the powers that the individuals who compose a polity would possess if there were no government. While each individual in the state of nature is morally entitled to use force to defend himself and to retaliate against those who initiate the use of force against him, no one is morally entitled to tax his neighbors, to "take" their land for the community's use, or to punish his neighbors for acts that do not constitute invasions of his person or his property.)

The existence of these powers authorizing government to violate the property rights of persons who themselves have not violated the rights of any specific individuals makes the task of securing those rights against government infringement vastly more difficult than the task of securing freedom of religion or speech, which are not subject to such powers. But the need to maintain these three powers has been an unquestioned assumption of Western political theory. In a polity which admits these powers, as all do, the securing of the rights of property is, of necessity, a matter of keeping their exercise within strict limits.

All three powers were indeed limited, in one way or

another, under the Constitution as the Framers intended it. Section 9 of Article I limits the federal taxing power by prohibiting "Capitation, or other direct, Tax[es] unless in proportion to the Census of Enumeration." [42] This had the effect of prohibiting the imposition of an income tax until the passage of the Sixteenth Amendment in 1913. Article I, Section 8 includes the clause, "The Congress shall have Power To lay and collect Taxes, Duties, Imposts and Excises, to pay the Debts and provide for the common Defence and General Welfare of the United States." This clause can be, and has been, interpreted to limit the purposes for which Congress may tax.[43] Indeed, a requirement that federal taxes be used only for the *general* welfare would clearly prohibit the use of tax money to provide transfer payments to private *individuals*, as the modern welfare state does. But this limitation is a dead letter today. As to the taxing powers of the individual states, neither the Constitution nor the Bill of Rights limits these in any way.

The Framers' intentions as to the limiting of the other two powers, eminent domain and police, require more effort to discern. There is the clause in the Fifth Amendment which states "nor shall private property be taken for public use, without just compensation." This limitation of the federal government's power of eminent domain is clear and unmistakable. But it is not, of course, the work of the Framers of the original Constitution, who had not intended to include a bill of rights. This does not mean, however, that the Framers intended this power to be altogether unlimited, for, according to William Stoebuck, the paying of compensation for eminent domain takings was already standard practice in the United States in 1789. Says Stoebuck, the paying of compensation was standard procedure in England at the time of the Revolution, and it had also been the rule in the American colonies, as well. (The formalizing of the "public use" limitation was an American innovation.)[44]

Two of the state bills of rights contained compensation clauses before the adoption of the federal Bill of Rights. And, as explicit written limitations in constitutions and bills of rights eventually became indispensable to the defense of individual rights in the early decades of the nineteenth century, every state constitution but one would come to include a compensation clause. The one state that never adopted such a clause, North Carolina, has always paid compensation as a matter of common law. So the Framers would have had reason to believe that the federal government's power of eminent domain would indeed be limited, even though they did not themselves explicitly limit it in the Constitution. (We can interpret the inclusion of the "takings" clause in the Fifth Amendment by the First Congress as an effort to solidify the position of this limitation by formally giving it constitutional status.)[45]

Ex Post Facto

As for the third of the "three powers," The Framers did include an explicit clause in the body of the Constitution which, had it been interpreted as intended, could have become a powerful limitation on the police-regulatory powers of the states. (The federal government does not possess police powers, *per se*, although the Commerce Clause has been interpreted—improperly—to authorize the exercise of vast police-like powers by the federal government.) Indeed, so powerful was the limiting potential of this clause that it contributed to the clause's evisceration as a limitation on legislative power the very first time the clause was used in court.

The clause in question reads, "No state shall ... pass any ... *ex post facto* Law." [46] (A similar clause, Article I, Section 9, limits Congress in the same way.) An *"ex post facto"* law is a law passed after the fact, a law which renders a person liable for an act that was legal at the time it was committed. It was common to interpret "*ex post facto* law" during the constitutional period to include laws which, after the fact, attached to some act which had been legal when it was committed what *amounted to* a penalty, even though the intent of the act was not to penalize. Such, for example, was Rhode Island's legal tender act of 1786. It required creditors to accept as payment for debts—contracted before the passage of the act—paper money of lower value than the money they had lent out. William Crosskey provides evidence that the term "*ex post facto*," as used in the Constitution, was indeed understood to apply to such legislation.[47]

There is evidence that at least some types of *ex post facto* laws were prohibited under the common law. Two members of the Constitutional Convention suggested as much in their objections to including the clause in the Constitution. Oliver Ellsworth, a future Chief Justice of the U.S. Supreme Court, said, "there was no lawyer, no civilian who would say that ex post facto laws were not void of themselves." James Wilson, a future Associate Justice of the same court, argued that inclusion of the clause would "bring reflections on the Constitution—and proclaim that we are ignorant of the first principles of Legislation." [48]

The 1792 Virginia Supreme Court case of *Turner v. Turner's Executrix*, among others Crosskey cites, also supports the contention that at least some types of *ex post facto* laws would have been prohibited regardless of whether there had been a clause to that effect inserted into the Constitution.[49] The Framers' explicit prohibition of such enactments by both the federal and the state legislatures suggests, therefore, that they considered the passing of *ex post facto* laws to be a serious problem that they were determined to bring to an end.

But in the Federal Convention, George Mason had raised an objection to including the clause in the Constitution which eventually would prove decisive for the effectiveness of the clause itself as a limitation on the power of legislatures to infringe property rights. "There never was nor can be a legislature but must and will make [ex post facto] laws, when necessity and the public safety require them." [50] Mason worried that the normal functioning of government required the passing of at least some *ex post facto* statutes, and that these inevitable violations of the Constitution's *ex post facto* clauses would provide a precedent for the eventual violation of other clauses. Mason proved prescient. The very first time a suit came before the U.S. Supreme Court on *ex post facto* grounds, in the famous case of *Calder v. Bull*, the Court interpreted the *ex post facto* prohibition to apply only to legislative enactments that would render a previous act *criminal*, and not to civil enactments such as would merely deprive an individual of legally acquired property rights. [51]

The evidence suggests that the Court's interpretation of *ex post facto* was incorrect, and that the clause should indeed have applied to legislative enactments that "punished" innocent people by dispossessing them of legally acquired property. [52] Thirty-one years after the *Calder* decision, for example, Justice William Johnson came to the question afresh, conducted an exhaustive review of the Court's reasoning in the *Calder* case, and concluded that their ruling was incorrect. Wrote Justice Johnson:

> This court has had more than once to toil up hill in order to bring within the restriction of the states to pass laws violating the obligation of contracts, the most obvious cases to which the Constitution was intended to extend its protection; a difficulty which it is obvious might often be avoided by giving to the phrase ex post facto its original and natural application. It is then due to the venerable men whose opinions I am combating, to believe that had this and many other similar cases which may occur and will occur, been presented to their minds, they would have seen that in civil cases, the

restriction not to pass ex post facto laws could not be limited to criminal statutes, without restricting the protection of the constitution to bounds that would import a positive absurdity.[53]

Crosskey argues that the reasons behind the Court's evisceration of the *ex post facto* clause in *Calder v. Bull* were political. But the opinion of James Iredell in that case suggests another motive at work, the one broached by George Mason in the Constitutional Convention. Wrote Iredell:

The policy, the reason and humanity, of the [ex post facto] prohibition, do not, I repeat, extend to civil cases that merely affect the private property of citizens. Some of the most necessary and important acts of Legislation are, on the contrary, founded upon the principle, that private rights must yield to public exigencies. Highways are run through private grounds. Fortifications, Lighthouses, and other public edifices, are necessarily sometimes built upon the soil owned by individuals. In such, and similar cases, if the owners should refuse voluntarily to accommodate the public, they must be constrained, as far as the public necessities require; and justice is done, by allowing them a reasonable equivalent. Without the possession of this power the operations of Government would often be obstructed, and society itself would be endangered. It is not sufficient to urge, that the power may be abused, for, such is the nature of all power,—such is the tendency of every human institution: and, it might fairly be said, that the power of taxation, which is only circumscribed by the discretion of the Body, in which it is vested, ought not to be granted, because the Legislature,

> *disregarding its true objects, might, for visionary and*
> *useless projects, impose a tax to the amount of nineteen*
> *shillings to the pound.* [54]

Iredell implies that to admit a civil application of *ex post facto* would be, in effect, to prohibit all legislation detrimentally affecting property rights. He suggests that such an interpretation would even threaten to extinguish the power of eminent domain. More broadly, it would have threatened the existence of all of the government's powers to override "private rights" for the sake of "public exigencies."

Carried to a certain level, the logic of *ex post facto* could have disallowed all legislative enactments that caused a diminution in the value of property, such as a zoning ordinance that prohibited the commercial use of a parcel of land especially suited to such use. It could have disallowed any regulation that required positive expenditures by property owners, such as a statute requiring the upgrading of buildings to accommodate the disabled, or a statute requiring the installation of certain safety features. Interpreted even more broadly, *ex post facto* could have disallowed any statute that deprived an individual of *any* right that had been legal when the act was passed. Followed to its logical conclusion, a civil application of the *ex post facto* prohibition could have meant the virtual abolition of the police power altogether. This was apparent to Justice Iredell, and it figured in his decision to eviscerate the clause by restricting its application to criminal matters.

The problem with the *ex post facto* limitation was that, in a polity committed to the proposition *"that private rights must yield to public exigencies,"* it was too powerful a defense of those rights. Iredell went on to say in his *Calder* opinion that "We must be content to limit power where we can, *and where we cannot, consistently with its use*, we must be content to repose a salutary confidence."[55] He held, in effect, that it was not possible to limit the government's police power in any way that would be consistent with its continued existence as a viable power of government. In the next section we shall encounter a jurist who did devise a way to limit the police power in a manner "consistent with its use." But first a final word on Iredell.

If one wishes to protect property rights from legislative infringement while, at the same time, admitting the validity of government "powers" to override those rights, then one must place well-defined limits on those powers, such as the Fifth Amendment requirement that compensation be paid for takings and that those takings be limited to public purposes. To admit government powers such as taxation and police without placing strict limitations on them is virtually to abolish the inviolabity of the rights of property. It is to render property rights merely statutory in nature, subject to the whims of the next legislature.

The question of the inviolablity of property rights has never been settled in the United States but has been the subject of debate from the beginning. Iredell was not only comfortable with the idea of an unlimited power of taxation, he used the absence of such limitation to support his argument for making other government powers, especially the police power, unlimited as well. His 1798 opinion in *Calder v. Bull* thus represents an early argument against rendering property rights constitutionally inviolable. He held that only those rights explicitly enumerated in the Constitution or the Bill of Rights should be judicially enforceable. But the Framers of the Constitution and the authors of the Bill of Rights had never intended their lists of rights to be exhaustive, and the Ninth Amendment attests to this. ("The enumeration in the Constitution, of certain rights, shall not be construed to deny or disparage others retained by the people.") Unenumerated rights are every bit as valid as enumerated ones. If the courts are to enforce the latter, then logic requires that they enforce the former as well. Property rights are, far and away, the most important of the "unenumerated" rights.

Kent

Eminent domain is the power of government to seize full title to private property. When the Framers met at Philadelphia in 1787, compensation for eminent domain takings was already established practice in England and America. The First Congress gave the compensation requirement constitutional status by including it in the Fifth Amendment of the Bill of Rights. Subsequently, in the early decades of the nineteenth century, all of the states but one inserted a compensation requirement into their own constitutions. Throughout U.S. history, governments at all levels have more or less consistently fulfilled the requirement that they compensate property owners for eminent domain takings, at least as far as outright takings of full title to property are concerned. To this extent, this requirement has proved to be the model of how to protect property rights against a "power" of government.

There is another kind of "taking," one that leaves the owner of private property in full physical possession of it but which damages the property or diminishes its value in some way. An example, common in the 1830s, would be a legislative authorization for a private party to erect a dam that caused his neighbors' farmland to be flooded. The owner of such permanently flooded land was in no way deprived of title to his property, but its utility, and value, were virtually destroyed. Such "partial" takings are effected not by the power of eminent domain but by that of police, though their similarity to eminent domain takings is manifest. The U.S. Supreme Court did not recognize the need to compensate such takings until 1871.[56] Yet this type of partial taking might well have been prohibited altogether had *Calder v. Bull* not limited the reach of the *ex post facto* clause. But we have seen why *ex post facto* was rejected as a limitation on legislatures' power to effect such partial takings. An alternative approach to limiting legislatures' power to effect such takings would have been to require that they be compensated. This was the approach championed by New York State Chancellor, James Kent

In 1823 James Kent began publishing his *Commentaries*, a series of law lectures delivered at Columbia University. These were destined to exert a great influence on the development of American Law. Kent was an eloquent advocate of the idea of property as a natural right.[57] Although he held that "private interests must be made subservient to the general interest of the community," he was, in the early decades of the Republic, the nation's most systematic champion of placing strict constitutional limitations on governments' powers of taxation, eminent domain, and police.[58] (That "private interests must be made subservient to the general interest of the community" flatly contradicts the logic of Lockean *individual* rights. But this premise underlies the existence of the three powers of government, and every effort to make property rights secure from the Revolution onward would have to contend with it, as Kent does here.)

As to the power of "regulation," as he referred to the states' police powers, Kent would have limited this largely to the existing common law of nuisance.[59] "Unwholesome trades, slaughter houses, operations offensive to the senses, the deposit of gunpowder, the application of steam-power to propel carts, the building with combustible materials, and the burial of the dead, may all be interdicted by law, in the midst of dense masses of population."[60] To restrict the police power to nuisance-abatement would, in itself, have proven a hugely important limitation on that power. Kent goes on to suggest that if a legislature exceeded this limitation "to the destruction of existing property values," then it should be required to compensate the owners, even though the "power" involved was not that of eminent domain but of police.[61] Said Kent, "If A. be the owner of a mill, and the legislature authorizes a diversion of the watercourse which supplies it, whereby the mill is injured, is that not a consequential damage to be paid for? The solid principle is too deeply rooted in law and justice to be shaken." [62]

The U.S. Supreme Court would adopt Kent's principle insofar as physical invasions of property were concerned in the 1871 case of *Pumpelly v. Green Bay Co.*[63] The Court would then extend the requirement for compensation to non-invasive "damages" consequent to police regulations in the 1922 case of *Pennsylvania Coal Co. v. Mahon.*[64] Since 1922, courts have adhered to Mahon only unevenly, although two recent cases hold promise of a more consistent application of the principle enunciated in *Mahon* that compensation may be required when police regulations go "too far" in diminishing an owner's rights in a property.[65]

Kent's proposed limitation on the taxing power is also worth noting here.

> *Every person is entitled to be protected in the enjoyment of his property, not only from invasions of it by individuals, but from all unequal and undue assessments on the part of government. It is not sufficient that no tax or imposition can be imposed upon the citizens, but by their representatives in the legislature. The citizens are entitled to require that the legislature itself shall cause all public taxation to be fair and equal in proportion to the value of property, so that no one class of individuals, and no one species of property, may be unequally or unduly assessed.*[66]

Kent fundamentally opposed the use of the taxing power to redistribute wealth in the interest of "equality."

The Doctrine of Vested Rights

The ideal outcome of Kent's efforts to limit the police power would have been the inclusion of explicit clauses in the federal and state constitutions limiting that power to enforcing the common law of nuisance and requiring compensation for all regulations that went beyond nuisance abatement and caused a "taking" by diminishing rights of ownership. But, although Kent's ideas on compensation for some types of "partial" takings would find their way into constitutional law, his idea of limiting the police power to nuisance abatement ran head-on into a counter movement that I shall describe later in this chapter. Meanwhile, though, the need for some sort of limitation on legislatures' regulatory powers had been apparent even as *Calder v. Bull* was eliminating the potential of the *ex post facto* clause as such a limitation. And by the time that Kent began publishing his *Commentaries*, there was already developing an alternative limitation.

With the *Calder v. Bull* decision, the police power was left virtually unlimited. This was not lost on Justice Samuel Chase, who dissented in *Calder*:

> *I cannot subscribe to the omnipotence of a state legislature, or that it is absolute or without control, although its authority should not be expressly restrained by the constitutional or fundamental law of the state.... The nature and ends of legislative power limit the exercise of it.... There are acts which the federal or state legislatures cannot do without exceeding their authority.... An Act of the legislature (for I cannot call it a law) contrary to the great principles of the social compact cannot be considered a rightful exercise of legislative authority.... A law that punished a citizen for an innocent action, or, in other words, for an act, which, when done, was in violation of no existing law;*

> *a law that destroys, or impairs the lawful private
> contracts of citizens; a law that makes a man a judge in
> his own case; or a law that takes property from A and
> gives it to B; it is against all reason and justice for a
> people to entrust a legislature with such powers; and
> therefore it cannot be presumed that they have done it.
> The genius, the nature, and the spirit of our state
> governments amount to a prohibition of such acts of
> legislation; and the general principles of our law and
> reason forbid them.*[67]

Chase argues here that legislatures do not have the power to infringe the legally acquired rights of individuals. He implies that courts ought to void enactments that infringe such rights, and he would base the courts' authority to do so on "the nature and ends of the legislative power," on "the great principles of the social compact," on "reason and justice," and on "the general principles of law and reason." He argues, in other words, for judicial limitation of legislatures' powers on a basis *other than* the explicit limitations contained in written constitutions. Edward S. Corwin interprets Chase's dictum as a call to the state legislatures to fill the void left by the evisceration of the *ex post facto* clause. Corwin reports that in the early decades of the nineteenth century the state courts responded with a flurry of decisions voiding legislative enactments infringing legally vested property rights, and often citing fundamental, extra-constitutional principles, as opposed to explicit constitutional clauses, as a basis for their decisions."[68]

This limiting of legislatures' powers began with the state courts' disallowing incursions by the state legislatures into matters essentially judicial in nature. Says Corwin,

> *Thanks to notions inherited from Colonial days, which
> were confirmed by the prevalent analogy between the*

State legislatures and the British Parliament, these bodies [the state legislatures] were prone during the early years of our constitutional history, and some of them for many years afterward, to all sorts of "special legislation" so called; enactments for revising or setting aside court decisions, for suspending the general law for the benefit of named individuals, for interpreting the law for particular cases, and even for deciding cases.[69]

Calder v. Bull, for example, had involved an act by the Connecticut Legislature overturning the decision of a probate court. Corwin says that following the *Calder* decision, the state courts actively set out to bring an end to the state legislatures' interfering in judicial matters, especially in cases involving property rights, and that the courts often cited such fundamental, extra-constitutional principles as Chase had mentioned in his dictum to justify their actions.[70]

According to Corwin, these early efforts by the state courts marked the beginning of a long-lived judicial doctrine aimed at protecting property rights from legislative interference. This body of judicial ideas Corwin christened the Doctrine of Vested Rights, which he termed "the most prolific single source of constitutional limitations of any concept of American constitutional law." He defines the doctrine as the idea that *"the effect of legislation on existing property rights was a primary test of its validity*; for if these were essentially impaired then some clear constitutional justification must be found for the legislation or it must succumb to judicial condemnation."[71] In other words, the courts' Doctrine of Vested Rights placed the burden on the government to justify, by strict constitutional standards, its own enactments any time they interfered with individuals' rights of ownership. Or, to put it differently, freedom to possess, use, and dispose of one's property would be the rule, and government infringement of that freedom would be the exception. The Doctrine of Vested Rights came into being around the beginning of the nineteenth century, and it continued as an active force in American jurisprudence into the New Deal.

The Contract Clause

The Doctrine of Vested Rights originated in the state courts and was grounded largely in general principles of reason, justice, and the separation of powers, as opposed to explicit clauses of the state constitutions and bills of rights. It eventually found its way into federal jurisprudence, this by way of a specific clause in the U.S. Constitution. During the Federal Convention, several of the Framers had argued, anticipating the *Calder v. Bull* decision, that *ex post facto* did not apply to civil matters, such as paper money laws, but only to criminal matters. Benjamin Fletcher Wright has argued that it was precisely in case *ex post facto* were ever held to apply only to criminal matters that the Framers included the Contracts Clause in the Constitution.[72] This clause prohibits the states from "impairing the Obligation of Contracts," and its potential application to the notorious legal tender laws of the Confederation period is obvious. The Contract Clause would become, in the early decades of the nineteenth century, an important limitation on the ability of the state legislatures to infringe property rights.

The primary moving force behind the judicial use of the Contract Clause to protect property rights was John Marshall, Chief Justice of the U.S. Supreme Court from 1801 to 1835. Marshall used the clause to block the Georgia Legislature from rescinding grants of land it had made to private individuals; he used it to block the New Hampshire Legislature from altering a corporate charter it had issued to Dartmouth College; and he used it to block the New York Legislature from aiding debtors at the expense of creditors through a retroactive bankruptcy law. The protection of property rights became, much through Marshall's efforts, a central concern of the federal courts, a development that would have occurred much earlier had the *Calder* court not eviscerated the *ex post facto* clause. More than any other individual, Marshall gave effect to the Framers' intention that the federal government act as a check upon the power of the state legislatures to infringe the property rights of individuals.

Another instrument that Marshall made use of in the interest of economic freedom was the clause in Article I, Section 8 of the constitution, which states that "The Congress shall have power ... To regulate Commerce ... among the several States." In the famous case of *Gibbons v. Ogden* Marshall, arguing that the clause meant that *only* Congress could regulate interstate commerce, used it to invalidate Robert Livingston's steamboat monopoly, which had been authorized and enforced by the state of New York. "The right of intercourse between State and State ... derives its source from those laws whose authority is acknowledged by civilized man throughout the world.... The constitution found it an existing right, and gave to Congress the power to regulate it." [73] In this instance Marshall used the commerce clause to abolish state-sponsored restrictions on freedom of commerce. [74] Marshall's idea of Congress's "promoting" commerce was thus limited to removing regulatory barriers to commerce, and he did not condone such *positive* measures as federal funding of "internal improvements" such as roads and canals.

It is interesting to compare Marshall's idea of "federalism" with that of Alexander Hamilton. Hamilton saw an invigorated federal government as a means to effecting certain *positive* goods, such as the transforming of the United States into an industrial power. Marshall, on the other hand, would limit the federal government's role in economic matters largely to effecting the *negative* good of preventing state interference with individuals' exercising of their rights, as he had done in striking down New York's state-sponsored steamboat monopoly. Where Hamilton favored a kind of alliance of commerce and government, Marshall advocated more of a separation of commerce and government. Marshall's conception was much more in line with that of the Lockeans among the Founders than was Hamilton's.[75]

Like Locke, Marshall believed that the right of private property rests on a foundation of "higher law." In the case of *Fletcher v. Peck*, in which the Supreme Court used the contract clause to void the Georgia Legislature's attempt to rescind grants of land it had authorized, Marshall identified the "rules of property" with "certain great principles of justice." He continued, "If any [limits to the legislative power] be prescribed, where are they to be found, if the property of an individual, fairly and honestly acquired, may be seized without compensation?"[76]

Marshall subscribed to Locke's idea that the right of private property originates in labor.

> That [the slave trade] is contrary to the law of nature will scarcely be denied. That every man has a natural right to the fruits of his own labor, is generally admitted; and that no other person can rightfully deprive him of those fruits, and appropriate them against his will, seems to be the necessary result of this admission.[77]

But Marshall goes Locke one better—by according natural right status to contracts as well. In *Ogden v. Saunders* he argued that the right of contract "results from the right which every man retains to acquire property, to dispose of that property according to his own judgment, and to pledge himself for a future act. These rights are not given by society, but are brought into it."[78]

The Bill of Rights

This concludes the discussion of the place of property rights in the Constitution itself. As for their place in the federal Bill of Rights, little need be said. The authors of this document did not include any mention of such abstract, *general* rights as life, liberty—or property. Instead, they focused their attention on such *specific* rights as freedom of speech, press, and religion; on certain specific limitations relating to the government's exercising of its proper powers of law enforcement, such as the prohibition against unreasonable search and seizure and the requirement for jury trial; and on a specific limitation of one of the three "powers" of government touching property rights, the power of eminent domain.

This last, of course, is the Fifth Amendment's requirement for compensation for "takings," and its limiting of such takings to "public purposes." As I have said, compensation was already standard practice in the states at the time of the ratification, and the inclusion of the takings clause in the Fifth Amendment is best understood as intended to solidify the status of this practice by making it a constitutional requirement.[79] The second of the three powers, the police power, was already limited under the Constitution's *ex post facto* clause (and the Contracts Clause), so it did not require limitation in the Bill of Rights. Likewise, the third of the three powers, taxation, was explicitly limited by the prohibition against income ("capitation") and other "direct" taxes. No one had yet conceived of any additional way to limit the taxing power. The Constitution and the Bill of Rights combined, therefore, contain all that was possible, in the context of the time, to complete the constitutional limitation of the three "powers" of government, and, thereby, to secure the rights of private property against government infringement.

Before drawing any further conclusions as to the place of property rights in the Constitution, I want to trace the development of judicial protection of property rights throughout the nineteenth century, because that development has a bearing on the conclusions to be drawn.

Democracy and the Police Power

At the time of the Constitutional Convention, property qualifications for voting and office-holding were common throughout the United States. The Framers chose not to alter these arrangements, deciding that the qualifications for voting for federal representatives would be whatever pertained in the respective states. But from the ratification onward, there was a steady lowering of these qualifications, affording an ever-larger proportion of the citizenry the opportunity to participate in voting and political office-holding at the state and federal levels. Samuel Eliot Morison says that with the defeat of the (Jeffersonian democratic) Republicans by John Quincy Adams in 1824, elements within the Republican Party expressly set about removing the few remaining barriers to voting by white males in the hope that an expanded electorate would tilt the balance both toward the Republican Party and away from federal dominance.[80] The result was a resounding victory for Andrew Jackson in 1828, and the beginning of a period of dominance by democratic elements within the states that would last almost unbroken until the Civil War.

One result of the newfound power of the democratic Republicans within the states was an increased demand for legislative activism, often at the expense of property rights. As Corwin notes, demands arose for free schools, and for "internal improvements," such as roads, canals, and railroads, all financed or aided by state legislative action.[81] During this period, governments and private concerns made liberal use of the power of eminent domain for such "public" endeavors as railroads and milldams. Also, notes Corwin, the era saw the rise of a number of moral crusades, including women's rights, abolitionism, and prohibitionism, all of which, to one extent or another, threatened vested property rights. Amid this growing demand for the use of legislative power, democratic elements within the states, reinforced by the growing "states' rights" movement, opposed the intervention of the federal courts to protect the property rights of individuals against the state legislatures.

Under the influence of Jackson-appointed Supreme Court justices, especially Chief Justice Roger Taney, James Kent's strictly limited "power of regulation" became the vastly expanded "Police Power," a power that gave the state legislatures complete territorial sovereignty within their own borders.[82] Said Taney in the famous *Charles River Bridge* case:

> *The object and end of all government is to promote the happiness and prosperity of the community by which it is established; and it can never be assumed, that the government intended to diminish its powers of accomplishing the end for which it was created.... While the rights of property are sacredly guarded, we must not forget that the community also have rights, and that the happiness and well-being of every citizen depends on their faithful preservation.*[83]

Under this formulation, the purpose of government becomes the achieving of a positive, "to promote the happiness and prosperity of the community" by positive intervention in its economic life, as opposed to the essentially negative purpose of punishing the use of force and fraud in order to enable individuals to pursue their own happiness and prosperity privately. As Corwin points out, this new conception of the police power signified:

> ...*that legislation affecting property rights detrimentally must nevertheless be judged from the point of view that the legislature intended thereby to promote the public interest, not to punish the holders of the said vested rights; and that in the absence of specific constitutional provision to the contrary the public interest was ordinarily entitled to prevail against such vested rights.*[84]

In the passage above quoted from Taney's *Charles River Bridge* opinion, there is also the novel conception of *collective* "rights," which stand in contrast to the Lockean conception of *individual* rights. And we have the corollary of such a conception, the idea that, as a general rule, individual rights must give way to "collective rights," except in those instances in which explicit constitutional clauses protect individual rights, as in the case of the "obligation of contracts." Taney's conception of the police power stands in sharp contrast to the Doctrine of Vested Rights, which held that the integrity of individual rights should be the rule, and their legislative impairment in the name of the public good the exception. The one represented largely unfettered legislative power, the other strictly limited legislative power. Taney makes clear, says Corwin, that the *only* limitation that the federal courts could enforce against the police powers of the state legislatures was contained in the Contract Clause. There would be no more limiting of that power on the basis of the social contract, natural rights, or any other of the unwritten principles of fundamental law upon which the state courts had erected the Doctrine of Vested Rights.

James Iredell had argued back at the time of *Calder v. Bull* that the only rights that courts could protect against legislatures are those that are explicitly enumerated in constitutions or bills of rights, much as the Fifth Amendment protects the right to compensation for eminent domain takings by the federal government. With his *Charles River Bridge* decision, Taney gave formal recognition to Iredell's viewpoint. But at the time that the First Congress was drawing up the Bill of Rights, it was not yet clear what role the explicit clauses of that document would play in securing men's rights. Specifically, it was not yet clear that the Supreme Court would eventually take it upon themselves to enforce the clauses of the Bill of Rights against Congress itself. Had this been clear, then the authors of the Bill of Rights might well have attempted a more exhaustive enumeration of the rights which the Supreme Court could thus protect.

But the authors of the Bill of Rights did not attempt such an exhaustive enumeration of individuals' rights, and they explicitly acknowledged this fact in the Ninth Amendment. When the Supreme Court later did assume the responsibility of securing individuals' rights as against Congressional enactments, it was important that the Court protect *all* the rights which individuals possessed, not just those enumerated in the Bill of Rights. The Doctrine of Vested Rights thus represented a constitutionally justified effort to afford judicial protection to certain fundamental, though unenumerated, rights, especially property rights.

Popular Sovereignty vs. Vested Rights

In its initial form, resting upon extra-constitutional, general principles, the Doctrine of Vested Rights reached the peak of its influence about 1830, says Corwin. But with the rise of Jacksonian democracy, a new, competing set of ideas dictated that the doctrine assume a new form. Corwin refers to this new set of ideas as the Doctrine of Popular Sovereignty. John Marshall had written of this phenomenon earlier. The Doctrine of Popular Sovereignty was the idea that

> *[T]he people alone were the basis of government. All powers being derived from them, might, by them, be withdrawn at pleasure. They alone were the authors of the law, and to them alone, must the ultimate decision on the interpretation belong. From these delicate and popular truths, it was inferred, that the doctrine that the sovereignty of the nation resided in the departments of government was incompatible with the principles of liberty.*[85]

Recall that Locke had held that in order to be legitimate a government must satisfy two criteria: it must secure the natural rights of its individual citizens, and it must enjoy the consent of those citizens; Locke held both to be indispensable. But there is a tension between these two principles when it comes to putting them into practice. The idea that the legitimacy of a system of government originates in the consent of the people can easily be interpreted to imply that "the people" should have an ongoing role in the political process, specifically in the form of voting. Lockeanism, therefore, tends in practice to entail a high degree of democratic participation in the political process. This brings it into conflict with the obligation to secure the rights of individuals from majority rule.

Locke himself, as I have said, was fully aware that pure majority rule is incompatible with the securing of the rights of individuals. He held that the people may, and indeed ought to, divest themselves to some extent of their rights of political participation and institute a less democratic form of government. This, incidentally, is approximately what the Americans did when they ratified their Constitution. They did vote to institute a decidedly less democratic form of government than had existed under the Articles of Confederation. Their new government could more effectively secure the rights of individuals while at the same time laying claim to enjoying the consent of the governed.

The ratification of the U.S. Constitution thus represented a commitment to maintaining Locke's twin bases of political legitimacy. But the idea that "the people" are sovereign is a potent one. Any polity that subscribes to it will experience an ongoing, powerful tendency to expand the suffrage to include an ever-broader segment of the population. To whatever degree a polity submits to this tendency, to that extent will it find that the principle of individual rights as a basis of political legitimacy is threatened.

According to Corwin, the idea of natural rights reached the high point of its influence during the Revolutionary period. Thereafter it entered upon a period of decline. By the 1820s it had all but disappeared from political discourse. It did continue to exert an influence among the judiciary throughout the nineteenth century, but this was its last refuge as an active force. It was inevitable, under these circumstances, that the idea of individual rights as a basis of political legitimacy would also decline in influence.

Corwin reports that by 1830 or so the idea that anything to which the people formally give their consent is thereby rendered legitimate had gained ascendancy over the principle of individual rights as a basis of legitimacy.[86] One consequence of this, says Corwin, was that the explicit clauses of written constitutions, as the embodiment of the express will of the people, achieved preeminence as fundamental law, much to the detriment of unwritten principles of natural law. In consequence of this, the Doctrine of Vested Rights, which up until 1830 had rested largely upon such unwritten principles, needed to find a home in some specific clause or clauses of the written constitutions and bills of rights of the nation.

Thomas McCaffrey

Substantive Due Process

The use of a certain pair of constitutional clauses to protect property rights had begun in the late eighteenth century. At that time, judges began to use the "law of the land" clause contained in some of the state constitutions and, later, its historical equivalent, the "due process" clause, which appears in the Fifth Amendment of the U.S. Bill of Rights, to protect vested property rights from state legislative encroachments.

The "law of the land" clause derives from Magna Carta. It appears in the Massachusetts Constitution of 1780: "No subject shall be arrested, imprisoned, despoiled, or deprived of his property, immunities, or privileges, put out of the protection of the law, exiled, or deprived of his life, liberty, or estate, but by the judgment of his peers or the law of the land."[87] Corwin points out that the "due process" clause derives from Chapter 3, Statute 28 Edward III of 1355: "No man of what state or condition he be, shall be put out of his land or tenements, nor taken, nor imprisoned, nor disinherited, nor put to death, without he be brought to answer by due process of law."[88]

Corwin notes, citing Coke, that the "law of the land" and the "due process" clauses had virtually the same meaning, which is that no one may be punished for breaking the law unless he is first tried and found guilty according to certain prescribed *judicial* processes. This was the meaning accorded the "law of the land" clause in a 1794 case in which the South Carolina Supreme Court voided as a violation of the "law of the land" clause of the state constitution a legislative enactment which permitted a municipal court to levy a fine without a jury trial.[89]

Corwin reports that the "law of the land" clause first came to the service of the Doctrine of Vested Rights in 1804. *North Carolina v. Foy* involved a legislative enactment rescinding an earlier grant of lands to a university. The Supreme Court of North Carolina voided the enactment as depriving the university of title to property in violation of the law of the land, i.e., without proper *judicial* proceedings. In this instance, we see associated with the "law of the land" clause the assumption that any *legislative* act that deprived someone of legally acquired property rights constituted a punishment, and that only a court could administer punishment.

Now, from the standpoint of justice, any legislative act which impairs legally acquired property rights is indeed the *functional equivalent* of a punishment, regardless of whether it is intended as punishment. It is entirely just, therefore, to require that no government impairment of vested property rights may occur without proper judicial proceedings. On the other hand, the "law of the land" and "due process" clauses historically had applied only to instances in which punishment was the *intention* behind the government's depriving an individual of his rights. They applied, in other words, only to instances in which an individual had been accused of violating the law, and they stipulated that he could be punished for it only if he were tried and found guilty according to specified judicial procedures, such as a jury trial. It was an extension of the "law of the land" clause beyond its historically established application to employ it in a case in which the government's deprivation of rights was not intended as punishment for a violation of the law.

The "due process/law of the land" manifestation of the Doctrine of Vested Rights took a huge leap in the 1856 case of *Wynehamer v. New York*. For one thing, the New York court was far more influential than that of North Carolina. But *Wynehamer* also broke important new ground. The statute in question prohibited the selling of most intoxicating liquors, including existing stocks thereof. It also prohibited the storage of any existing stocks in any place but a residence. The statute also called for the summary destruction of such stocks upon discovery, and it provided for the trial and punishment of violators of these prohibitions. On the basis of the statute's effect upon existing stocks of liquor, the New York Supreme Court voided the statute as a violation of "due process."

Now, the summary destruction of private property without a judicial proceeding is indeed a violation of due process. But the justices objected to more than this element of the statute. For one thing, they judged the statute's diminishing of the *value* of existing stocks of liquor as a violation of due process as well. This objection applied to people who had not been prosecuted under the law, and who therefore had not been accorded due process in a judicial setting, but who nevertheless had had the value of their property severely diminished. They had indeed been deprived of most of a vested right without due process.[90] But this was the first time that the due process clause had been used to void a statute that did not divest individuals of *full* title to their property, but merely diminished its value. (Recall that Kent had argued that "indirect" and "consequential" damages to property from legislative enactments ought to be compensated.)

But even beyond this the *Wynehamer* judges held, in effect, that the statute's provision for trial and punishment for the possession of liquor that had been legally acquired also amounted to a violation of due process. They held, in other words, that an individual tried under this statute and accorded full judicial process before being deprived of his rights nevertheless had *not* been accorded due process if the property rights in question had been legally acquired before the passage of the statute.[91] Under this interpretation, "due process" would require not only that prescribed judicial proceedings occur before a vested right could be taken away, but also that the statute in question be ascertained not adversely to affect property rights legally acquired prior to the passage of the statute. *Wynehamer* interpreted "due process" to have both a procedural component and a substantive one, the one requiring that certain judicial procedures be followed, the other that certain preexisting rights not be adversely affected by the statute in question; thus the term "substantive due process" to describe this expanded conception of the clause.

In a dissenting opinion, Judge T.A. Johnson said that "It might be urged with precisely the same pertinency and force, that a statute which prohibits certain vicious actions and declares them criminal deprives persons of their liberty and is therefore derogatory of the constitution." [92] His point was that the *Wynehamer* interpretation of "due process" could be turned to the protecting of the rights of liberty as well as those of property, thus making it impossible to pass any statute declaring *any* act illegal which had been legal before the passage of the statute. This would amount, Corwin notes, to a virtual abolition of all legislative power to pass new laws of any kind.[93] We hear echoes here of the objection that Justice Iredell had raised against the *ex post facto* clause in *Calder v. Bull*.

Wynehamer was a controversial decision when it came down. But Corwin reports that "In less than twenty years from the time of its rendition the crucial ruling in *Wynehamer v. the People* was far on the way to being assimilated into the constitutional law of the country."[94] As a result, by the last decades of the nineteenth century, the judicial protection of vested property rights, which at the start of the century had rested largely upon unwritten principles of natural law and the separation of powers, was on its way to securing a new home in a pair of explicit clauses of the federal and state constitutions. This development was facilitated by the existence of due process or law of the land clauses in almost all of the state constitutions.

From *Wynehamer* to the New Deal

Until well after the Civil War, the development of "substantive due process" as a judicial means of protecting property rights against legislative encroachment occurred entirely in the *state* courts. Although the Fifth Amendment to the U.S. Constitution contains a "due process" clause, this could not form a basis for the *federal* courts to employ "substantive due process" against state legislation because, until after the Civil War, the U.S. Bill of Rights applied only to Congress and not to the state legislatures. But 1868 brought the ratification of the Fourteenth Amendment, which did apply to the state legislatures, and which contained its own "due process" clause. Ratification of the amendment thus cleared the way for the adoption by the federal judiciary of the "substantive due process" doctrine.

But it would be a full generation before the Supreme Court would adopt it. Corwin attributes the delay to the fact that members of the Court feared that the Reconstruction program, and especially the Fourteenth Amendment, threatened to upset the Constitutional balance between federal and state power.[95] Corwin argues that until a later Court decision limited the scope of Congress' power under Section 5 of the amendment in an important respect, the Court had reason to believe that the amendment would render Congress much more powerful in relation to the states than the Constitution had ever contemplated.

Among the Court decisions that limited Congress's power under Section 5 of the Fourteenth Amendment were the *Civil Rights Cases* of 1883. The Civil Rights Act of 1875 had made it a federal offense "for innkeepers, common carriers, and theater managers to refuse admission or accommodation to persons 'on account of race, color, or previous condition of servitude.'"[96] The Court overturned the act on the grounds that the Fourteenth Amendment authorized Congress to prohibit only *positive* state enactments mandating such discrimination, and not a state's passive allowance of such discrimination by private individuals.[97]

Another important development on the road to federal adoption of "substantive due process" came in the *Slaughterhouse Cases* of 1873. These involved a Louisiana statute that established an animal-slaughtering monopoly in New Orleans. Those butchers who were deprived by the statute of their rights to practice their trade in New Orleans brought suit, arguing that they had been divested of these rights without due process of law. The Supreme Court rejected their argument, but the case is notable for Justice Bradley's dissent, which would become influential later on.

> *Rights to life, liberty, and the pursuit of happiness are equivalent to rights of life, liberty, and property. These are the fundamental rights which can only be taken away by due process of law, and which can only be interfered with or the enjoyment of which can only be modified by lawful regulations necessary and proper for the mutual good of all.... This right to choose one's calling is an essential part of that liberty which it is the object of government to protect; and a calling, when chosen, is a man's property and right. Liberty and property are not protected where these rights are arbitrarily assailed.... [A] law which prohibits a large class of citizens from adopting a lawful employment previously adopted, does deprive them of liberty as well as property without due process of law.*[98]

In 1884 the Court revisited the New Orleans slaughterhouse monopoly, this time to sustain a Louisiana statute that impaired the monopoly established by the earlier statute.[99] Justice Bradley submitted as a now-assenting opinion his earlier dissent in the previous *Slaughterhouse Cases*. Corwin reports that Bradley's opinion now became, as an assenting opinion, far more influential than it would have been as a dissenting one. Says Corwin, "Just as after the decision in *Calder v. Bull* the State judiciaries took over the task of defending vested rights against unjustifiable "retrospective" legislation, so now they took up the gauge in increasing numbers in behalf of "liberty of pursuit" or, as it soon came to be called, "liberty of contract," especially in the field of labor relations." [100]

It would not be until 1897, however, after the Civil War generation of justices had left the Supreme Court, that the Court would take up the doctrine of "substantive due process" in earnest. In that year, in the case of *Allgeyer v. Louisiana*, the Court employed Bradley's assent from the 1884 slaughterhouse case to set aside a Louisiana statute that made it illegal for citizens of that state to contract for marine insurance covering property in Louisiana with an out-of-state insurer not licensed to do business in Louisiana.

Then in the famous 1905 case of *Lochner v. New York* the Court set aside, as a violation of freedom of contract, a statute limiting working hours in bakeries to ten per day and sixty per week. For a generation following *Lochner*, federal courts would involve themselves in determining whether state statutes constituted a violation of individuals' rights to liberty and property. Sometimes the courts would decide in favor of individual rights, sometimes in favor of the police power. It was not a period of complete *laissez-faire*, but it was a period when courts at all levels were actively engaged in scrutinizing legislation in the name of liberty and property.

In 1936, the Supreme Court voided a New York minimum wage statute for women and minors.[101] Then in early 1937, hoping to defeat the Court opposition to his New Deal, President Roosevelt asked Congress to authorize him to appoint to the Court one additional justice for each sitting justice over age seventy. This effort to "pack" the Court failed, but on March 29, 1937 the Supreme Court, in an act of capitulation to Roosevelt, upheld a Washington State minimum wage law, overturning a 1923 decision. This decision, says Corwin, marked the beginning of the end of "substantive due process" as a limitation on the power of state legislatures.[102] Two weeks later, in *National Labor Relations Board v. Jones and Laughlin Corp* the Court upheld Congress's authority to compel companies to permit labor to organize and bargain collectively.

Prior to this decision, "liberty" had meant the absence of coercion, and governments secured liberty by punishing those who initiated the use of coercion. A labor contract was something that either party could end at any time for any reason (in the absence of explicit provisions to the contrary), as it still is today on the side of the employee. Thus, if an employer fired an employee for engaging in labor organizing, he was not initiating the use of force against that employee, so he was not violating his rights in any way.

NLRB v. Jones and Laughlin Corp. transformed "liberty" into something that an employer could violate without ever resorting to the use of coercion against an employee, but simply by ending his contract with the employee. It put the government in the position, in effect, of forcing an employer to remain in a contract with an employee against the employer's will, all in the name of protecting the employee's "liberty." Such coercion by government of one group of people in order to make other people "free" was something the Supreme Court had ruled, back when it voided the Civil Rights Act of 1875, that Congress could not do. *NLRB v. Jones and Laughlin Corp.* marked the end of "substantive due process." For two generations thereafter, the U.S. Supreme Court virtually ceased altogether to protect economic liberties against legislative infringement. The fundamental reasons for this abrupt reversal form the subject of the next chapter.

Thomas McCaffrey

Property Rights and the Constitution:
Conclusion

An essential purpose of the U.S. Constitution was to make property rights secure. The means of accomplishing this was not to list property rights in a federal bill of rights and then to expect the courts to protect those rights against government encroachment. To begin with, a federal bill of rights would apply only to the federal government, whereas the primary threat to property rights to date had been the state legislatures. As to enforcing such a bill of rights against the federal government, the U.S. Supreme Court would not even claim the authority to do so until 1808, and it would not actually exercise that authority until a half century after that.[103] Furthermore, many of the states had included property in their own Revolutionary era bills of rights, but this had failed notoriously to make property rights secure. One reason for this failure was that the state courts had lacked clear authority to enforce the state bills of rights as against the acts of their respective state legislatures. For another, it would have been difficult to enforce such generalized rights as "life, liberty, and property" against legislative enactments. But the main reason the states had failed to secure the rights of property was that they had lacked the constitutional structure needed to maintain those rights against the forces of democracy.

Rather than employ a bill of rights strategy, therefore, the Framers of the Constitution employed a primarily architectural one. They concentrated on devising a structure of government, and a distribution of powers within that structure, that would be capable of maintaining the security of property rights against the kinds of pressures likely to assail them. *The very existence of the Constitution, and the specific structure of government delineated by it, are an expression of the Framers' intention to make property rights secure.* Indeed, the Framers were more intent on protecting property rights than they were on protecting such other rights as freedom of religion, speech, and press. These latter rights gained explicit constitutional protection only as an afterthought, when the First Congress appended the Bill of Rights to the newly ratified Constitution.

But did the Framers intend to render property rights constitutionally inviolable in the way that freedom of religion, speech, and press, have since become (relatively) inviolable? The constitutional inviolability of individual rights, as we understand it today, did not exist in 1787, and it only developed gradually over time. The dominant political tradition during the American founding era was the one embodied in the unwritten British Constitution. Under that tradition, there existed bills of rights, but these did not have the force of law in the sense that there existed a formal, institutional mechanism for enforcing them as against acts of Parliament. Certain traditional rights were widely considered to be inviolable, but they were only *politically* inviolable, in that no Parliament would dare risk the political uproar that would ensue following any Parliamentary attempt to abrogate those rights. Such rights were not, in fact, *legally* inviolable. Legally, Parliament had the authority to legislate such rights out of existence.

Under such a system of "legislative supremacy," the integrity of individual rights depended on the discretion of Parliament and, ultimately, upon the electoral process. The British people did not object to this system of Parliamentary supremacy in large part because, historically, Parliament had been the protector of the Rights of Englishmen against assaults by the British Crown.

This was the reigning model available to the Framers as they set about devising a constitution for the United States. Their first departure from the British model was the very act of devising a *written* constitution. This was itself a fundamentally Lockean undertaking, a people contracting among themselves to devise a frame of government from nothing. (Or almost from nothing; the state governments that would be an integral part of this constitutional structure were already in existence.)

This first departure from the British model, a written constitution expressly ratified by the governed themselves, would have profound consequences for the legal status of individual rights in the new American polity. As a fundamental expression of the political will of the sovereign entity in this new polity, the people themselves, the written Constitution would acquire the force of law, fully enforceable by the courts. Thus the rights identified in the Constitution and its amendments would also become legally enforceable in a way that rights had never been enforceable under the British Constitution. But the full implications of such conceptions as written constitutions and popular sovereignty, including the judicial enforceability of constitutional rights, would only become clear over time; they were not fully apparent to the Framers in 1787.[104]

As for property rights, the key to securing them against legislative encroachment would prove to be the placing of strict, judicially enforceable limits on the "powers" of taxation, eminent domain, and police. When the Framers met at Philadelphia in 1787, the idea of natural rights was only a century or so old. The Americans would be the first people in history to attempt to give institutional form to this new idea. No one anywhere had, as yet, grasped that to give effect to natural rights, short of abolishing the "three powers" outright, men would have to place certain limitations on those powers.

The English had pioneered the way toward limiting one such power, that of eminent domain, when they evolved the practice of compensating eminent domain takings. And in the mid-1600's the Dutch natural law theorist, Hugo Grotius, advocated compensation for takings and a public use limitation, as a matter of fundamental justice. But not until James Kent began publishing his *Commentaries* in the 1820s would anyone conduct a full, systematic analysis of the "three powers" that touch property rights and the limitations needed to be placed on them in order to make property rights as secure as their status as natural rights merited. Specifically, until the Framers gathered in Philadelphia that summer of 1787, no one had as yet recognized a need to, and proposed a way to, limit either the taxing or the police powers in accordance with the requirements of natural rights.

In the Contract and *Ex Post Facto* clauses, the Framers would make the first attempt in history to place limitations on the police power, although they likely did not think of what they were doing in those terms. It is doubtful they ever conceived of a need to limit something called the "police power," per se. More likely, they thought they were prohibiting only certain specific manifestations of legislative power, such as legal tender, stay, and moratory laws. But the *ex post facto* clause had the potential to develop into a formidable limitation on the police power, as George Mason and James Iredell recognized.[105]

The Americans achieved a momentous advance when they prohibited the exercise of the police power in the realms of religion, speech, and press. As worded, this prohibition was a complete one. And, as America's constitutional history has unfolded, this prohibition has proven to be a model of how to limit a "power" of government. But the Framers' pioneering attempt to limit the police power as it touches *property rights* would prove short-lived. The reason, as I have said, was that it had the potential, via the *Ex Post Facto* Clause, to become as complete a prohibition as the one shielding religion, speech, and press. But from the beginning of the Republic, there never has been sufficient agreement among American lawmakers and jurists about the need to limit the government's power to regulate private property in the way that the power to regulate the dissemination of ideas was limited.

Thomas McCaffrey

Had the science of securing individual rights been more advanced in 1787, things might have developed differently. Kent's idea of limiting the police power primarily to the traditional common law of nuisance, and then awarding compensation for all legislative encroachments upon private rights that went beyond nuisance abatement, would have limited the police power without abolishing it altogether. Had this idea been available to the Framers in 1787, then perhaps this limitation would have developed into as solid and long-lasting a defense of property rights as the comparable limitations on the eminent domain power did.

"Freedom of religion," as it is enunciated in the First Amendment, was a relatively new concept in 1790. The amendment states that "Congress shall make no law respecting an establishment of religion." Neither the Constitution nor the Bill of Rights prohibits an establishment of religion by the state governments; indeed, state establishments of religion continued for some time after the ratification of the Constitution. But in passing the First Amendment the United States adopted what has proven to be the *principle* of religious freedom. Widespread acceptance of this principle eventually lead the remaining states with established churches to disestablish them. Over time, individual states added religious freedom clauses to their own constitutions. The current effort to ban prayer from public schools is a logical extension of this principle of religious freedom, however unintended such a ban was on the part of the authors of that clause.[106] The acceptance of the principle of religious freedom has, since 1791, worked inexorably toward effecting a complete separation of church and state.

Likewise, the idea of a right of private property, though much older than the idea of religious freedom, was still far from fully realized in 1787. But just as the principle of religious freedom, once accepted, would lead to an ever-widening separation of church and state, so the principle of the inviolability of private property would continue to develop, as we have seen, throughout the nineteenth century. For example, the Fifth Amendment formally recognized the principle that "takings" of private property should be compensated. Following the ratification, state after state added similar clauses to their own constitutions.

In the beginning, only physical takings of full title to property were compensated. But the *principle* underlying the takings clause inarguably requires the compensation of *all* takings, whether full or partial, and whether effected by the power of eminent domain or by police regulation. Impelled by the logic of the principle, the U.S. Supreme Court accepted the need to compensate partial takings involving physical invasion in 1871, and in 1922 it accepted the need to compensate partial, "regulatory takings" that involved no physical invasion of property at all but merely a diminution in value. In the development of constitutional enforcement of freedom of religion and the right to compensation for takings, we see how the initial commitment to a correct principle issued eventually in applications of the principle, and a consequent expansion of liberty against government not contemplated by the Founders.

Likewise, the *Ex Post Facto* Clause, construed to apply to civil as well as criminal matters, also enunciated a principle, that legislation should not adversely affect existing rights (especially property rights, as the Framers intended the clause). (This principle, especially as it applies to property rights, did not depend entirely on the *Ex Post Facto* Clause for its place in the Constitution; it was, as I have shown, an important reason for the very existence of the Constitution and the form of government laid out therein.)

Even after the Supreme Court limited the reach of *ex post facto* to criminal matters only, the principle underlying that clause continued to manifest itself in increased security of property rights against legislative infringement. In the first half of the nineteenth century, for example, courts ended the practice of legislative interference in judicial matters, especially those involving property rights. In a related development, the courts also ended "special" legislation that applied to the property of specific, named parties. In the second half of the century, courts enlisted the doctrine of substantive due process to do what the *Ex Post Facto* Clause should have been able to do in the first place. Substantive due process, however, is best understood not as a net advance for property rights, but, rather, as a holding action against a burgeoning police power.

In the case of freedom of religion, speech, and press, this salutary process of growth has continued to the present day. But the corresponding development of the rights of property was cut short, beginning about a hundred years ago. In principle, property rights ought to be every bit as inviolable today as are freedom of religion and freedom of speech. Indeed, freedom of religion, freedom of speech, and private property form an integrated whole, the first representing, in a general sense, the freedom to think as one chooses, the second the freedom to communicate one's thoughts freely, and the third the freedom to act upon those thoughts as one sees fit. To secure the freedom to think and the freedom to communicate one's thoughts, without also securing the freedom to act, is, ultimately, to leave all rights vulnerable to government infringement.

But that is the contradictory course the United States has pursued for most of the past century. Several factors conspired to ensure that the development of property rights would be cut short, whereas that of such rights as freedom of speech and religion would continue to this day. One such factor is that, for the reasons I have already identified, property rights were never explicitly endorsed as inviolable rights in America's two formative documents, the Declaration of Independence and the U.S. Constitution cum Bill of Rights. This placed property rights at a distinct disadvantage when, by the 1830s, courts found themselves limited to protecting only those rights that were explicitly identified in constitutions or bills of rights.

A second reason that the constitutional securing of property rights never reached full development in America has been, on the one hand, the inherent difficulty of securing them in the face of the three "powers" of government, taxation, eminent domain, and police, and, on the other, the ambivalence toward making property rights inviolate to which the existence of these powers gave rise. But the single factor most responsible for cutting short the development of constitutional protections for property rights has been the opposition they faced from the democratic conception of "popular sovereignty," which derives from the Lockean idea of the consent of the governed. It was, in part, expressly to bring an end to assaults upon property rights issuing from "an excess of democracy" that the Framers gathered in Philadelphia in 1787 and devised the specific form of government they did.

In the 1905 case of *Lochner v. New York*, Justice Oliver Wendell Holmes averred:

> This case is decided on an economic theory which a large part of the country does not entertain.... [A] constitution is not intended to embody a particular economic theory, whether of paternalism and the organic relation of the citizen to the state or of laissez-faire.... I think the word "liberty" in the Fourteenth Amendment is perverted when it is held to prevent the natural outcome of a dominant opinion, unless it can be said that a rational and fair man necessarily would admit that the statute proposed would infringe

fundamental principles as they have been understood by the traditions of our people and our law.[107]

Holmes argues here that the passing of legislation that infringes, or even abolishes, the property rights of individuals is entirely consistent with the letter and spirit of the U.S. Constitution, *if* that legislation is reflective of a "dominant opinion." He argues, in other words, that under the U.S. Constitution the securing of the property rights of individuals is not a matter of fundamental, inviolable right, but that it is, rather, a matter of majority vote.

But one of the purposes of the U.S. Constitution, as stated in the Preamble, is to "secure the Blessings of Liberty to ourselves and our Posterity." If it were possible to secure liberty without reference to a "particular economic theory," then Holmes might have been correct when he said that the Constitution does not embody any particular economic theory. But there is *in fact* only one "economic theory" that is compatible with the securing of liberty, and that is the economic theory based on private property. The Framers of the U.S. Constitution were fully aware of this connection between liberty and property. History and experience had taught them that unrestrained democracy was antithetical to the securing of property rights and, thus, to liberty. The triumph of Holmes' counter-principle, which would be accomplished through the efforts of the Progressive movement, would thus represent the undoing of the Constitution in a fundamental respect that the Framers had considered essential to their purposes. The details of this undoing form the subject of the next chapter.

One final point, though. One of the greatest threats to private property rights today comes not from the states but from the federal government. It comes in the form of environmental legislation, such as the Clear Air and Clean Water Acts. The constitutional grounds for this legislation is not any one of the three powers of government, but Congress's Constitutional power "to regulate commerce ... among the several states." As I have said, the original purpose of this clause was to lodge in the federal government the *sole* authority to regulate interstate commerce, thereby prohibiting the states from erecting the kind of interstate barriers to trade that had been common during the Confederation period. The purpose of the Commerce Clause, in other words, was to e*liminate* the regulation of interstate commerce—and most assuredly not to create a new, federal power to interfere in interstate commerce, much less *intra*state commerce.

As I have argued, the Commerce Clause functioned as intended for almost a century, until the Progressives began to distort its meaning. It has since evolved into a plenary power to regulate any economic activity even remotely related to interstate commerce, a power such as the Framers would never have imagined. One might argue that the Progressives have simply twisted the Commerce Clause to mean something it was never intended to mean in just the same way as the advocates of substantive due process once twisted the due process and law of the land clauses to suit their purposes. But those nineteenth century jurists were reacting to changes in the interpretation of the Constitution that the Framers could never have foreseen, and in light of those changes they were attempting to re-interpret certain clauses of the Constitution in order to serve the Framers' original intention to make property rights secure against legislative infringement.

The Progressives, on the other hand, in reinterpreting the Commerce Clause to mean exactly the opposite of what the Framers intended, have been working to *defeat* the intentions of those Framers. The Constitutional remedy for the metastasizing of the Commerce Clause into a plenary power to regulate all things economic between and within the states is simply to interpret the Commerce Clause as it was originally intended to be understood.

Notes

1. Aristotle, *The Politics of Aristotle*, trans. Ernest Barker (London: Clarendon Press, 1946), 3, 1280B.
2. John Adams, *The Works of John Adams*, ed. Charles Francis Adams (Boston, MA: Little, Brown, and Company, 1850-56), 3, 470.
3. Ibid., 6, 206-7.
4. Ibid., 4, 553.
5. Ibid., 4, 554.
6. John Adams, Letter to Rev. De Walter, October, 1797. Quoted in C. Bradley Thompson, *John Adams and the Spirit of Liberty* (Lawrence, KS: University of Kansas Press, 1998), 119.
7. Adams, *Works*, 5, 453-4.
8. Declaration of Independence.
9. Thomas Jefferson, Letter to Isaac McPherson, August 13, 1813. Quoted in *Thomas Jefferson: Writings*, ed. Merrill D. Peterson (New York, NY: The Library of America, 1984), 1291.
10. Thomas Jefferson, Letter to DuPont de Nemours, April 24, 1816. Quoted in Peterson, ed., *Writings*, 1387.
11. Jean Yarbrough, "Jefferson and Property Rights," in *Liberty, Property and the Foundations of the American Constitution*, ed. Ellen Frankel Paul and Howard Dickman (Albany, NY: State University of New York Press, 1989), 68.
12. Thomas Jefferson, Letter to James Madison, Oct. 28, 1785. Quoted in *Papers of Thomas Jefferson*, ed. Julian Boyd, (Princeton, NJ: Princeton University Press, 1950), 8, 682.
13. Emphasis added.
14. Thomas Jefferson, *Prospectus* recommending the political economist DeStrutt de Tracy, copy sent to Joseph Milligan, April 6, 1816. Quoted in *The Writings of Thomas Jefferson*, ed. Andrew A.

Lipscomb & Albert Ellery Bergh (Washington, D.C.: The Thomas
Jefferson Memorial Assocation, 1903), 14, 456. Emphasis in
original.

15. Locke, *Two Treatises*, II, 138, 3-8.

16. William Blackstone, author of *Blackstone's Commentaries* and *the*
authority on English law among the Americans, held a similar
view. William Blackstone, *Commentaries on the Laws of England*, ed.
Wayne Morrison (London and Sydney: Cavendish Publishing,
Ltd., 2001), 2, 3-11.

17. Ibid., 340.

18. Preamble to the Constitution of the United States.

19. *The Roots of the Bill of Rights*, ed. Bernard Schwartz (New
York, NY: Chelsea House, 1980) 2, 234.

20. Gouverneur Morris, *Records of the Federal Convention of 1787*,
ed. Max Farrand (New Haven, CT: Yale University Press, 1966),
1, 536. Emphasis in original.

21. Farrand, *Records*, 1, 536-7, 541-2.

22. Farrand, *Records*, 1, 541.

23. See *The Politics of John Dickenson*, ed. Paul Leicester Ford
(Cambridge, MA: Da Capo Press, 1970), 44 and George Athan
Billias, *Elbridge Gerry: Founding Father and Republican Statesman*
(New York, NY: McGraw Hill, 1972), 548, note 33.

24. William Patterson, *Vanhorne's Lessee V. Dorrance*, 2 U.S. 304
(1795).

25. Hugh Williamson, "Remarks on the New Plan of
Government" in *Essays on the Constitution of the United States
Published during Its Discussion by the People*, ed. P. Ford (Brooklyn,
NY: Historical Printing Club, 1892). Quoted in Michael
Kammen, "The Rights of Property, and the Property in Rights:
The Problematic Nature of Property in the Political Thought of
the Founders and the Early Republic," Paul and Dickman, *Liberty*,
1.

26. Chief Justice John Marshall, *Ogden v. Saunders*, 25 U.S. (12

Wheaton) 213, 354-55 (1827).

27. "When in the Course of human events, it becomes necessary for one people to dissolve the political bands which have connected them with another ..."

28. William Winslow Crosskey, *Politics and the Constitution* (Chicago, IL: The University of Chicago Press, 1953), 298-392, 304-307.

29. S.E. Morison, H.S. Commager, and Wm. E. Leuchtenburg, *The Growth of the American Republic* (New York, NY: Oxford University Press, 1969), 240.

30. Crosskey, *Politics*, 965-66.

31. George Bancroft, *History of the United States* (New York, NY: D. Appleton and Company, 1885), 6, 171.

32. Ibid., 172.

33. Ibid., 173.

34. Farrand, *Records*, I, 48.

35. James Madison, Federalist No. 10, *The Federalist Papers*, ed. Clinton Rossiter (New York, NY: New American Library, 1961), 79.

36. Ibid., 78.

37. Ibid., 84.

38. James Madison, Letter to James Monroe, Oct. 5, 1786 , *The Papers of James Madison* , ed. William T. Hutchinson and William M.E. Rachal (Chicago, IL: University of Chicago Press, 1975), 9, 140-41.

39. Ernst Freund, *The Police Power: Public Policy and Constitutional Rights* (Chicago, IL: Callaghan & Co., 1904), 11.

40. Ibid., 21.

41. Less harm is done by police regulations that require a complaint from a harmed property owner to trigger enforcement of a nuisance ordinance. But police regulations today typically prohibit, for example, sending smoke from a factory across neighboring lands regardless of whether the owners of those lands

object. In this way, police regulation has moved away from its common law concern with protecting the rights of individuals and has moved toward a notion of protecting the "rights" of the public considered collectively.

42. A direct tax is a tax placed on an individual. Until the passage of the Sixteenth Amendment, income taxes were classed as direct taxes. As for "apportioning" such a tax, if Rhode Island, for example, had 5% of the population of the U.S., then the amount collected from Rhode Islanders under a direct tax would have to equal 5% of the total amount collected by the federal government.

43. *United States v. Butler*, 297 U.S. 1 (1936). Cited in Richard Epstein, *Takings: Private Property and the Power of Eminent Domain* (Cambridge, MA and London, England: Harvard University Press, 1985), 296. Edward Corwin reports that what constituted a "public purpose" had by 1880 become a question reserved to the courts in "probably…every State in the Union." Edward Corwin, *Liberty against Government* (Baton Rouge, LA: Louisiana State University Press, 1948), 79.

44. William Stoebuck, "A General Theory of Eminent Domain," *Washington Law Review* 47, 4 (August, 1972): 554.

45. The 2005 *Kelo* decision, which abolished the public use limitation on the power of eminent domain, is a symptom of the deterioration of the U.S. legal system.

46. U.S. Constitution, Article I, Section 10.

47. Wrote "Lycurgus," for example, in *The New-Jersey Gazette* on November 1, 1784, "*ex post facto* laws to lessen the right of the creditor [were] incompatible with the station [America] ha[d] taken among the nations, and [were] inconsistent with foreign commerce." Crosskey, *Politics*, 326. Brackets in Crosskey.

48. Farrand, *Records*, 2, 376.

49. The case involved an act by the Virginia Legislature that was intended to clarify the meaning of an earlier act by the same body,

but which was intended to apply to actions which had occurred after the passage of the first act but before passage of the second. The act in question, the second one, dated from 1787, which placed it before the adoption of the U.S. Constitution, so the *ex post facto* prohibition in the Constitution was not a factor. The Virginia constitution of the time did not contain an *ex post facto* prohibition. The judge in the case was Edmund Pendleton, who had earlier presided over the Virginia state convention that had ratified the U.S. Constitution. Said Pendleton, "They [the legislature] may amend as to future cases, but they cannot prescribe a rule of construction, as to the past. For a legislative interpretation, changing [property] titles founded upon existing statutes, would be subject to every objection which lies to *ex post facto laws*, as it would destroy rights already acquired under the former statute, by one made subsequent to the time when they became vested." *Turner v. Turner's Executrix,* Va. (4 Call) 237 (1792).

50. Farrand, *Records*, 2, 640.

51. *Calder v. Bull* involved a case in which an enactment of the Connecticut Legislature overturned the finding of a probate court that a will was invalid. The party aggrieved by the Legislature's action appealed to the U.S. Supreme Court on the grounds that the legislative enactment depriving him of legally vested property rights amounted to an *ex post facto* law.

52. See Crosskey, *Politics*, chapter 11.

53. Note appended to Johnson's opinion in *Satterlee v. Matthewson*, 27 U.S. (2 Peters) 380, 416n. (1829).

54. *Calder v. Bull*, 3 U.S. (3 Dallas) 386, 388-89 (1798).

55. Emphasis added.

56. *Pumpelly v. Green Bay Co.*

57. Although Kent denied the validity of Locke's idea of a state of nature, he did subscribe to the concept of property as a natural right. "The sense of property is inherent in the human breast and

the gradual enlargement and cultivation of that sense from its feeble force in the savage state to its full vigor and maturity among polished nations forms a very instructive portion of the history of civil society. Man was fitted and intended by the author of his being for society and government and for the acquisition and enjoyment of property. It is, to speak correctly, the law of his nature…." Echoing Madison, he said, " A state of equality as to property is impossible to be maintained, for it is against the laws of our nature…." And, sounding like John Adams denying the virtue of Spartan asceticism, Kent said, "No such fatal union (as some have supposed) necessarily exists between prosperity and tyranny or between wealth and national corruption in the harmonious arrangements of Providence." James Kent, *Commentaries on American Law* (Boston, MA: Little, Brown, and Company, 1896), 2, 318 ff. Quoted in Corwin, *Liberty*, 77-78.

58. Kent, 2, 340.

59. Corwin, *Liberty*, 81.

60. Kent, *Commentaries*, 2, 340.

61. Corwin, *Liberty*, 81.

62. Kent, *Commentaries*, 2, 339, note a.

63. The Court awarded compensation for land permanently flooded consequent to the building of a statute-authorized dam.

64. Mahon owned a home whose deed reserved the underground mineral rights to the Pennsylvania Coal Co. The deed explicitly absolved the company from liability should their mining beneath the house cause damage to the house. Pennsylvania's Kohler Act, passed long after the deed took effect, required the company to leave enough coal in place under the house to prevent damage to the house. Although the police regulation did not involve any sort of physical invasion of the coal company's property, the Court ruled that the company were due compensation from the State for the "taking" of the coal they were required to leave in place under Mahon's house.

65. *Nollan v. California Coastal Commission* and *Lucas v. South Carolina Coastal Council*.

66. Kent, *Commentaries*, 2, 331.

67. *Calder v. Bull*, 388-89.

68. Corwin, *Liberty*, 71-72, 75.

69. Ibid., 70.

70. Ibid., 73.

71. Ibid., 72. Emphasis in the original.

72. Benjamin Fletcher Wright, *The Contracts Clause of the Constitution* (Cambridge, MA: Harvard University Press, 1938), 10.

73. *Gibbons v. Ogden*, 22 U.S. (9 Wheaton) 1, 211 (1824).

74. Both James Kent and Joseph Story, another champion of property rights on the Supreme Court, advocated judicial protection of the kind of state-sponsored monopoly involved in the *Gibbons* case. A fair reading of the opinions of both men would disclose that they were not at all advocates of the kind of union of commerce and state that Alexander Hamilton had promoted. Rather, they labored under the economic premise, common at the time, that without the incentive of state-sponsored monopolies entrepreneurs would lack the motivation to risk their wealth on such "public" endeavors as turnpikes, bridges, and ferry lines. It is worth noting that such monopolies tend to promote *private* ownership and management of such facilities.

75. In *Gibbons v. Ogden*, Marshall achieved this "negative" good by asserting the power of the federal government. The question whether a powerful federal government has ever been the best means of securing individual liberty in the United States is beyond the scope of this book. My purpose is only to show that for the Framers of the Constitution, and for such Federalists as Marshall, the protecting of individuals' property rights was a central purpose of government, and the establishing of a powerful federal

government, as a check upon the state legislatures, was a primary means to this end.

76. *Fletcher v. Peck*, 10 U.S. (6 Cranch) 87, 133, 135 (1810).

77. *The Antelope*, 23 U.S. (10 Wheaton) 66, 120 (1825).

78. *Ogden v. Saunders*, 25 U.S. (12 Wheaton) 213, 346 (1827).

79. It is also worth noting that a logically consistent application of the principle underlying the compensation requirement, as well as literal interpretation of the clause itself, would require that *all* "takings" of private property be compensated, whether full or partial, direct or "consequential," and whether effected by eminent domain or police regulation.

80. Samuel Eliot Morison, *The Oxford History of the American People* (Oxford, England: Oxford University Press, 1965), 421.

81. Corwin, *Liberty*, 85.

82. Crosskey, *Politics*, 696.

83. *Charles River Bridge Co. v. Warren Bridge Co.*, 36 U.S. (11 Pet) 420, 547-48, 552 (1837).

84. Corwin, *Liberty*, 88. Emphasis in the original.

85. John Marshall, *Life of Washington*, (Philadelphia, PA: Crissy & Markley, and Thomas, Cowperthwait and Company, 1850), II, 281-82.

86. Corwin, *Liberty*, 84-85.

87. Ibid., 90-91.

88. Ibid., 91.

89. *Zylstra v. Charleston*, 1 S.C.L. (1 Bay) 382, 384 (1794). See Corwin, *Liberty*, 91-92.

90. Crosskey, *Politics*, II, 1148.

91. Corwin, *Liberty*, 102-03.

92. *Wynehamer v. The People of New York*, 13 N.Y. 378, 468 (1856). Quoted in Corwin, *Liberty*, 104.

93. Corwin, *Liberty*, 104.

94. Ibid., 114.

95. Ibid., 118.

96. Ibid,. 134, quoting the Civil Rights Act of 1875 (18 Stat. 335-337).

97. Since the U.S. Supreme Court itself already possessed the power to prohibit *positive* state actions, this decision eliminated the danger that a branch of the federal government would acquire a *new* power over the states, the power to prohibit passive tolerance of private discrimination by state and local governments.

98. *The Slaughterhouse Cases*, 83 U.S. (16 Wall) 36, 116, 122 (1873).

99. *Butcher's Union v. Crescent.*

100. Corwin, *Liberty*, 136.

101. *Morehead v. N.Y. ex rel. Tipaldo.*

102. Corwin, *Liberty*, 159.

103. *Dred Scott v. Sandford.*

104. As I have said, the federal courts did not even assert the authority to enforce rights against the federal legislature until 1808, and they did not actually exercise it until a half century later. (*Marbury v. Madison, Dred Scott v. Sandford.*) The authority of the federal courts to enforce the federal Bill of Rights against the *state* legislatures was not tested (and denied, the first time around) until forty-three years after the ratification of the Constitution. (*Barron v. Baltimore,* 1833)

105. The contract clause could also have developed into an important defense of freedom of contract had the courts not limited its application to already-existing contracts. William Crosskey, for one, argued that the clause should indeed be interpreted to ban *all* legislation "impairing the obligation of contracts," not just acts that affect contracts entered into before the legislation in question was enacted. Crosskey, *Politics*, Chap. 12.

106. Just as no citizen should be forced, through taxation, to support the dissemination of religious ideas with which he

disagrees, so, and for the identical reason, should no citizen be forced to support the dissemination of *any* ideas with which he disagrees. Since schools are in the business of disseminating ideas, the real problem with prayer in tax-supported schools is not the prayer but the existence of the schools themselves. To sponsor praying in them is simply to compound a wrong.
107. *Lochner v. New York,* 198 U.S. 45, 75-76 (1905).

Thomas McCaffrey

3: Progressivism

THE AUTHORS OF the U. S. Constitution intended to limit democracy at the state and federal levels in order to render individuals' rights of private property secure. The Constitution functioned largely as intended in this respect for just about a century or so. In the late nineteenth century, however, a movement arose that would severely erode the constitutional protections surrounding property rights. Progressivism aimed explicitly to make American governments at all levels more democratic and to give those governments vastly increased powers to control the use and disposition of private property.

In its early years, Progressivism drew heavily upon the philosophical movement called Pragmatism. Three Americans composed the core of the Pragmatic movement in philosophy, Charles S. Peirce (1839–1914), William James (1842-1910), and John Dewey (1859-1952). Among the services these men provided to progressivism was a redefinition of what constitutes truth.

The Pragmatists

To men who could write "We hold these Truths to be self-evident," the meaning of the word "truth" is clear; a statement is true if it corresponds with the facts of reality. This is the correspondence theory of truth, and it goes back to Aristotle. "To say of what is that it is not, or of what is not that it is, is false; to

say of what is that it is, or of what is not that it is not, is true." [1]
This definition of truth is based on the assumptions that something
exists, that man can know this something, and that truth is a
correspondence between what a man knows and what this
something in fact is. [3] According to this conception of truth, for
example, the statement, "The earth is spherical," is true because
the object known as the earth is *in fact* shaped like a sphere, as that
term is commonly defined. When the American Founders wrote,
for example, that "All men are endowed by their Creator with
certain unalienable Rights," they meant that it is an objective fact
of reality that men have such rights.

If a person wanted to undermine this idea of unalienable
rights, one approach might be to argue that it is *untrue* that men
possess such rights by nature. Another approach, pursued by the
pragmatic philosophers, would be to redefine the meaning of the
word "truth" altogether.

The Pragmatists rejected the correspondence theory of truth.
For them, the meaning of an idea "lies exclusively in its
conceivable bearing upon the conduct of life; so that ... if one can
define accurately all the conceivable experimental phenomena
which the affirmation or denial of the concept would imply, one
will have therein a complete definition of the concept." [2]
Consider, for example, the statement, "It is raining." For the
pragmatist, this means that if I were to go outside, I would *see*
drops of water falling from the sky, *feel* drops of water hitting my
skin, *hear* drops of water hitting the ground, *taste* water on my
tongue, and undergo myriad other experiences that I would come
to associate with the statement, "It is raining." The sum total of
these experiences *would be* the meaning of "It is raining" for the
pragmatist. "It is raining" would not mean that actual drops of
water are actually falling from an actual sky. For the pragmatist,
the "meaning" of a proposition is not some objective fact of reality
but, rather, the *subjective* experiences that men come to associate

with that proposition.

James

Accordingly, pragmatism gives rise to distinctly subjectivist conceptions of knowledge and truth, in contrast to the objectivist conception of the Aristotelian, "common sense" school, as William James referred to it. Charles Peirce, for example, writes that "There is absolutely no difference between a hard thing and a soft thing so long as they are not brought to the test." [3] The objective view of reality, on the other hand, is that the difference between a hard thing and a soft thing is an independently existing fact of reality that pertains *whether or not* it is "brought to the test." Consider the following from James:

> Thus a room is a physical thing considered from the point of view of the metal or wood that makes it up. Considered from the point of view of my relations to it, as part of my life-history, it is an experience of mine and so is mental. The basic stuff is only one kind of thing—not two as in the mind-matter metaphysics.... But the basic stuff ... is in itself neither [mind nor matter], although it may be taken as either. [4]

What James is saying here, as E.C. Moore has pointed out, is that there is no such thing as an objective reality that exists apart from and independently of human consciousness. They are both, mind and matter, simply the same basic stuff as seen from different perspectives. [5] (James called it "pure experience.") According to Moore, Pierce agreed with James on this point. [6] In the world of the pragmatist, subject and object become one.

In the objective, Aristotelian universe, knowledge and truth are *discovered*. During the Renaissance, for example, the Europeans discovered the truth that the earth is spherical. But truth in the subjectivist universe of the pragmatists is otherwise: "The truth of an idea is not a stagnant property inherent in it. Truth *happens* to an idea. It *becomes* true, is *made* true by events." [7] From the Aristotelian perspective, of course, this is false. The correspondence between the meaning of "The earth is spherical" and the fact of reality to which it refers was itself a fact of reality before Columbus ever set sail.

So, what is truth for a pragmatist? It is *not* a correspondence between an idea and an independently existing fact of reality. It is, rather, as William James put it, an idea's "working," its leading to the perceptual experience that the idea "predicts." Consider again the statement, "It is raining." If this statement is true, then I should *feel* drops of water striking my skin, *hear* drops of water striking the ground, *see* drops of water falling from the sky, and *taste* water on my tongue. If someone were to say "It is raining," and then I did experience all these perceptions, the statement, "It is raining," would indeed have "worked." It would be "true" in the pragmatic sense. The correspondence here is, in effect, between a statement and a series of experiences, and *not* between a statement and a fact of reality. Pragmatism does not concern itself with facts of reality that exist independently of the perceptions I experience.

This notion of truth as an idea's "working" seems at least to preserve some connection between truth and objective reality. The statement "It is raining" does *imply* that I will feel drops of water on my skin, hear rain striking the ground, etc. But the real meaning of "working" is revealed by James's handling of the "working" of more abstract ideas. [8] And it discloses the "flexibility" of his subjectivist conception of truth. "Such is the large loose way in which the pragmatist interprets the word agreement [between idea and reality].... He lets it cover any process of conduction from a present idea to a future terminus, provided only it run prosperously .... It is ... *as if* reality were made of ether, atoms or electrons, but we mustn't think so literally. The term "energy" doesn't even pretend to stand for anything 'objective.'" [9] James is suggesting here that a scientific theory can be "true" without making any attempt at describing in any literal sense a fact of reality. Under this conception, it is not necessary that an idea agree with reality; it is only necessary that it not disagree with experience.

Dewey

Indeed, under John Dewey's later conception, in order that an idea be true it must *not* be known with certainty to correspond with the facts of reality. Quoting Peirce, Dewey wrote:

> *Truth is that concordance of an abstract statement with the ideal limit towards which endless investigation would tend to bring scientific belief, which concordance the abstract statement may possess by virtue of the confession of its inaccuracy and one-sidedness, and this confession is an essential ingredient of truth."* [10]

Dewey says that any statement we can make with certainty, such as what we had for breakfast this morning, cannot be true at all, for the concept "true" applies only to those statements that we do not know with certainty, such as certain scientific hypotheses. "The judgment that my friend is in Constantinople would, if I am sure that he is there, not be a judgment at all but a "truism," "tautology," or "triviality." So, since it would not be a judgment or a proposition to say that my friend is in Constantinople, it would not be true, since only judgments or propositions can be true, and to say that a proposition is true is to imply that it might not be true. The statement that my friend is in Constantinople is only a proposition *if* I am not entirely certain he is there but I have reason to believe that he is." [11]

So truth, for the pragmatist, becomes at best a hypothesis, on the order of Columbus' hypothesis that the earth is spherical. And as soon as the question is settled and all doubt is removed as to the shape of the earth, then the statement "The earth is spherical" ceases to be a "proposition" and, therefore, can no longer be "true." George Orwell could not have devised a more subversive conception of truth.

The analogy to scientific hypothesis goes to the heart of pragmatism. Scientific investigation was, for the pragmatists, the model of how all knowledge should be acquired (or, rather, made). The active, hands-on approach of the scientist, his use of experiment, and his employment of mathematical measurement where possible, epitomized the process of "inquiry," as Dewey called knowledge-seeking.

Dewey would turn this idea of truth-as-hypothesis directly against many of the truths that had animated the American Founders. "If the pragmatic idea of truth has itself any pragmatic worth, it is because it stands for carrying the experimental notion of truth that reigns among the sciences, technically viewed, over into political and moral practices, humanly viewed." [12] Dewey meant to apply the scientific method quite literally to moral and political questions. Just as the scientist is not hindered in his investigations of non-human life and of inanimate nature by abstract moral prohibitions, so Dewey believed that social scientists and moral and political leaders should not be hindered by such abstract principles as unalienable rights and private property from conducting moral, social, and political experiments. To allow such hindrance to occur, he believed, would be unscientific.

Dewey's philosophy is about as antithetical to the ideas that animated the American Founders as a philosophy could be. Whereas the Founders predicated their institutions upon a highly determinate human nature grounded in reason and free will,

Dewey saw a creature controlled by habits, a creature who, to a great extent, could be molded into whatever his environment influenced him to be. Because man, for Dewey, is not fundamentally self-directing, there is no reason that political philosophy should treat him as an autonomous being.

Whereas the Founders placed the highest value on the liberty of the individual to think for himself and to act accordingly, Dewey restricted liberty primarily to the freedom to think for oneself, and he treated the corresponding freedom to act accordingly as though it were entirely superfluous. "The democratic ideal of freedom is not the right of each individual to *do* as he pleases, even if it be qualified by adding "provided he does not interfere with the same freedom on the part of others."... The basic freedom is freedom of *mind* and whatever degree of freedom of action and experience is necessary to produce freedom of intelligence." [13]

"Freedom of mind," for Dewey, was the kind of freedom contemplated by such Constitutional rights as freedom of religion, speech, and press. Such freedom must be secured, said Dewey, so that the individual could participate in the all-important process of democratic decision-making. Because under Dewey's system the individual was to be governed by such collective decision-making, Dewey could not very well advocate individual freedom of *action* as he did freedom of thought. Freedom of action therefore becomes a nullity under Dewey's conception, which called, quite logically, for the socialist abolition of private property. Indeed, Dewey equated private property with coercion, and socialism with liberty—exactly inverting the thinking of John Locke and the American Founders on this crucial point.

Pragmatism, and notably its subversive conception of truth, placed Dewey and other Progressives in a position to argue that unalienable rights and private property, among other key ideas, were not *true* in any timeless, Aristotelian sense. They were not grounded in the facts of reality. They were simply experimental hypotheses; they worked well for a time, but circumstances had changed. Industrialism, especially, had shown that these hypotheses no longer worked, so it was time for new hypotheses and new social and political experiments.

William James' conception of truth would become especially influential in America. The idea that that which "works" is, *ipso facto*, the right thing to do has long since become synonymous with acting pragmatically. This interpretation of James as meaning that any solution to a problem that accomplishes one's purpose, however short-range the solution or immoral the purpose, James himself always denied. But he unquestionably set out to undermine men's confidence in the idea of truth as correspondence with objective reality. "Truly objective truth ... is nowhere to be found," he wrote.[14]

And when we combine James' assault on objective truth with Dewey's denigration of abstract moral, social, and political principles in favor of "scientific" experimentation, we see that the equating of "pragmatic" with "unprincipled" is not unjustified. That unprincipled pragmatism is today the ruling "principle" of our political life, and that such abstract principles as individual rights and private property have been mutilated almost beyond recognition by legislative and regulatory experimentation, are due in no small part to the work of Peirce, James, and Dewey. At the end of the nineteenth century, the American idea would come under assault from many directions. Pragmatism was an important element in this assault. It was one of the intellectual solvents that would eventually eat away, like acid, at what Eric Goldman called "the steel chain of ideas" that had kept America moored to its founding principles for a century.[15]

Thomas McCaffrey

The Jurists

Holmes

In *The Common Law*, published in 1881, Oliver Wendell Holmes wrote:

> *The life of the law has not been logic: it has been experience. The felt necessities of the time, the prevalent moral and political theories, intuitions of public policy, avowed or unconscious, even the prejudices which judges share with their fellow-men, have a good deal more to do than the syllogism in determining the rules by which men should be governed.*[16]

If Holmes's elevation of "experience" at the expense of logic sounds a bit like the pragmatism of Peirce, James, and Dewey, it is no accident. During the 1870s a small group of men who called themselves "The Metaphysical Club" met periodically in Cambridge, Massachusetts to discuss philosophical issues. This group included Charles S. Peirce, William James, and—Oliver Wendell Holmes. The pragmatic influence in Holmes' thought is unmistakable: "The substance of the law at any given time pretty nearly corresponds, so far as it goes, with what is then understood to be convenient."[17]

One of the purposes of the U.S. Constitution enumerated in the Preamble is to "establish Justice." The ancient Greeks defined justice, in its broadest sense, as treating a man as he deserves. In the narrower, legal sense justice is (or ought to be) the holding men accountable for violating other men's rights. Justice, in this sense, cannot exist without law; law provides an objective basis for administering justice. This is the import of the expression, "a government of laws and not of men." But in Holmes's *Common Law* one will search in vain for a discussion of justice; it is the kind of abstract, *a priori* principle that pragmatists eschew.

Justice as the Greeks and the American Founders conceived of it has a timeless quality to it; it does not fundamentally change from time to time or from place to place. Justice thus conceived tends to invest the law, as well, with an aura of timelessness and permanence. But this view of law conflicts with the pragmatic idea that all laws, like the laws of science, are mere hypotheses that need updating from time to time. (William James was fond of pointing out that the expression, "*the* law," was premised on a convenient fiction.) There was no place in Holmes's legal universe for the idea of a fundamentally unchanging law grounded in an abstract conception of justice.

Indeed, Holmes took pains to show that law is *not* consistently based on abstract principles of any sort. He pointed out, for example, that because "early forms of legal procedure were grounded in vengeance," they at least concerned themselves with personal blame.[18] But modern law, Holmes argued, dispenses with the idea that the law ought to punish only those who are morally guilty of some violation of the law.[19] To illustrate his point, he adduced the principle that "Ignorance of the law is no defense." Because a defendant who is in fact ignorant of the law cannot be morally guilty, Holmes reasoned that the law is therefore not consistently grounded in principles such as morality (or justice), but, rather in "convenience" and "the felt necessities of the time."

In accordance with his implicit denigration of the idea of justice and his pragmatic indifference to abstract principle, Holmes denied the validity of the idea of natural rights. "A legal right is but a *permission* to exercise certain natural powers.... Just so far as the aid of the public force is given a man, he has a legal right, and this right is the same whether his claim is founded in righteousness or iniquity." [20] Not only did Holmes deny the validity of natural rights, he insisted that society's needs must take precedence over those of the individual.

> *No society ever admitted that it could not sacrifice individual welfare to its own existence. If conscripts are necessary for its army, it seizes them, and marches them, with bayonets in the rear, to death. It runs highways and railroads through old family places in spite of the owner's protest, paying in this instance the market value, to be sure, because no civilized government sacrifices the citizen more than it can help, but still sacrificing his will and his welfare to that of the rest.*[21]

This quotation illustrates a type of argument that Holmes used often in his court opinions. As a product of many minds working under various historical circumstances, law can come to embody contradictory principles. The U.S. government, for example, may not force a man to attend church, but it may force him to march off to his death in a war. There are different ways one might respond to such a contradiction; one might conclude that if it is wrong to force a man to attend church, then perhaps it is also wrong to force him to fight in a war. Or one might conclude that if it is permissible to force a man to fight in war, then perhaps it ought to be permissible to force him to attend church as well. The one response would expand liberty at the expense of state power, the other would expand state power at the expense of liberty. As he does in the passage quoted above, Holmes consistently used such contradictions to argue for doing the latter, to argue for expanding the sphere of government power and for diminishing the sphere of individual autonomy.

James Kent provides an instructive contrast to Holmes in this regard. Kent insisted, just as Holmes did, on the need for government to override the rights of individuals under certain circumstances. But Kent consistently worked to erect well-defined limitations around government's powers to do so. Holmes, on the other hand, though he acknowledged the need for some limitations, chose to emphasize time and time again government's power to sacrifice the will and welfare of the individual to those of the collective. Where, for example, Kent tried to solidify important limitations on the police power, "to [Holmes] we are indebted for a definition of the police power which is one of the most sweeping ever to have been enunciated by the Court, a definition which was to be characterized by Harold J. Laski ... as 'the modern charter of the federal state.'" [22] Due largely to his efforts to unshackle the police power, Holmes played a crucial role in bringing about the final demise of the Doctrine of Vested Rights, the doctrine that courts have an obligation to protect the property rights of individuals against legislative infringement.

Examples abound of Holmes opinions that extended (or would have extended, had Holmes not been in the judicial minority) legislative power at the expense of property rights. In the famous 1905 case of *Lochner v. New York*, for example, Holmes would have allowed a statute limiting bakery workers to ten hours per day and sixty hours per week. This was the case that marked the adoption by the Supreme Court of the doctrine of substantive due process, and Holmes went on record as opposing that doctrine. In the 1908 case of *Adair v. United States*, to cite one other example, Holmes would have permitted Congress the power to prohibit railroads from firing workers who joined unions.

Taken as a whole, Holmes judicial decisions were not pragmatic in the sense of lacking a unifying principle. For most of his judicial career he argued much more often than not in favor of governance by majority rule, on the one hand, and judicial deference to legislative enactments, on the other. The guiding "principle" of his jurisprudence was that "The first requirement of a sound body of law is, that it should correspond with the actual feelings and demands of the community, *whether right or wrong.*" [23]

Up until 1919, when he turned 78, Holmes' position on the sanctity of individual rights as against legislative power was similar to that which James Iredell had held back in the early decades of the Republic. Like Iredell, Holmes would have denied judicial protection to "unenumerated" rights, and the protection he would have accorded even to enumerated rights such as free speech was as limited as what judges accorded it in Iredell's time. But Holmes began to formulate an expanded conception of the right of free speech in 1919, when he came under the influence of the arch-Progressive, Louis Brandeis. (This development brought Holmes into compliance with John Dewey's idea that in a democracy the right to *think* freely must be held sacrosanct, though there can be no right to *act* accordingly—as that right is given effect in the judicial protection of private property.)

Holmes was not himself a Progressive. Indeed, his private opinions ran contrary to important elements of the Progressive program. But his judicial espousal of a substantially unfettered democracy, and his willingness to permit the resulting diminution of the rights of private property, created the widespread impression that Holmes was sympathetic to the Progressive agenda.

And he made an inestimable contribution to the Progressive cause with his idea that jurists should interpret the Constitution not as having a fixed meaning that is embodied in the plain language of its articles, but as being a "living" law that each generation is entitled to interpret afresh according to its own "felt necessities" and the "prevalent moral and political theories" of its time.

When we are dealing with words that are a constituent

act, like the Constitution of the United States, we must realize that they have called into life a being the development of which could not have been foreseen completely by the most gifted of its begetters. It was enough for them to realize or to hope that they had created an organism; it has taken a century and has cost their successors much sweat and tears to prove that they created a nation. The case before us must be considered in the light of our whole experience and not merely in that of what was said a hundred years ago. . . . We must consider what this country has become in deciding what that Amendment [the Tenth] has reserved.[24]

Brandeis

Just as in the field of philosophy William James helped pave the way for the Progressive, John Dewey, so in the field of jurisprudence Oliver Wendell Holmes paved the way for the Progressive, Louis Brandeis. Brandeis was an activist liberal lawyer in the modern sense. In 1916 he took his progressivism to the U.S. Supreme Court, where he remained until 1939. He set out early to defeat the Framers' constitutional strategy that would limit American democracy so as to secure property rights against popular majorities. For Brandeis, democracy was primary.

We Americans are committed not only to social justice in the sense of avoiding things which bring suffering and harm, like unjust distribution of wealth, but we are committed primarily to democracy.[25]

When democracy reigns supreme, there can be no place for unalienable rights that are judicially enforceable against the people's legislatures. Thus Brandeis rejected the idea that individuals possess rights which are in any sense prior to government. Rights were, for Brandeis, as they were for Holmes, whatever the government said they were. "All rights are derived from the purposes of the society in which they exist," wrote Brandeis.[26] "Rights of property and the liberty of the individual must be remolded, from time to time, to meet the changing needs of society." [27]

To be sure, Brandeis would allow judicial protection against legislative enactments that constituted "a clear, unmistakable infringement of rights secured by fundamental law."[28] What he meant by "rights secured by fundamental law" were rights that are explicitly enumerated in constitutions or bills of rights, as freedom of religion, speech, and press are enumerated in the U.S. Bill of Rights. Any statute that did not infringe such rights, on the other hand, should be presumed to be constitutional by the courts. Brandeis did not consider property rights to be "secured by fundamental law," so statutes infringing those rights were entitled to a presumption of constitutionality by the courts, he believed. In limiting judicial protection only to enumerated rights and in denying it to property rights, Brandeis intended to abolish altogether the Doctrine of Vested Rights, the century-old idea that property rights merited judicial protection as fundamental constitutional rights.

Brandeis was not just a Progressive; he was also a Pragmatist, and his Pragmatism was part and parcel of his Progressivism. We hear from Brandeis, for example, the same Pragmatist argument in favor of political experimentation that John Dewey had used. Just as Thomas Edison engaged in "mechanical" experimentation, so should American legislatures be free to engage in "social" (i.e., political) experimentation:

> *In any or all of this legislation there may be economic and social error. But our social and industrial welfare*

> demands that ample scope should be given for social as
> well as mechanical invention. It is a condition not only
> of progress but of conserving that which we have.
> Nothing could be more revolutionary than to close the
> door to social experimentation.[29]

To promote political experimentation, legislatures needed to be given as free a hand as possible. Unless it infringed enumerated rights, experimental social legislation needed to be evaluated not according to some fundamental constitutional principle, such as the right of free contract, but according to the idiosyncratic "facts" of the case. This was Brandeis's pragmatism in its clearest expression.

The signal case in this regard was *Muller v. Oregon*. The case involved an Oregon ten-hour law for women workers. Brandeis argued the case for the State of Oregon before the U.S. Supreme Court. Brandeis's strategy in the *Muller* case epitomized the pragmatic mind at work. A jurist of an earlier era might well have argued that *on principle* a person ought to be free to contract with another person on any terms that are agreeable to both— regardless of the peculiarities of the case in question. But to Brandeis the pragmatist, an abstract principle such as freedom of contract was meaningless; all that mattered were the facts of the case in question—in *Muller* the "fact," among others, that long hours of work are bad for a woman's health.

Accordingly, in his *Muller* brief Brandeis devoted just two pages to the legal argument, "merely summarizing the rules applicable to the case," [30] and an unprecedented one hundred-plus pages to statistics and other facts purporting to show that the Oregon Legislature had acted reasonably in enacting the statute. (It was not even necessary, Brandeis argued elsewhere, that the facts on which legislatures based their enactments be true. All that was needed was that the legislatures could reasonably be understood to have believed the "facts" on which they based their determinations.)[31]

Brandeis did not intend to suggest that courts should go into the business of sifting through the mountains of facts upon which legislatures base their enactments. Indeed, precisely because courts are not equipped to evaluate such quantities of facts, he believed, they ought not try; they should leave such matters to the legislatures. Courts should limit their own determinations to those cases that clearly involve matters of law, such as cases involving the infringement of enumerated rights (or other clearly identified constitutional principles such as the separation of powers or the distribution of power between the states and the federal government.) Implicit in Brandeis's *Muller* brief was the admonition that if the Court intended to intervene in matters essentially legislative in nature, then they should do so on the same terms that legislatures deal with such matters, that is, factually, and not in terms of abstract moral or constitutional principles, such as freedom of contract.

We see Brandeis's pragmatism exemplified in the case of *Duplex Printing Press Co. v. Deering*. The management of Duplex had refused to unionize. In retaliation, other unions refused to work on installing Duplex printing presses. Duplex sought an injunction against the boycott, even though the boycotters had not violated the rights of Duplex or anyone else. The U.S. Supreme Court (wrongly) sided with Duplex, in effect forcing the boycotting unions to perform work against their will. Brandeis rightly dissented, pointing out that the boycott was permissible both under the Clayton Act and under the common law.

But the common law, because it tends to embody timeless principles of law and liberty, and because it stands prior to and, to one extent or another, beyond the reach of legislative enactments, is precisely the sort of *a priori* legal principle that Pragmatic Progressives eschew. Accordingly, Brandeis "pointedly disclaimed any purpose to attach constitutional or moral sanction" to the boycotters' actions.[32] In other words, even though the boycotting unions had not violated anyone's rights, and even though their actions had been legal under the Clayton act and fundamentally protected by the common law, Brandeis "pointedly" refrained from arguing that the unions had had either a moral or a constitutional right to engage in their boycott, choosing instead to defend the boycott on the grounds of social policy (the policy being that labor needed all the weapons that legislatures could give them in their struggle against capital).

Like his pragmatic philosophical counterpart, John Dewey, Brandeis embraced democracy at the expense of property rights, and not merely out of indifference to property rights. Rather, in true Progressive fashion, Brandeis positively opposed the natural economic order, the economic order, that is, which is based on *voluntary agreements*. To the Progressive mind, in general, and to Brandeis's mind, in particular, the problem with allowing the economic order to develop along purely voluntary lines was that the resulting arrangement would tend to be unjust and even dangerous to democracy. The injustice would be that some individuals wound become very wealthy while others would not. And the danger to democracy would result from the *bigness* of industrial and commercial enterprises left to grow freely. "There develops within the State a state so powerful that ordinary social and industrial forces are insufficient to cope with it." [33]

Accordingly, Brandeis would "democratize" the economic realm:

> *The end for which we strive is the attainment of rule by the people, and that involves industrial democracy as well as political democracy.... The problems of his business, and it is not the employer's business alone, are the problems of all in it.... You must create something*

Thomas McCaffrey

akin to a government of the trade before you reach a
real approach to democratization.[34]

There are many examples of court cases in which Brandeis argued for the exercise of the legislative power at the expense of economic rights. Two cite just two: in *Stettler v. O'Hara* he defended Oregon's minimum wage statute before the U.S. Supreme Court, and in *Burns Baking Co. v. Bryan* he argued to let stand a Nebraska statute requiring minimum and maximum weights for loaves of bread.

Brandeis was an important judicial spokesman for the twin Progressive goals of increased democracy and decreased judicial protection of economic rights. He influenced American jurisprudence through his testimony before Congressional committees, through his participation as a justice in majority decisions that affirmed legislative (and executive) power to the detriment of property rights, and through his dissents in a great many more cases that affirmed the sanctity of property rights. But even in dissent, his was the voice of the future. His *Muller* strategy won official sanction and succeeded over time in helping to divert judicial attention away from considerations of principle and directing it toward the "facts" of individual cases. After Franklin Roosevelt was through refashioning the Supreme Court, pragmatic Progressives in the Brandeis mold would dominate the Court for generations.

Pound

Perhaps nothing contributed so much to create and foster the [popular] hostility to courts and law and constitutions, which was conspicuous at the end of the nineteenth century and at the beginning of the present century, as the conception of the courts as guardians of individual rights against the state and against society,

> *of the law as a final and absolute body of doctrine declaring these natural rights, and of constitutions as declaratory of common-law principles, which are also natural-law principles, anterior to the state and of superior validity to enactments by the authority of the state, having for their purpose to guarantee and maintain the natural rights of individuals against society and all its agencies.*[35]
> —Roscoe Pound

Roscoe Pound's *Spirit of the Common Law* is remarkable for the sharp, clear lines it draws between the jurisprudence of the nineteenth century and that of the later, Progressive era. It is not what Progressive jurists actually achieved in the first two decades of the twentieth century, so much as what they aspired to, that so distinguishes them from their nineteenth century colleagues. They eventually would achieve virtually all that they aspired to, of course, but not until later, during the New Deal, the Fair Deal, the New Frontier, and the Great Society. So the break between the two eras occurred gradually, thus masking the sharpness of the divide that separated the governing principles of the earlier era from the ideology of the later one. But, as the quotation at the top of this page suggests, the differences were profound.

Whereas the earlier period valued individual rights, the later one valued the prerogatives and interests of society collectively. Whereas the earlier era saw the law that secured those rights as final and absolute, the later one saw it as fluid, evolving, and as relative to time and place. And whereas nineteenth century jurists considered the common law, which they believed to be coextensive with natural law, to be superior to the enactments of legislatures, Progressives considered the enactments of the people, represented by their legislatures, to be the highest form of law. Pound's short classic makes clear that progressivism represented a genuine revolution in American jurisprudence.

If Oliver Wendell Holmes's *Common Law*, published in 1881, anticipated the central themes of Progressive jurisprudence, Pound's book, published in 1921, was able to report substantial progress by those ideas in winning acceptance in American courts. Pound identified the central contradiction at the heart of American law: "While the lawyer as a rule still believes that the principles of law are absolute, eternal, and of universal validity, and that law is found, not made, the people believe no less firmly that it may be made and that they have the power to make it." [36] This is the contradiction, which I have traced back to Locke himself, between the idea of unalienable rights, on the one hand, and the democratic principle of consent, on the other. Pound's solution: "Each of these absolutes must be given up." [37] In practical terms, this meant a dramatic diminution in the status of property rights as judicially enforceable checks upon the power of legislatures.

Like his fellow Progressives, Pound denied the validity of the idea of natural rights.

> *A legal system attains its end by recognizing certain interests, individual, public and social; by defining the limits within which these interests shall be recognized legally and given effect through the force of the state, and by endeavoring to secure the interests so recognized within the defined limits. It does not create these interests. There is so much truth in the eighteenth-century theory of natural rights.* [38]

Thus does Pound reduce natural rights to the status of mere "interests." Which of these interests government ought to secure is a matter of expediency. "For legal rights, the devices which the law employs to secure such of these interests as it is expedient to recognize, are the work of the law and in that sense the work of the state." [39]

Pound identified the most pressing problem in American law at the turn of the century as the stifling of much-needed "social" legislation by a too-strict judicial enforcement of individual rights, especially the rights of property. He argued that fundamental change was needed. And, by means of his analysis of the history of legal development in Europe and the United states, he was able to portray such change as part of an ongoing, natural process of legal evolution. In Pound's account, the 1700s were a time of change and liberalization in English law under the influence of the ideas of natural law and natural rights. This stage culminated, in America, in the great age of constitution-making and in our bills of rights. With the end of this period of change and liberalization, there then followed a period of consolidation and stabilization, during which there was little fundamental change, and during which time courts tended to administer the law with increasing inflexibility. This period lasted, in America, until about the end of the 1800s.

On the basis of this historical account, Pound was then able to portray progressivism as a natural, timely, and much-needed return of a period of change and liberalization in American law analogous to that which had ushered in the natural law and natural rights conceptions of an earlier era. (The application of the idea of evolution to a field other than biology, in this case the history of law, reflected the influence of Darwin's theory of evolution and was characteristic of much of Progressive thought.)[40]

The historical process of legal evolution, as Pound portrayed it, is not a story of forward progress to an ever more just world. Pound's account is remarkable for its non-judgmentalism as to the relative merits of the respective stages of legal development that he described. The Progressive era, for example, was not more moral or more just that the era of natural rights which preceded it. The reforms of progressivism were simply more attuned to the needs of an urban, industrial society than were the individualist conceptions of the earlier era.

> *Thus we may think of the task of the legal order as one of precluding friction and eliminating waste; as one of conserving the goods of existence in order to make them go as far as possible ... so that where each may not have all that he claims, he may at least have all that is possible.*[41]

The end of law, per Pound, is not justice, not the *moral* problem of securing to each that which he deserves, but the purely *practical* matter of securing to each "all that is possible," given the competing claims of other individuals and of society collectively. Pound does not eschew the use of the concept "justice" altogether. "It is true the world wide movement for socialization of law, the shifting from the abstract individual justice of the past century to a newer ideal of justice, as yet none too clearly perceived, is putting a strain on law everywhere." [42] "Justice" here is relative; there is no such thing as absolute justice, there are only those conceptions of justice which each age finds expedient.

Indeed, in *The Spirit of the Common Law* Pound makes the remarkable assertion that the rise of progressivism marks the reemergence in the common law of its feudal element, which had lain largely dormant throughout the century-long dominance of the natural rights idea. Whereas the nineteenth century's individualist conception of law dealt with a man based on what he did, the feudal conception dealt with him based on what he was. Feudal law treated individuals in terms of groups and relations, such as landlords and tenants, with attendant "rights, duties and liabilities arising not from express undertaking, the terms of any transaction, voluntary wrongdoing or culpable action, but simply and solely as incident of a relation." [43] Progressive-era statutes that require an employer to compensate an employee for an on-the-job injury not because the employer is in any way at fault, but simply because he is the employer, are an example of this feudal element in Progressive jurisprudence.

Feudal law, in Pound's estimation, is no less (nor more) just than is law grounded in natural rights. Sir Henry Maine had asserted in the nineteenth century that the development of law is a progress from status to contract. Pound was able to report that Sir Henry had obviously been incorrect, given that by 1921 the direction of legal evolution had clearly turned back toward status. [44] Pound's book is commendable for the frankness with which it portrays the rise of progressivism as a departure from the dynamic individualism of the nineteenth century and as a return to ideas more characteristic of the stasis of the feudal era.

Pound also evinced the Pragmatism that we have seen was so prominent a part of the thinking of Holmes and Brandeis. But whereas Brandies had argued that courts should not second-guess legislatures on social questions because only the legislatures were equipped to deal with the mass of factual data needed to fashion such legislation, Pound boldly recommended that the courts should indeed begin to equip themselves to process the great quantities of facts needed to validate social legislation. [45] The days of courts' deciding constitutional questions primarily on the basis of fundamental law and abstract principles were coming to an end.

As we have seen, this movement away from deciding constitutional questions on the basis of long-established constitutional and legal principles and toward deciding them on a case-by-case basis depending on the "facts" of the matter was integrally related to the effort to undermine the status of property rights as judicially enforceable checks on the power of legislatures.

In *The Spirit of the Common Law*, Pound identified "[e]ight noteworthy changes in the law in the present generation, which are in the spirit of recent ethics, recent philosophy and recent political thought." Four of these involved diminutions in the judicial protection afforded property rights. These included diminutions of owners' rights to use property for "anti-social" purposes, such as for building of "spite fences"; limitations on the freedom of contract, such as minimum wage laws; diminutions of the rights of creditors or injured parties to win satisfaction, such as homestead exemption statutes; and the revival of the idea of liability without fault, as exemplified by worker's compensation statutes. A fifth such change involved the exerting of public control over resources that had previously been managed privately; Pound expressed it as "a very marked tendency in judicial decision to regard the social interest in the use and conservation of natural resources." [46]

By the early 1940s, the Doctrine of Vested Rights had ceased to exist in American jurisprudence. For the next two generations, the courts would offer virtually no protection to property rights against legislative and executive infringement. The success of the effort to undermine the status of property rights as judicially enforceable checks upon the power of democratic pluralities and their legislatures was the crowning achievement of the Progressive movement in American jurisprudence. To a great extent, this movement owed its success to Oliver Wendell Holmes, Louis Brandeis, and Roscoe Pound.

The Moralists

In *Rendezvous with Destiny* Eric Goldman describes how in the late nineteenth century American Protestantism joined in the call for Progressive reform. Among the writings that influenced Protestant reformers were those of the Englishmen, John Ruskin and William Morris.

<u>Ruskin</u>

John Ruskin (1819-1900) was a writer and an art critic. In an 1860 series of magazine articles titled "Unto This Last" he identified the fundamental contradiction between England's dominant, Christian ethics and an economic system grounded in the pursuit of private profit.

> *I know no previous instance in history of a nation's establishing a systematic disobedience to the first principles of its professed religion. The writings which we (verbally) esteem as divine, not only denounce the love of money as the source of all evil, and as an idolatry abhorred by the Deity, but declare mammon service to be the accurate and irreconcilable opposite of God's service: and, whenever they speak of riches absolute, and poverty absolute, declare woe to the rich, and blessing to the poor.*[47]

Capitalism, on the other hand, gives rise to "economic man," who consults his own rational self-interest and makes his decisions on that basis. But Ruskin disagreed that the term "economic man" either does or ought to describe an actually existing person. He argued that the force of the human soul "enters into all of the political economist's equations, without his knowledge, and falsifies every one of their results."[48] Man, believed Ruskin, neither is nor ought to be either primarily self-interested or entirely guided by coldly calculating economic rationality. "The largest quantity of work will not be done by this curious engine for pay.... It will be done only when ... the will or spirit of the creature, is brought to its greatest strength by its own proper fuel; namely, by the affections."[49]

In opposition to "economic man," Ruskin posited Christian man. Ruskin's Christian was, above all else, unselfish in economic matters. Comparing the moral status of the merchant to that of lawyers, physicians, clergymen, and generals, Ruskin found the merchant lacking because he was "presumed to act always selfishly." [50] Ruskin's great objective was to Christianize commerce; he was out to persuade the merchant to place his Christian obligation to serve others ahead of his own selfish interests. "[T]he merchant's function ... is to provide for the nation. It is no more his function to get profit for himself ... than it is the clergyman's function to get his stipend." [51] And just as the general must risk his life for his countrymen, so also must the merchant risk "any form of distress, poverty, or labour" rather than diminish the quality of his product or charge unfair prices for it. [52]

Central to Ruskin's conception of Christian commerce was the idea of the "just price." To his way of thinking, all exchange is, in the final analysis, an exchange of labor; the proper recompense for any quantity of labor is an equal quantity of labor of equal quality, or an equivalent thereof. If it takes an hour's labor to make a pair of shoes, and if shoemaking is roughly equivalent to carpentry, then a pair of shoes would be "just" remuneration for an hour's worth of carpentry. Everything offered in exchange, in this view, has a single, just price. To take advantage of a distressed buyer by demanding a higher price, or of a distressed seller by offering a lower price, would be unjust, immoral, and un-Christian. The only just profit is that which derives from the creation of value where none existed before—the farmer who plants one bushel of corn and harvests two is entitled to the "profit" of the second bushel, for example. But profit that arises purely as a result of changes in supply and demand is unjust.

To implement the "just price" idea would require the repeal of the law of supply and demand, a fact of which Ruskin was aware. "I believe the sudden and extensive inequalities of demand, which necessarily arise in the mercantile operations of an active nation, constitute the only essential difficulty which has to be overcome in a just organization of labor."[53] To overcome them Ruskin recommended, for one thing, that employers pay lower wages for steady work, rather than paying higher wages during periods of high demand and then laying off workers when demand falls. Prices come about, of course, from the *competition* of market participants for scarce goods and services. But competing is an un-Christian way of relating to other men, said Ruskin, who would have substituted cooperating instead. He also counseled men to be satisfied with their station in life, whatever it might be.

Ruskin viewed the production and accumulation of wealth as a zero-sum proposition; in order for one man to become wealthy, others must become poor. And indeed, as Ruskin saw it, a man who would become wealthy *has an interest in beggaring others*; for one thing, if all a man's neighbors were as wealthy as he, then they would only drive up the prices he must pay for scarce resources. For another, the only way a wealthy man could hope to hire the help he needed would be if others were poor enough to need the employment.

All of these conceptions—the "just price," the criticism of economic competition, the recommendation that men stay put in whatever social stratum they happen to find themselves, and the conception of wealth production and accumulation as a zero-sum endeavor—point to a *static* economy as Ruskin's ideal; the social dynamism of capitalist industrialism was a major object of his antipathy. Ruskin believed that wealth afforded some men power over others. He did not object in principle to one man's ruling another; indeed, he advocated it. "My continual aim has been to show the eternal superiority of some men to others, sometimes even of one man to all others; and to show also the advisability of appointing such persons or person to guide, to lead, or on occasion even to compel and subdue, their inferiors according to their own better knowledge and wiser will." [54] What he objected to in capitalism was the *kind* of men who tend to become wealthy—and, as he saw it, powerful. "[I]n a community regulated only by the laws of demand and supply, but protected from open violence, the persons who become rich are, generally speaking, industrious, resolute, proud, covetous, prompt, methodical, sensible, unimaginative, insensitive, and ignorant." [55]

To his credit, Ruskin favored the Christianizing of the British economy primarily not by force of legislation but by rational persuasion. Another Englishman, arguing from similar Christian premises, favored a more coercive means of Christianizing English economic life.

<u>Morris</u>

William Morris (1834-1896) was an English artist and writer. A person of refined esthetic sensibility, Morris was repelled by modern industrial manufacturing, especially by what he judged to be the poor quality of manufactured goods and by the displacement of skilled craftsmen by what he considered to be demoralized factory workers. In an effort to restore the place of crafts and craftsmen, he helped to start in 1861 what later became famous as Morris and Company. The company used skilled craftsmen to turn out high quality goods such as carvings, stained glass, carpets, wallpaper, and chintz, and it helped give rise to the Arts and Crafts movement with which Morris's name is associated. In 1891, Morris published a utopian novel, *News from Nowhere*, which contributed to the rise of progressivism in the United States by encouraging opposition to capitalist industrialism.

News from Nowhere is the story of a late nineteenth century Londoner who finds himself mysteriously transported two hundred years into the future. The people of this future England have long ago become disenchanted with industrialism and they have overthrown the entire capitalist order. As they see it, private property is the root of all evil. Before their great revolution, private property had given rise to an idle ruling class and an enslaved working class—social and economic inequality were self-evidently evil in Morris' view. Private property also gave rise to crime, by creating a permanently deprived class of people. It thus necessitated the instituting of police, courts, and prisons, i.e., government, in order to protect the property of those who have it from those who do not. (Morris agreed with Locke that the primary purpose of government is to secure the rights of private property. But for Morris, in contrast to Locke, the society based on private property and freedom of contract is fundamentally a society grounded in coercive relations among men, rather than in voluntary cooperation.) Property also gave rise to competitive relations among men, which Morris would have eschewed in favor of cooperative relations.

In *News from Nowhere*, Morris's protagonist is a self-proclaimed communist. In his utopian England of 2100 AD, there is no private property. People live where they choose, often communally, with whomever they choose. There is no industrialism. People produce what they need mostly by hand or with simple machinery, and they take great pride in their workmanship. There is no money. People work simply for the joy and spiritual reward of working, which Morris sees as the great well-spring of human motivation. And there is no government, because the absence of private property has obviated the need for it.

But the dominant theme throughout *News from Nowhere* is esthetic. Morris' utopia is, above all else, esthetically pleasing. The people of England have become universally attractive, both in body and in spirit. They are joyful and friendly and generous. Even their clothes are attractive, clothes being a subject of some importance to Morris. Their architecture is equally attractive, with the exception of what remains of the era from the

Renaissance through the industrial age. (Morris idealized the pre-Renaissance Middle Ages.) Then there is the land, the hills and fields and forests and rivers. Morris' utopian England has become a garden. Much of London and the entirety of the Midland industrial cities have been leveled. Most persons have taken to living in small villages, much as people did in the Middle Ages. Salmon even swim in the Thames. (Morris's ideas prefigured a great deal of what we now call environmentalism, including the high priority assigned to esthetic considerations.)

One purpose of Morris's portrait of a happy, healthy people living in peace and cooperation in a garden-like landscape of abundance and beauty was to undermine men's confidence in the *morality* of the economy based on private property and of its most prominent achievement, industrialism. If indeed private property was the cause of so much misery and ugliness as Morris alleged, and if indeed the simple abolition of property could produce such happiness and beauty as Morris claimed, then the moral case for property rights would be seriously undermined. According to Eric Goldman, Morris's Christian critique of capitalist industrialism, and that of John Ruskin before him, contributed to the rise of an energized, reform-minded Protestantism in the United States.[56] Among the Americans who took up the Christian critique of capitalist industrialism propounded by Ruskin and Morris was Edward Alsworth Ross.

Ross

Edward Alsworth Ross (1866-1951) grew up in Marion, Iowa, attended tiny Coe College, did advanced study in Germany, completed his studies "among the German-trained rebels at Johns Hopkins," and became one of America's pioneer sociologists.[57] He spent the bulk of his career, from 1906 to 1937, in the Progressive hothouse at the University of Wisconsin. In 1907, he published *Sin and Society: An Analysis of Latter-Day Iniquity* which would prove influential in the quickening development of progressivism. The book's introduction, a letter from Theodore Roosevelt, attests to the influence of *Sin and Society*.

We have seen how, influenced by Darwinian ideas about evolution, Brandeis and Pound set out to replace static conceptions of justice and law with fluid ones, ones that evolve over time. In *Sin and Society* Ross applied this same idea to certain classes of sin. He argued that a new age characterized by large-scale, corporate enterprise required new, expanded conceptions of what constitute moral transgressions. He believed that, whereas people learned long ago the degree of moral opprobrium they ought to attach to such violent crimes as robbery, assault, and murder, the modern economy had brought into being a host of new offenses that Americans had not yet learned to deplore to the degree Ross thought they should. The list of sins and sinners that Ross nominated for increased popular–and legal–condemnation was a long one.

One category included bribery, graft, ballot-fraud, illicit rebating, and embezzlement of public funds. These are the kind of offenses that tend to proliferate as government begins to insert itself into private economic affairs. Ross criticized, for example, "tariff-protected businesses, the railroads, the public utility corporations, telegraph, telephone, express, lumber, coal, oil, insurance, and the various trusts" for exerting, either licitly or illicitly, a too-great influence over the government's regulation of their respective fields of endeavor.[58] On the one hand, Ross censured politicians and regulators who took advantage of the rapidly expanding powers of government for purposes of self-enrichment. On the other, he demonized those captains of industry who tried to counteract, by legal or illegal means, the new forces of regulation that assailed them.

From either perspective, this degree of government regulation of economic life was new in America, and it certainly did give rise to immense opportunities for abuse, among regulators and regulated alike. These were indeed new "sins" in the American landscape, and Ross' purpose in *Sin and Society* was to persuade a slow-to-catch-on public that these government-related sins were as evil as the older, traditional sins.

Other "sins" that Ross wanted to raise in the public consciousness were "monopoly," offenses against consumers (worthless patent medicines, medical quackery, fly-by-night construction), offenses against occupational safety (unsafe mines, unfenced machinery in factories), offenses against labor (child labor, night labor for women, labor contracts that prohibit union membership), improper tenement housing, and countless more of this ilk. In these cases, Ross' task was more difficult, since most of these offenses did not involve the violation of anyone's rights in the sense that Americans were accustomed to think of such violations. Nowhere in *Sin and Society* did Ross present a reasoned argument as to why these new kinds of offenses should have been classified as sins, much less why they should have been made illegal. He simply treated it as self-evident that the *need* of the "victims" should take precedence over the *rights* of the "offenders." The alleged need of laborers, for example, to have a labor union should trump the rights of employers to hire or fire whomever they chose.

In giving precedence to the needs society's less "powerful" classes, Ross was being true to the altruist premise inherent in America's dominant, Christian morality. That premise had long stood in contradiction to the individualist morality that underlay the principle of individual rights, but it had lain dormant as a force in American political life. Before the late nineteenth century, there had never been an organized movement to give political expression to the altruist impulse inherent in American Christianity. The genius of the Progressive revolution was its tapping into this moral vein, and Ross' work was characteristic in this respect.

Consider the following from *Sin and Society*:

> *"Conservatism!" piled on top of inertia and the strangle-hold of sinister interests, in a tumultuously changing society, where an evil condition may be rapidly worsening while we speechify and procrastinate! Here is growing evil,—so much blood of brakemen on cars and rails. Give heed, ye legislators! No impression. The legislator removes his cigar long enough to sneer, "hot air," "mawkish sentimentality," "they take the risks." So, on with the slaughter. Let the wheels redden until the totals are formidable. "Now will you act?" No, "interference" would "undermine individual responsibility," or be "unconstitutional." So let the mangled pile up.*[59]

If railroad employees were being injured or killed in accidents, then regardless of whether the railroad owners were at fault, the government needed to intercede and force changes upon the owners. The workers' need for relief trumped the rights of the railroad owners to run their businesses as they saw fit.

As the passage above makes clear, Ross understood that his effort to vilify the employers placed him in opposition to fundamental American politico-economic principles, which he dismissed with a rhetorical wave of the hand, "individual responsibility" and "unconstitutional[ity]" amounting to so many rationalizations. But elsewhere in *Sin and Society* Ross argued that he opposed neither individual rights nor private property. Whereas the socialists advocated outright public ownership of the means of production, for example, Ross advocated leaving ownership in private hands, though he called for transferring the *control* of property to the public, i.e., the government.

Thomas McCaffrey

If we recall that the definition ownership includes the rights to possess, to use, and to dispose of a thing, whereas the socialist would have transferred all three to the public, Ross would have left the first, the right to possess a thing, in private hands and would have transferred only the second and third to the public. (Private "ownership" shorn of the right to control the use to which property is put is every bit as much socialism as is outright public ownership.) In an effort to avoid being identified with outright socialism, Ross would have preserved what is, in effect, merely the outward appearance of capitalism. He called his new hybrid "transfigured individualism." [60]

Despite his protestations in favor of capitalism and private property, Ross's "transfigured individualism" evinced a deep-seated distrust of profit-seeking:

> [T]he active producers, such as farmers and workingmen, think in terms of livelihood rather than of profit, and tend therefore to consider the social bearings of conduct. Intent on well-being rather than pecuniary success, they are shocked at the lenient judgment [which society accords] the commercial world. Although they have hitherto deferred to the traders, the producers are losing faith in business men's standards, and may yet pluck up the courage to validate their own ethics against the individualistic, anti-social ethics of commerce.[61]

Ross' elevating of "livelihood" and "well-being" above profit and "pecuniary success," and his similar treatment of farmers and workingmen at the expense of "traders" and their "individualistic, anti-social ethics" bespeak the limited range within which he considered the pursuit of self-interest to be morally acceptable. They suggest that his 'transfigured individualism" was a good deal closer to the ideas of the socialists he criticized than it was to the ideas of the men who drafted the U.S. Constitution. In this respect, his thinking was typical of an important moral strain within Progressive thought.

Rauschenbusch

The most systematic treatment of these moralist themes came from the pen of an American Baptist clergyman. Walter Rauschenbusch (1861–1918) worked among the German immigrants in New York's "Hell's Kitchen." In 1902 he became professor of church history at Rochester Theological Seminary. Through his books, *Christianity and the Social Crisis* (1907) and *Christianizing the Social Order* (1912), Rauschenbusch became a leader in the effort to enlist Christianity in the Progressive reform movement.[62]

In *Christianizing the Social Order*, Rauschenbusch observed that in the past Christianity had focused on saving individual souls, and saving them not for any benefits that might accrue for this life on this earth, but for benefits that would follow in another life. Influenced by John Ruskin, among others, Rauschenbusch advocated that Christianity radically alter its focus and devote itself to improving life for persons in this life. More to the point, he argued that Christianity should set about transforming the economic life of America to accord with the requirements of Christian morality.

> *The chief purpose of the Christian Church in the past has been the salvation of individuals. But the most pressing task of the present is not individualistic. Our business is to make over an antiquated and immoral economic system; to get rid of laws, customs, maxims, and philosophies inherited from an evil and despotic past; to create just and brotherly relations between*

*great groups and classes of society; and thus to lay a
social foundation on which modern men individually
can live and work in a fashion that will not outrage the
better elements in them. Our inherited Christian faith
dealt with individuals; our present task deals with
society.*[63]

Rauschenbusch identified the great, fundamental
contradiction between America's dominant, Christian morality of
self-sacrifice and an economic system grounded in private gain.
"We had long been living a double life, but without realizing it.
Our business methods and the principles of our religion and of our
democracy have always been at strife."[64] Christian morality
teaches that one should love thy neighbor as thyself. To an
important extent, it consists of service to others. America was
finally rediscovering her Christian morality, said Rauschenbusch.
"A new sense of duty ... a new capacity of self-sacrifice"
characterizes the "moral adolescence" through which America is
passing, he said, as she awakens to the "evil" of the reigning
economic order.

Rauschenbusch echoed John Ruskin on "the five great
intellectual professions": "The Soldier's profession is to defend
[society]; the Pastor's is to teach it; the Physician's is to keep it in
health; the Lawyer's to enforce justice in it; the Merchant's to
provide for it."[65] Each of these professions, in others words, is
conceived of in terms of a *duty* to society. The merchant who
places private profit above his duty to provide for society is
immoral. "If the relation of the Merchant to the nation is founded
altogether on selfishness, and has no sacrificial qualities in it, he
may get money, but he will not get honor and love, and the
Business which he fashions and rules will remain an un-
Christianized portion of the social order."[66]

To understand Rauschenbusch's meaning here, we need to

understand his criticism of the profit motive. Like Ruskin, he opposed the placing of profit ahead of other considerations, such as the welfare of one's countrymen. Just as the soldier should be prepared to die for his country, so also should the merchant be prepared to risk financial ruin before, say, raising his prices to cover rising costs.

Rauschenbusch distinguished between what he called earned profits and unearned profits. "Work and service must become the sole title to income," said Rauschenbusch.[67] The workman who is justly remunerated for the work he performs is entitled to his "profit." So is the manager who brings exceptional knowledge and experience to his job entitled to a higher "profit" than an unskilled workman. In certain circumstances, even the capitalist investor is entitled to some return on his investment to compensate him for the risk he has incurred. These are all examples of "earned" profits. But should a temporary labor shortage enable a worker to demand more than the usual going rate for his type of work, then for him to so would be to exploit the employer's misfortune. This would violate the Christian imperative to love one's neighbor as oneself. To "drive a hard bargain" in this way is to take money out of another's pocket to which one has no just claim. Rauschenbusch called it extortion.

Another example of "unearned profit" would be the income of a corporation founder's great grandson who merely sits back and watches the value of his stock appreciate. Likewise the profit of the landowner whose land value increases due to factors for which he can claim no credit. In each case, he who receives the profit has no moral claim on it because he has not given equal value in exchange for it. The effect of Rauschenbusch's criticism of the profit motive was to morally sanction the pursuit of economic gain only to the extent that it benefits others equally as it benefits oneself. Even contracts freely entered into were not morally legitimate if the exchange agreed upon was not equal

Thomas McCaffrey

according to the concept of the just price. Rauschenbusch thus delegitimized whole categories of profit, morally eviscerating capitalism in the process.

Rauschenbusch's critique of profit is based on the concept of the "just price," which prevailed in the relatively static economic life of medieval times, when Christian morality dictated economic law. The idea resurfaces today every time we have a natural disaster, such as a hurricane, that causes shortages in certain commodities; these shortages tend to cause prices to rise briefly, and the rising prices often evoke declamations of "price gouging," which is a violation of the just price principle. The idea of the just price is utterly unworkable in a dynamic capitalist economy.

The Christian critique of capitalism manifests itself today in many other ways. Besides obstructing the normal functioning of the law of supply and demand, the just price principle also has the effect of de-legitimizing large, private concentrations of wealth, *per se*, since any individual who limited himself to accumulating wealth according to Christian principles could never become vastly more wealthy than the people around him. From the Christian perspective, a man has less of a moral claim on the second $100,000 of his annual income than he has on the first $100,000. Thus today's graduated income tax, which was, not coincidentally, an important element of the Progressive agenda.

Consider the following declaration agreed upon by the thirty three Protestant denominations represented in the Federal Council of Churches of Christ in December, 1908. Rauschenbusch wrote that this declaration represented "the common sense of the Protestant churches of America" at the time. "We deem it the duty of all Christian people to concern themselves directly with certain practical industrial problems," says the dedication.[68] Among the items for which "the churches must stand" are the following:

- for the right of workers to some protection against the hardships often resulting from the swift crises of industrial change;
- for the protection of the worker against dangerous machinery, occupational disease, injuries, and mortality;
- for the abolition of child labor;
- for such regulations of the conditions of toil for women as shall safeguard the physical and moral health of the community;
- for the gradual and reasonable reduction of the hours of labor to the lowest practicable point, and for that degree of leisure for all which is the condition of the highest human life;
- for the release from employment one day in seven;
- for a living wage as a minimum in every industry, and for the highest wage that each industry can afford;
- for the most equitable division of the products of industry that can ultimately be devised;
- for the suitable provision for the old age of the workers and for those incapacitated by injury;
- for the abatement of poverty.[69]

With one exception, every one of these recommendations incorporates, in one way or another, the Christian principle that one person's *need* constitutes a moral claim on someone else.[70] The need, for example, of women not to work long hours, and the need of the community that they not do so, morally justify depriving employers (and women employees) of the right to contract for whatever hours of work they judge best. Before the end of the twentieth century, every one of these recommendations, and countless others like them, would become law as need would take its place alongside rights as a moral foundation of law in the economic realm.

According to the principle of Christian selflessness, virtually any kind of need on the part of one man can constitute a reason for curtailing or expropriating the property rights of another man. If morality requires that one treat one's neighbor as oneself, then no one with more property than he needs to survive is morally entitled to the entirety of it whenever someone else is in need. On this principle the whole of the welfare state is erected; the welfare state is Christian charity made law. No one who subscribes to the Christian principles of selflessness, self-sacrifice, and duty is in a position to object in *on moral grounds* to the welfare state, or to any of the countless other need-based infringements of property rights that Progressive liberalism has given us. The Christian critique of capitalism thus helped pave the way for an open-ended expansion of government power to intervene in the economic life of America.

This vast expansion of government power at the expense of the economic rights of individuals occurred hand in hand with a dramatic expansion of the political power of the demos, the people at large. Just as the Framers of the U.S. Constitution had sought to make American government less susceptible to the forces of democracy in order to make economic rights more secure, so the Progressives sought to tilt the constitutional balance back towards democracy precisely in order to extend the power of government over the property of individuals. And just as the Christian morality of selflessness helped to justify this infringing of property rights, so also would Christianity's history of governance-by-coercion find itself repeated in the modern guise of progressivism.

Rauschenbusch argued that Christian selflessness is integrally related to political democracy.

The idea of the Kingdom of God was filled with democratic spirit, but it had come down from despotic times and was cast in monarchical forms. The Messiah was expected as a king, and his followers hoped to rule as his courtiers. Jesus flatly contradicted such expectations and laid down the law of service as the fundamental law of his kingdom. He himself had not come to be served, but to serve to the death, and all greatness in the kingdom would have to rest on the same basis. Modern democracy is destined to establish the same principle and to abolish all lordships that cannot show their title on that basis.[71]

Thus did Rauschenbusch enlist Christianity in the Progressive effort to make America more democratic. Along with most Progressives, Rauschenbusch advocated popular participation in party primaries, direct legislation (today's "initiative"), recall, and the direct popular election of U.S. senators.[72]

But Rauschenbusch went far beyond advocating just the expansion of *political* democracy. He objected to what he saw as capitalism's tendency to subject the many who labor to the few who own the means of production.

The moral objection [to industrialism] lies, not against the size and complexity of the modern system, but against the fact that this wonderful product of human ability and toil with its immense powers of production has gravitated into the ownership and control of a relatively small class of men.[73]

Rauschenbusch opposed this subjugating of the many by the few because "Such power on the one side and such weakness on the other constitute a solicitation to sin to which human nature ought never to be subjected. None of us is good enough to hold the lives of his fellow men in the hollow of his hand." [74] He argued that such concentrating of power in the hands of a few contradicted not only Christian morality but also the American political principle that power ought always remain widely dispersed. [75] So he called for an economic democracy to complement the political democracy that Progressives aspired to erect. "Political democracy without economic democracy is an uncashed promissory note ... a form without substance." [76]

Pure democracy, which means pure majority rule, is, of course, nothing less than rule by brute force. When democratic majorities are able to legislate *without limitation*, which is what pure majority rule means, there can be no such thing as unalienable rights of life, liberty, or property. As I discussed in Chapter 2, the Framers of the U.S. Constitution were acutely aware of the danger that democratic majorities pose in this respect to the rights of individuals. For this reason, they established a constitutional republic that placed important limitations on the political power of the demos.

Neither Walter Rauschenbusch nor Progressives generally advocated the instituting of pure democracy in the United States, so neither advocated complete rule by brute force. But to the extent that they did advocate the tilting of the constitutional balance in favor of democracy *specifically as a means of extending government control over economic matters*, to that extent they advocated the introduction into American governance of rule by brute force, because constitutionally protected property rights are the primary means of guarding individual liberty against arbitrary mob rule. But as a Christian Rauschenbusch, drew upon a heritage rich in governance by force.

Under the Lockean conception of rights, the primary evil is the use of force in human affairs; force is evil because it prevents individuals from acting according to their own rational judgment.

Government exists to protect private persons from the use of force by other private persons (which it accomplishes by retaliating against those persons who initiate the use of force against other persons). But the government itself must be strictly prohibited from initiating the use of force against its own citizens. There is only one enforceable commandment under Lockeanism, and that is the negative commandment, which applies to private individuals and governments alike, not to commence the use of force.

In contrast to Lockean societies (of which the United States is the prime example), throughout much of Christianity's history its adherents have shown a marked predilection to employ force to get men to obey God's commandments. During medieval times in Europe, this tendency manifested itself in economic matters.

> When the Church gained its wonderful ascendancy over the social life of the Teutonic nations, it set itself against those economic tendencies which have since resulted in capitalism. It prohibited the breeding of money by interest. It discountenanced the unlimited accumulation of wealth. It condemned monopoly profit and taught the doctrine of the "fair price" as a just reward for labor cost. To take advantage of a man's necessity to raise the price was regarded as extortion. The Jews, who were not under the law of the Church, were free to practice the principles of capitalism centuries before its time. That made them rich; it also made them hated and despised.[77]

With the coming of the Renaissance, Christianity's hold on the social and economic life of Europe began to loosen. The beginnings of capitalism soon followed. By virtue of his advocating the separation of church and state, John Locke helped to end government enforcement of Christian morality in the economic sphere. With their state and federal constitutions and bills of rights, the Americans, by institutionalizing the separation of church and state, also institutionalized the separation of their economic life from a government-imposed Christian morality.

What Rauschenbusch and the reform Protestants—and the Progressives—were calling for was, in effect, a restoration of government-imposed morality on the economic life of the West, a restoration of something that had begun to disappear from Europe four hundred years earlier, and that had not existed to any great extent in the United States since she had come into being. The enormity of this attempt to reverse four hundred years of history cannot be exaggerated. Even Rauschenbusch failed to appreciate the true significance of this reversal, though he wrote that "All this is nothing less than a political revolution. It is a second war of independence." [78] It was that and much, much more.

It is ironic that today American liberals decry attempts by the religious right to "impose their morality on America," when, in fact, the very essence of modern-day liberalism in the United States—the single idea that, more than any other, defines modern liberalism—is the attempt to impose the morality of Christian selflessness on the American polity, a polity which in 1789 became the first nation in history implicitly to draw a line of separation between its economic life and *any* government-imposed religious morality.

The Economists

What Holmes, Brandeis, and Pound did in the realm of law, Richard T. Ely and Thorstein Veblen did in economics. Whereas the jurists undermined the idea of permanent, timeless principles of law, so did Ely and Veblen undermine the idea of permanent, universally valid economic principles and institutions.

Ely

Richard T. Ely (1854-1943) was born at Ripley, New York, and educated at Dartmouth and Columbia. He took his Ph.D. at Heidelberg, Germany, and went on to become a highly influential teacher of economics at Johns Hopkins, Wisconsin, and Northwestern. Ely helped found the American Economic Association, and his *Outlines of Economics* was a standard college text for years.

In *The Past and Present of Political Economy*, which he published in 1884, Ely contrasted the older, classical political economy of the late eighteenth and nineteenth centuries with the newer political economy of the Historical School, which he encountered at Heidelberg. As Ely observed, the older political economy, originated by Adam Smith and elaborated by David Ricardo and John Stuart Mill, among others, depended heavily upon deductive reasoning from certain premises. Among these premises was the idea that self-interest is the "animating and overwhelmingly preponderating cause of economic phenomena," and the idea that the pursuit of self-interest by private individuals tends to redound to the good of society as a whole.[79] This last is, of course, Adam Smith's conception of the "invisible hand." [80] The classical political economists considered ideas like these to be laws of nature, analogous to the natural law doctrine of John Locke.

If it accords with the law of nature for men to pursue their self-interest, and if it is a further law of nature that the pursuit of their self-interest by private individuals will tend to benefit society as a whole, then it stands to reason that governments should, as a rule, refrain as much as possible from interfering in the economic activity of their citizens. Thus the injunction, *Laissez-faire, laissez-passer* (Let things alone, let them take care of themselves), which came to epitomize the view of the classical economists on the subject of government intervention in economic matters.

But Ely denied that there can exist natural laws of political economy that are universally valid in the way that, say, Newton's laws of physics are universally valid. He argued instead for a science of political economy not as "something fixed and unalterable, but as a growth and development, changing with society." [81] Germany's Historical School, which Ely advocated as the preferable, newer alternative to the classical economists, treated political economy as an organic outgrowth of a people's past, as something peculiar to a certain time and place, much the way, say, that classical Greek culture was peculiar to the Attic Peninsula in the period of 400 B.C.

According to this view, there can be no single set of politico-economic principles that are proper to all people everywhere at all times. Rather, which principles are proper to a given people at a given time are a function of their history, their culture, and even their geography. It would be wrong, for example, to say that all societies should adopt the institution of private property. According to the Historical School, that institution would only be proper to certain peoples at certain times and under certain circumstances. Under Ely's idea of politico-economic truth, the American Founders' conception of economic liberty had no more claim to universal validity than did their fashions in clothing or furniture.[82] (There is a family resemblance between Ely's conception of truth and William James's conception, which would appear some twenty years after Ely's; both were products of the same general revolt against Enlightenment rationality.)

Having thus disposed of classical political economy's truth that all governments everywhere should "Let things alone, let them take care of themselves," Ely argued that government ought to intervene in economic matters whenever "experience" indicates that such intervention is needed (another family resemblance to Jamesian pragmatism). On the basis of America's experience with industrialism, for example, Ely advocated a wide range of interventions, such as the regulating of female labor, pro-union labor legislation, and publicly funded education. Ely's characterization of the nature of government power anticipated the aura of mystical omnipotence that government would come to possess in the twentieth century. "[The new economists] recognize, therefore, a kind of *divine right* in the associations we call towns, cities, states, nations, and are inclined to allot to them whatever economic activity nature seems to have designed for them, as shown by careful experience."[83]

Ely is noteworthy as well for using political economy to import Christian ethics into economic life. He decried capitalism's "attempt to take one great department of social life, namely, the economic, entirely outside the range of ethical obligation."[84] He wanted to break down the barrier separating the economic sphere from the influence of Christian morality. The fundamental tenet of Ely's social ethics was the Christian imperative to love thy neighbor as thyself. He called this "the social law of service."[85] Morally, he argued, we have a responsibility for our neighbors when they are in need. Indeed, "We are responsible to a certain extent for all the poverty and sin and suffering about us."[86] And we are required to share at least our surplus wealth with our neighbors whenever they are in need of it. No Christian, said Ely, can justify spending his wealth on luxuries when others are in need.[87]

Ely believed there is a fundamental conflict between the propertied and the un-propertied, a conflict that could only be resolved if the propertied would temper their self-interestedness with consideration for the un-propertied.[88] He considered it a perversion of the idea of private property to view it as sanctioning the unrestrained pursuit of self-interest.[89] He held that private property is *not* so fundamentally grounded in self-interest as to be inconsistent with a significant degree of altruism in economic life. What political economy should aim at, he said, "is the union of self-interest and altruism in a broad humanitarian spirit."[90] By way of specifics, he quoted John Wesley: ""We cannot," said Wesley, "consistently, with brotherly love, sell our goods below the market price; we cannot study to ruin our neighbor's trade in order to advance our own; much less can we entice away or receive any of his servants or workmen whom he has need of. None can be gained by swallowing up his neighbor's substance without gaining the damnation of hell." [91]

To justify government's involving itself in making men moral, Ely appealed to the highest of philosophical authorities: "The only limit to the functions of the State is that laid down by Aristotle; the general principle cannot be better stated than he said it: "It is the duty of the State to do whatever is in its power to promote the good [i.e., virtuous] life." Any other principle is false to the fundamental principles of Protestantism, both ancient and modern." [92]

Not only did Ely oppose the idea of constitutional limitations on the power of government to infringe economic rights, he also opposed, in effect, the principle of the separation of church and state. "The distinction of ecclesiastical and profane laws can find no place among Christians. The magistrate himself is holy and not profane, his powers and laws holy, his sword holy." [93] "Let all Christians," said Ely, "work ... to change the constitution in so far as this may stand in the way of righteousness. The nation must be recognized fully as a Christian nation." [94] In Ely's view, the science of political economy tells us that the state has a sacred obligation to do the Lord's work in the economic realm, and, in consequence, there should be virtually no limit on government power to intervene in economic matters.

Veblen

Thorstein Veblen (1857-1929) came into this world the son of poor Norwegian farmers in Wisconsin. He studied at Carleton College and Johns Hopkins, and he earned a PhD. at Yale. He published his best known book, *The Theory of the Leisure Class*, in 1899. Veblen's ideas were slow to catch on, but he came into his own, says Eric Goldman, when the Crash of 1929 seemed to make the case for reform imperative. Thereafter, says Goldman, Veblen would be widely considered "the most penetrating thinker of modern American dissidence."[95]

Veblen's "leisure class" consists of "the non-productive." It includes not only the "idle rich," but also those men of wealth who hold positions of leadership in government, the military, and the churches. It also includes the "captains of industry." Veblen's leisure class, in other words, are the wealthy. In the great dichotomy between capital and labor, the leisure class are capital. If the reader detects a certain bias, an implicit value judgment, in Veblen's characterizing of capitalists as, most fundamentally, non-productive people, then he is on his way to understanding what Veblen was about. His style is dryly academic, his tone that of the dispassionate scientist. He would have us believe that his business is merely to report the facts.

> [I]t is by no means here intended to depreciate the economic function of the propertied class or of the captains of industry. The purpose is simply to point out what is the nature of the relation of these classes to the industrial process and to economic institutions. Their office is of a parasitic character, and their interest is to divert what substance they may to their own use, and to retain whatever is under their hand.[96]

Thus did Veblen, the scientific socio-economist, dryly report the plain fact that the wealthy are parasites.

As Veblen saw it, the leisure class made its appearance in history when society evolved from peaceful savagery to "consistently warlike" barbarism. In order for a leisure class to

161

come into being, "the community must be of a predatory habit of life," meaning that men "must be habituated to the infliction of injury by force and strategem." [97] In other words, the capitalist-class-to-be came into being precisely when men introduced coercion as an integral element into human affairs. At that stage of socio-economic development, honor accrued to those who were most successful at plundering their enemies in war. Mundane productive labor, by contrast, became dishonorable and was left largely to women and slaves. In this way the leisure class developed early on an aversion to productive labor, an aversion that would last into modern times. (Throughout *Theory of the Leisure Class* Veblen made little attempt to support his sweeping generalizations with specific facts drawn from the histories of specific peoples.)

It should come as no surprise that with the appearance of this predatory "leisure class" private property also came into being, according to Veblen. The earliest form of ownership was the ownership of women by men. The ability to afford owning a woman demonstrated a man's success at plundering; she became a status symbol. Accordingly, marriage and property both originated in predation. And both owe their existence to that fundamental human motivation which, with the exception of the instinct for self-preservation, is "the most alert and persistent of the economic motives." [98] That motive is the "invidious" desire to impress the neighbors, and it manifests itself most consistently in the form of "conspicuous consumption" (and "conspicuous leisure").

Indeed, according to Veblen, it is *not* the need to own the product of one's labor—which is indispensable to the ability to sustain one's life—that gives rise to the institution of private property. It is, rather, the desire to keep up with (or surpass) the Joneses. "The possession of wealth confers honor; it is an invidious distinction. Nothing equally cogent can be said for the consumption of goods, nor for any other conceivable incentive to acquisition," said Veblen. "Ownership began and grew into a human institution on grounds unrelated to the subsistence minimum." [99] Veblen did not deny in *Theory* that private property is necessary in some degree for subsistence. But he insisted that above and beyond what is needed for bare subsistence, the primary purpose to which men in fact put their wealth is ostentation.

So, as Veblen had it, property originated in predation and the desire to impress one's neighbors, and, beyond the bare minimum necessary for subsistence, its ongoing function is ostentation. The effect of Veblen's "analysis" of the history of the leisure class was to call into question both a man's right to the entire product of his labor, and the status of this right as a universally valid principle indispensable to man's survival. If indeed the primary purpose of all wealth above and beyond the subsistence minimum is to impress the neighbors, then a man's claim to that "surplus" portion of his wealth can hardly be considered a pressing one. As if this weren't enough to undermine the reigning economic order, Veblen went on to deny, in a manner reminiscent of James and Dewey, Holmes and Brandeis, and Richard T. Ely, that *any* economic institution could be valid for all times and all places. [100]

Economic institutions such as private property are nothing more than "prevalent habits of thought." [101] And habits of thought change continually as "population increases, and men's knowledge and skill in directing the forces of nature widen." [102] But, as economic institutions are always a product of *past* circumstances, they are always out of sync with current conditions. The process of "selective adaptation" by which economic institutions (i.e., habits of thought) adjust to ever-changing circumstances can never quite catch up with the conditions of the present. In addition, men tend to hold onto old, established ideas. "This is the factor of social inertia, psychological inertia, conservatism." [103]

So anyone who would argue for preserving established institutions, such as private property and contract, is exhibiting "social" and "psychological inertia." The leisure class, i.e., the wealthy, tend to be insulated from the forces that bring about economic evolution, said Veblen. Quite understandably, then, they are the last to recognize the need to discard old economic institutions and to embrace new ones. "The effect of the pecuniary interest [i.e., the leisure class] and the pecuniary habit of mind upon the growth of institutions is seen in those enactments and conventions that make for security of property, enforcement of contracts, facility of pecuniary transactions, vested interests." [104] But such "enactments and conventions" as these are "derivatives ... of the ancient predatory culture." [105] They are out of joint with the times.

And so, according to Veblen, the plain, scientific facts of economics are that private property and the entire regime of economic freedom that rests upon it are outdated. As "habits of mind" they derive from a time when the strong preyed upon the weak. In the modern, enlightened era that Veblen heralded, it was time for new habits of mind and new economic institutions.

The Historians

As we have seen, William James and John Dewey undermined the U.S. Constitution by calling into question both the possibility of truth, *per se*, and the validity of timeless, abstract principles. Oliver Wendell Holmes, Louis Brandeis, and Roscoe Pound added to this undermining by casting doubt upon the validity of abiding, abstract principles in the realm of jurisprudence. Richard T. Ely and Thorstein Veblen effected a similar undermining in the sphere of political economics. Edward S. Ross and Walter Rauschenbusch attacked the morality that implicitly underlays the Constitution and promoted in its place an alternative, antithetical morality grounded in selflessness. All of these thinkers contributed to the undermining of the Constitution by attacking it *indirectly*. The following two thinkers focused their attacks directly upon the Constitution itself.

Beard

Charles A. Beard was born in 1874 near Knightsbridge, Indiana. He graduated from DePauw University and then spent two years in Europe. In England, according to Eric Goldman, Beard became especially enamored of the ideas of John Ruskin.[106] After returning to America he enrolled for graduate work at Columbia University, where, after taking his degree, he joined what would become one of the most Progressive faculties in the land. In 1913 Beard published one of the signal works of the Progressive era, *An Economic Interpretation of the Constitution of the United States*.

In his *Economic Interpretation* Beard attacked the Constitution on two fronts; he attacked the *motives* of the Constitution's Framers and ratifiers, and he attacked the validity of the process by which the Constitution was ratified.[107] As to motives, Beard portrayed the movement to replace the Articles of Confederation with a more consolidated form of central government as fundamentally motivated by a desire for economic gain. He argued:

- that speculators in western lands subject to Indian troubles stood to gain from an invigorated federal government;
- and that Southern slaveholders saw in a strengthened central government increased security against slave revolt;
- and that owners of money had a direct interest in a federal government that would put an end to states' inflationary and debtor-relief legislation;

- and that holders of public securities, whose values were problematic under the government of the Articles, hoped for a central government committed to honoring those securities at par;

- and that manufacturers wanted protection against competition from British manufactured goods;

- and that shippers desired retaliation against British mercantilist regulations.

These economic groups represented *capital* in the newly-independent United States, and Beard argued that it was largely these groups who favored a new, more powerful federal government and a corresponding diminution in the power of the state governments.

Pitted against the owners of capital, in Beard's analysis, were the owners of *land*, mainly the small farmers, and the "mechanic" class. "Not one member [of the Constitutional Convention] represented in his immediate personal economic interests the small farming or mechanic classes." [108] The owners of capital tended to favor the limiting of democracy in the interest of protecting their rights of property, while the owners of land and the mechanic class desired a less fettered democracy and the prerogative to resort to such legislative expedients as legal tender and debt-relief laws.

To support his contention that the motives of the Federalists were primarily monetary, Beard analyzed the financial interests of the individual Framers at Philadelphia. He reported that of the fifty-five members of the Constitutional Convention, "no less than forty appear on the Records of the Treasury Department" as owning federal securities. (Ratification of the Constitution promised—and delivered—a full payoff of the federal and the states' Revolutionary War debts.) "[W]ith the exception of New York, and possibly Delaware, each state had one or more prominent representatives in the Convention who held more than a negligible amount of securities." [109] In addition, at least fourteen members of the Convention had money invested in lands for speculative purposes. At least twenty-four loaned money for interest (and were, therefore, creditors, rather than debtors). A least eleven members owned interests in mercantile, manufacturing, or shipping concerns. And at least fifteen owned slaves. [110]

Concluded Beard, "It cannot be said, therefore, that the members of the Convention were 'disinterested.'" [111] Indeed, "enough has been said to show that the concept of the Constitution as a piece of abstract legislation reflecting no group interests and recognizing no economic antagonisms is entirely false. It was an economic document drawn with superb skill by men whose property interests were immediately at stake; and as such it appealed directly and unerringly to identical interests in the country at large." Beard argued, in other words, that the Constitution was not a frame of government devised by disinterested men seeking solely to erect the best possible government for the new American polity. "As a group of doctrinaires," he said, "they would have failed miserably." [112] The Constitution was, instead, a frame of government devised by selfish men looking out for their own and their friends' monetary interests.

In the same manner as he impugned the motives of those who drafted the Constitution, Beard called into question the motives of the men who ratified it. "The state conventions do not seem to have been more "disinterested" than the Philadelphia convention; but in fact the leading champions of the new government appear to have been for the most part, men of the same practical type [as the Framers], with actual economic advantages at stake." [113] For example, in Massachusetts the vote in favor of ratifying the Constitution was close enough that, said Beard, had all the towns sent delegates to the ratifying convention that were entitled to do so, the margin of victory might have been only a half dozen votes. He pointed out that twenty-two of the thirty four delegates from Boston and Suffolk Counties, the Federalists' stronghold around Boston, held public securities, and all but two of these "probably benefited from the appreciation of the funds which resulted from the ratification." [114]

In Connecticut, 65 of the 128 delegates who voted for ratification held public securities at or about the time of the adoption of the Constitution.[115] Likewise, sixteen of the thirty who voted for ratification in New York.[116] In Pennsylvania, it was nineteen out of forty-six.[117] And so on throughout the ratifying states. Beard concluded, "Inasmuch as so many leaders in the movement for ratification were large security holders, and inasmuch as securities constituted such a large proportion of personalty, this economic interest must have formed a very considerable dynamic element, if not the preponderating element, in bringing about the adoption of the new system." [118] As in the case of the Framers of the Constitution, Beard implied that the men who ratified it in the respective state conventions were at least as interested in enriching themselves and their friends as they were in adopting the best possible frame of government for the young United States.

Whether the battle lines during the debate over the Constitution were indeed small farmers and mechanics against the owners of capital is beyond the scope of this discussion. That there was a clash of economic interests (as the participants perceived those interests), and that this clash did indeed, to one degree or another, pit a propertied and anti-democratic few against a less-propertied and democratic many is indeed true, as I have described in Chapter Three of this book. I have no dispute with Beard on this score, except that he tends to state the argument in stronger terms than I would.

Nor do I disagree with the conclusion that he drew from this clashing of economic interests against each other in the struggle over the Constitution. "The Constitution was essentially an economic document based upon the concept that the fundamental rights of property are anterior to government and morally beyond the reach of popular majorities." [119] Although, according to my own analysis of the Constitution, Beard overstated the Constitution's grounding in the idea of a *natural* right of private property, he was substantially correct in characterizing it as very much an economic document. Individual liberty, as I have argued, is above all else economic. But it is precisely because of its economic basis that Beard indicted the Framers' Constitution. Beard (implicitly) challenged the idea that liberty is essentially economic, and that without a secure right of private property no other rights are possible. This challenge, as I have said, and as Beard's *Economic Interpretation* evinces, is a defining characteristic of progressivism.

The other defining characteristic of Progressivism is the idea that a democratic polity is morally superior to a constitutionally limited republic. It is no surprise, then, that the other principal basis on which Beard indicted the Constitution is the less-than-entirely-democratic process by which it was framed and ratified. For example, given that the Framers met in Philadelphia with a mandate from their respective state legislatures to do nothing more than propose amendments to the Articles of Confederation, the whole idea of drafting an entirely new frame of government was never put to a vote of the people beforehand, either in the form a direct vote on the question itself or in the form of an election of state legislators committed one way or the other on the question. Indeed, in the election of the state legislators who selected the delegates to the Philadelphia Convention, a large proportion of the adult population was altogether denied the right to vote by the property qualifications upon which the suffrage was based in the Confederation era.

When it came time to ratify the new Constitution, Beard asserted, about three-quarters of the adult males failed to participate in the election of delegates to the state conventions, either from indifference or because of property qualifications. Only New York permitted manhood suffrage in the election of delegates to their state ratifying convention.[120] In Boston, only 760 out the 2700 who were entitled to vote for delegates to the state convention did so. Throughout the eleven states that approved ratification, "The Constitution was ratified by a vote of probably not more than one-sixth of the adult males."[121]

Beyond this, Beard argued, it is quite possible that the voters of New York, Massachusetts, New Hampshire, Virginia, and South Carolina, all of which ratified the Constitution, actually elected a majority of delegates to their respective conventions who were opposed to ratification, but that a number of these delegates sufficient to alter the outcome changed their minds after hearing the arguments of the generally wealthier, better-educated, and better-organized Federalists in their respective conventions.

Also, several states, notably New Jersey and Delaware, both of whose conventions voted unanimously for ratification, acted so quickly on the question that their citizenry had little time to consider the issue, and the potential opposition had little time to organize. And in Pennsylvania, where opponents of ratification tried to absent themselves from the legislature in order to deny a quorum when the question of authorizing a state ratifying convention came up, the advocates of ratification forcefully escorted these members of the legislature to that body's chamber in order to establish the necessary quorum and then hold the vote to authorize the calling of the Constitutional Convention. "[I]t seems a safe guess that not more than 5 per cent of the population in general ... expressed an opinion one way or the other on the Constitution."[122] Beard concluded that "The Constitution was not created by "the whole people" as the jurists have said" (and as the document itself suggests).[123]

Beard is suggesting here that because only a minority of the adult population participated in ratifying the Constitution, the process itself was not morally valid. He is assuming two things: that an individual may not rightfully be subjected to a government to which he does not consent, and that the consent requirement is satisfied if the individual participates in the process of adopting a new frame of government by voting on it. But even a person who voted on the adopting a new constitution might not consent to that constitution if the vote did not go his way. Now, a government that violated the rights of an individual would, to that extent, be morally culpable (just as an individual who did so would be). But a government that exercised authority over an individual who did not consent to it would not be morally culpable, simply because no government can expect to enjoy the unanimous consent of all who fall within its jurisdiction (Locke's idea of "tacit consent" notwithstanding).

In a properly functioning polity, the government enjoys the consent of the vast majority of its citizens as a matter of course; most people lead moral lives and obey the laws voluntarily because those laws comport with their moral principles. The people refrain from raping and pillaging not because those things are against the law, but because they are contrary to their morality. Governments tend to lose the consent of the governed when the morals of the people and the actions of the government fall out of sync, especially when the government begins to expand its role beyond a narrow focus on prohibiting things like raping and pillaging, upon which the vast majority of a citizenry tend to agree.

If the morals of the people become deficient, as in many of our inner cities, but the government continues to act justly, then, even if the morally deficient come to see the police as an illegitimate, occupying force, the government is no less legitimate. (The American Civil War provides an example of this phenomenon. Although the southern states explicitly withdrew their consent from the government in Washington, the North was morally justified in forcibly subduing the southern states because they were systematically violating the natural rights of a sizable portion of their populations.)

If, on the other hand, the people continue to live morally but the government begins to abandon its obligation to secure their rights, either by ineffective enforcement of the laws or by an unjust expansion of its powers at the expense of the rights of the law-abiding, then the people might withdraw their consent. Nevertheless, the government's legitimacy would be diminished not by virtue of this withdrawal of the consent of the governed, but because of the government's failure to secure the rights of the citizenry. *Pace* Locke, there is only one basis of political legitimacy, and that is the securing of individuals' rights. The consent of the governed is a practical consideration—it is difficult to govern without the consent of the governed, but not a moral one.

If political legitimacy depends above all else upon a government's securing of individuals' rights—to which the securing of the rights of private property is indispensable—then it does not matter whether Beard is correct either that the Framers and ratifiers of the Constitution were motivated by narrow self-interest or that the process of adopting the Constitution was less than entirely democratic. The standard for judging the legitimacy of the Constitution is the degree to which it succeeded in securing the rights, and *especially* the property rights, of individuals. On this score, the Framers' Constitution succeeded admirably.

Charles Beard presumed that democratic participation is the foundation of political legitimacy, and he ignored individual liberty. The next historian took a different approach to the relationship of democracy and liberty; he equated the one with the other.

Smith

James Allen Smith was born in Missouri in 1860 to homesteaders from Virginia. He attended the University of Missouri and earned a law degree there in 1887. He practiced law for nearly a decade in Kansas City before leaving to study political economy at the University of Michigan, where he developed an affinity for the German historical school of political economy. Smith went on to have an influential career at the Progressive University of Washington. In 1907 he published *The Spirit of American Government*, in which he placed the blame for America's alleged political ills squarely on the shoulders of the Framers of the U.S. Constitution.

In *The Spirit of American Government* Smith took issue with two major characteristics of the Constitution, first, that it sharply limits the power of democratic majority rule, and, second, that in so doing it puts individuals' rights of property and contract largely beyond the reach of democratic control. Smith's objections, in other words, go to the heart of the kind of government the Constitution was intended to establish. Smith argued that the government of the Constitution is destructive of human freedom. To understand his argument, we must understand his idea of freedom.

Smith viewed the development of human freedom in the world as a halting progression from rule by the few to rule by the many. Until *Magna Carta*, England was ruled by one man. But on that day at Runnymede in 1215, the English nobles extorted a measure of political power from King John, thus initiating a movement away from one-man rule. *Magna Carta* was only a beginning, since the sharing of power it effected between the Crown and the nobility excluded the great mass of the English people. Nevertheless, the power-sharing it began would eventually culminate, in the American Revolution, in the people's winning full control of their government for the first time in history. But the victory of freedom that the American Revolution represented was short-lived, in Smith's view, since the adoption of the U.S. Constitution cut short the reign of the people and returned control of the state to a wealthy minority. From 1789 onward, the continuing efficacy of the Constitution represented, for Smith, the on-going subjection of the non-wealthy many by the wealthy few.

Smith is entirely correct, of course, that the Constitution sharply limited the political power of democratic majorities, and that it did so for the purpose of making secure the rights of property and contract against infringement by such majorities. In his 1907 book he provided an excellent analysis of the Constitutional devices which the Framers used to limit the political power of democratic majorities. His analysis mirrors much of what I wrote on the subject in Chapter Two.

Under the Constitution as originally drawn up, what political power the people possessed they exercised mainly through the House of Representatives. But there were, observed Smith, a number of elements of the Constitution intended to keep the power of this body, and thus, of the people, within strict bounds. First, the House of Representatives could not enact legislation without the concurrence of the Senate, which was structured along distinctly non-democratic lines. The Framers intended that, to one degree or another, the Senate should be composed of men of wealth who would operate as a check upon any redistributive tendencies that might emanate from the House. Also, noted Smith, the fact that all the states wield equal power in the Senate meant that a minority of the people concentrated in the smaller states could, through their senators, block legislative enactments favored by a majority of the national population. In this way also the Senate functioned as an anti-majoritarian device.

Another check on the power of the democratic House, noted Smith, is the powerful chief executive. Under the original Constitution, the President was elected non-democratically, like the Senate. Through his use of a qualified veto power, he was expected to block any democratically motivated legislation harmful to property rights that might get past the Senate. Smith noted that in the heyday of American democracy, under the Articles of Confederation, the Continental Congress governed with no chief executive at all, and that at the state level during that period most governors were largely figureheads, with real power residing in the legislatures themselves. For this reason, said Smith, the powerful chief executive of the Constitution marked a clear departure from the democracy that predominated before the adoption of the Constitution.

A third device for checking the power of the people, acting through their House of Representatives, is, of course, the U.S. Supreme Court. With their life-time tenure and their total independence from popular election the justices represent, in Smith's opinion, a throwback to the English Crown. He argued that to empower these neo-monarchists to veto enactments of the duly elected state and federal legislatures was especially offensive to genuine freedom. Smith noted that no such judicial power existed under the democratic regime that governed before the adoption of the Constitution (though the first stirrings of such a power were becoming evident).

Finally, Smith argued that the amending process stipulated by the Constitution works to thwart the forces of democracy. By requiring that two thirds of both houses of Congress or three fourths of the states ratify any potential amendments, the Constitution makes it virtually impossible for the people ever to amend the Constitution at all. (Smith noted that all the amendments that existed in 1907 came either from the founding era or from the period immediately following the convulsions of the Civil War.)

In all these ways, Smith argued, the architecture of the U.S. Constitution limits the power of the majority of the American people to influence the workings of their government—at both the federal and the state levels. He noted that by 1907 the rest of the Western world had surpassed America in the degree to which their peoples exercised democratic control of their governments.[124] Smith clearly presupposed that democracy is the morally preeminent form of government, but nowhere did he validate this premise. He simply treated it as self-evidently true. "It is a fundamental principle of free government," he says, "that all legislative power should be under the direct control of the people."[125]

The Framers believed that unrestrained democracy would mean the end of liberty. What, then, did Smith have to say about liberty? Democracy *is* liberty, said Smith. "In fact, true liberty consists, as we have seen, not in divesting the government of effective power, but in making it an instrument for the unhampered expression and prompt enforcement of public opinion."[126]

The old, eighteenth-century conception of liberty, which gave rise to the checks and balances of the U.S. Constitution and the Bill of Rights, defined liberty as freedom from government, said Smith. (Actually, it defined liberty as freedom from *coercion*, whether from government or from other individuals.) Smith believed that such a conception of liberty quite usefully helped protect a politically powerless majority when government was controlled by a powerful minority. "It was but natural under the circumstances [in the pre-Revolutionary 1700s] that the people should seek to limit the exercise of political authority, since every check imposed upon the government lessened the dangers of class rule."[127] But with the ascendance of the people themselves to political power, as occurred in the American Revolution, there was no longer any need for such an idea of liberty. "In so far as government had now passed into the hands of the people there was no longer any reason to fear that it would encroach upon what they regarded as their rights." [128]

The key to understanding Smith here is to understand his redefining of the concept "liberty." Whereas the old conception of the Founders had been a *negative* liberty grounded in the *absence of* coercion, Smith conceives of the new "liberty" of democracy as a *positive* liberty. "We have a new conception of liberty. We see a tendency ... to reject the old passive view of state interference as limited by the consent of the governed and take the view that real liberty implies much more than the mere power of constitutional resistance—that it is something positive, that its essence is the power to actively control and direct the policy of the state." [129] *Liberty is political power.* If this is so, then clearly, to limit political power, as the Framers did, was to limit "liberty."

In Smith's view, once the people come to power, and especially with the rise of the industrial mode of production, there is no longer any need for such a thing as *individual* liberty. Before the rise of industrialism, when a great many men were independently employed on their own farms or in their own shops, it was true that individual initiative predominated. But with the rise of the industrial corporation, individual initiative became largely a thing of the past, in Smith's view.

> *It is manifestly impossible to restore to the masses the right of individual initiative. Industry is too complex and too highly organized to permit a return to the old system of decentralized control. And since the only substitute for the old system of individual control is collective control, it appears to be inevitable that*

government regulation of business will become a fixed
policy among all democratic states. [130]

"Government regulation of business," of course, must come at the expense of property rights. Smith emphasized that the Framers' intention to contain the forces of democracy grew out of their fear that an insufficiently restrained democracy would pose a threat to property rights. "They foresaw that [unrestrained democracy] would mean the abolition of all private monopoly and the abridgement and regulation of property rights in the interest of the general public." [131] He acknowledged that private property was indissoluably linked with liberty under the Framers' individualist conception of liberty. "Liberty, as the Framers of the Constitution understood the term, had to do primarily with property and property rights." Obviously, if individual liberty was to be abolished under Smithian democracy, so also must property be virtually abolished, just as the Framers feared it would be should democracy ever prevail in America. If the Framers' individualist conception of liberty was invalid, then their defense of property as indispensable to liberty would be moot.

Smith held that in fact the Framers served only their own narrowly selfish interests by building into the Constitution an elaborate defense of property rights. "It was the purpose of the Constitution as we have seen to establish the supremacy of the so-called upper class." [132] As evidence, Smith argued that the Framers sought in the Constitution not to evenly *balance* the political power of democratic House of Representatives with that of the Senate, the President, and the Supreme Court; they sought, rather, to render these anti-democratic institutions constitutionally *superior to* the House, according to Smith.

But Smith himself, on the other hand, would not have rectified the Framers' alleged transgression by establishing a proper *balance* between these opposing elements. As we have seen, he advocated absolute superiority for the democratic majority. What, then, of the danger, which so preoccupied the Framers, that the majority might oppress the minority? The question never arose in Smith's book. The reason it never arose seems to be that once the majority came to power there would be no minority to oppress. As Smith saw it, there were only two alternatives; either the few would rule the many, or the many would rule.

> *All governments must belong to one or the other of two classes according as the ultimate basis of political power is the many or the few. There is, in fact, no middle ground. We must either recognize the many as supreme, with no checks upon their authority except such as are implied in their own intelligence, sense of justice and spirit of fair play, or we must accept the view that the ultimate authority is in the hands of the few.*[133]

When the propertied few rule the many—which is how Smith characterized capitalism—then the few have separate and distinct interests from the many. It is therefore necessary for the propertied few to hold the power of the many in check—thus the system of checks and balances.

What happens to the propertied few when the many come to power? Smith does not say so, but the logic of his argument seems to be that they will cease to exist as a propertied few. Under true democracy, there will no longer be two classes of citizens with two separate and distinct sets of interests. This seems to be the reason that Smith saw no need to discuss the possibility that under democracy the many might oppress the propertied few. There won't be any propertied few to oppress; they will have been reduced to membership in the many. All of the citizenry would then be members of the majority, and therefore all would have the same economic interests.

At the time Smith published *The Spirit of American Government* in 1907, America was already becoming more democratic. Indeed, the United States began the process of democratizing almost as soon as the Constitution was ratified. It began with the expanding of the general electorate as property qualifications for voting and office-holding began to be eliminated. Then, under the democratic influence of the Jackson presidency, there was an increase in the power of government to infringe property rights, in the form of an expanded police power. By Smith's time, the Electoral College had ceased to function as a "filter" between the general populace and the presidency.

But America in 1907 was still a long way from the highly democratized polity it would become by the end of the twentieth century. Among the specific measures that Smith (and many other Progressives) advocated were the initiative, referendum, and recall, all of which are ways for the people to exercise political power directly; the direct party primary, which took the selection of party candidates out of the hands of party leaders and put it into the hands of the people; and the direct election of U.S. Senators, which went a long way toward turning the Senate into just a high-class House of Representatives. All of these measures would eventually become law, and they would have the desired democratizing effect. [134]

The error in Smith's reasoning was his equating of political power, which has coercion at its disposal, with economic "power," which does not. A factory owner might offer a worker poor wages and dangerous working conditions, but he lacks the power to *force* the worker at gunpoint to accept either one, and this is the all-important difference between political power and economic power.

The Critic

Herbert Croly had a radical pedigree. His mother was one of the first professional newspaperwomen in the United States, and the very first to be syndicated. She helped found the Woman's Press Club of New York City, and she also founded the first important women's club in the city. Croly's father, David, was editor of the *New York Daily Graphic* and the founder of *Modern Thinker*, "An Organ for the Most Advanced Speculation." Both mother and father were devout followers of the French positivist, Auguste Comte. Herbert was born in 1869 in New York City and raised a Comtean. He graduated from Harvard and worked as an editor for the *Architectural Record*. In 1909 he published his *Promise of American Life*, one of the signal works of the Progressive era.

At the time of the American founding, said Croly, the promise that America held out to native and immigrant alike was the opportunity to achieve financial well-being. All understood this promise to be inextricably linked to America's free political institutions. As long as she maintained those institutions, America would continue to offer every citizen a chance at prosperity and economic independence. To one extent or another, wrote Croly, economic opportunity continued to be a part of the promise of American life ever since.

But soon after the Revolution, said Croly, an additional element associated itself with that promise. Following the electoral success of Jefferson and his party and, later, of the Jacksonians, "The Land of Freedom became in the course of time also the Land of Equality. The special American political system ... was made explicitly, if not uncompromisingly, democratic." [135] Democracy thus became an added element in the promise of American life; in Croly's view, it became the *essential* element.

One would expect an advocate of American democracy to be an admirer of Thomas Jefferson. But Croly was a bitter critic of Jeffersonism. In his discussion of the Hamilton-Jefferson

181

controversy, Croly came down squarely on the side of Hamilton. His preference for Hamilton's ideas over Jefferson's was grounded in a fundamentally different conception of democracy from Jefferson's. Jefferson the democract was a sincere champion of individual liberty. Indeed, in light of the importance that he accorded to individual rights, Jefferson's democratism makes sense as a way of limiting political power by dispersing it. As a champion of individual liberty, Jefferson was a staunch opponent of concentrated political power.

Croly's democratism, in contrast, called for just such a concentrating of political power. Whereas, for example, American democrats since Jefferson's time had jealously guarded the powers of the states against encroachment by the federal government, Croly advocated, in the name of democracy, a dramatic expansion of federal powers, often at the expense of the states. Croly advocated, for instance, enlarging the federal power to regulate commerce, and he advocated authorizing the federal government to oversee the organization of labor and to exercise "control over property in the public interest." [136] As far as the federal-state balance of power was concerned, Croly was much closer to the anti-democratic Hamilton than he was to his fellow democrat, Jefferson. [137]

Progressivism, as I have argued, advocated a return to a pre-Constitutional level of democracy in the United States. The democracy that had prevailed under the Articles of Confederation had been true to one of Locke's twin principles of political legitimacy; it had indeed rested upon the consent of the governed. But, among other failures, it had not met Locke's second test of legitimacy; it had failed to secure the rights, and especially the property rights, of individuals. Democrat though he was, Jefferson genuinely subscribed to *both* of Locke's principles of legitimacy. He believed that a legitimate government must enjoy the consent of the governed *and* it must secure their individual rights.

It is precisely because of Jefferson's loyalty to the second of Locke's principles that Croly so bitterly opposed the ideas of his fellow democrat. In a critical vein, Croly wrote, "Neither is there any doubt as to which of these ideas [liberty or democratic equality] Jefferson and his followers really attached the more importance. Their mouths have always been full of the praise of liberty; and unquestionably they have really believed it to be the cornerstone of their political and social structure." [138]

In Croly's view, the Jeffersonian conception of democracy rested on the premise that if men were left free to pursue their own private interests as they saw fit, then the American promise of freedom for all and prosperity for those who could earn it would continue to find fulfillment automatically and indefinitely. But the problem with this idea, Croly believed, was that if men were left free to pursue their own private interests, then some would become much wealthier than others. These wealthy few would then wield great "power" over their fellow citizens.

Now Croly, democrat though he is, was not at all opposed to having some individuals exercise great power over others. Indeed, he considered it indispensable to a democracy that the more intelligent, energetic, and "efficient" citizens should possess extensive authority to order the lives of their less ambitious countrymen. But the problem with Jeffersonian democracy was that it afforded no adequate mechanism for removing individuals from economic power once they ceased to exercise that power with intelligence, energy, and efficiency. Indeed, left to their own devices, the wealthy usually try to consolidate their power and to block any attempts by others to deprive them of it. In this way, inequalities of wealth and economic power come to enjoy a degree of permanence that is more characteristic of a European monarchy than of a proper democracy. The non-wealthy majority who are thus excluded from power "lack the first essential of individual freedom when they cannot escape from the condition of economic dependence." [139]

American democracy had failed, argued Croly, not because, as some Progressives claimed, it had been betrayed by unscrupulous individuals, but because it had been too faithful to its Jeffersonian premise of individual rights.

> *[T]he political corruption, the unwise economic organization, and the legal support afforded to certain economic privileges are all under existing conditions due to the malevolent social influence of individual and incorporated wealth; and it is equally true that these abuses, and the excessive "money power" with which they are associated, have originated in the peculiar freedom which the American tradition and organization have granted to the individual.* [140]

The unequal distribution of wealth was, in Croly's view, the fatal flaw in their democracy as Americans learned it and had faithfully practiced it since the time of Jefferson. The solution was for the American state to make "itself responsible for a morally and socially desirable distribution of wealth." [141] Favoring high compensation for the exceptionally "efficient," Croly was not opposed to economic inequality, *per se*. But he was opposed to "over-compensating" the exceptionally ambitious and successful, as Jeffersonism tended to allow. [142] And he was also opposed to extending compensation beyond the time during which the owners of wealth employed that wealth efficiently and in the service of the public interest.

To prevent wealth from remaining in the possession of "inefficient" owners, Croly advocated confiscatory inheritance taxes. [143] The question of who should be compensated how much became, for Croly, a function of the public good. "While preserving at times an appearance of impartiality so that its citizens may enjoy for a while a sense of the reality of their private game, it [the state] must on the whole make the rules in its own interest. It must help those men to win who are most capable of using their winnings for the benefit of society." [144]

In Croly's hands, democracy became not a means of dispersing political power, as it had been for Jefferson, but of enlarging and concentrating it. Croly did not shy away from the term "socialism" to describe his system, but he preferred the term "nationalism." His nationalism was the most prominent feature of Croly's brand of progressivism. But in his system, the nationalizing of American democracy referred to something much more radical than simply shifting power from the states to the federal government. At root, Croly's nationalism was as much a moral conception as a political one.

For Croly, the moral was that which was motivated not by self-interest, but by "disinterest." "[D]emocracy, as a political and social ideal, is founded essentially upon disinterested human action." [145] A nationalized democracy would be a democracy of *selfless* individuals, individuals who had learned to place the collective interests of the nation ahead of their own private interests. Americans, argued Croly, had enjoyed too much individual liberty. People who are free to pursue their own private interests inevitably neglect the collective, public interest. Croly disdained the priority that Jeffersonism accords to the satisfaction of individuals' private needs and desires. For Croly, the good of the nation must take precedence over the interests of individuals. He quoted John Jay Chapman approvingly:

> It is thought that the peculiar merit of democracy lies in this: that it gives every man a chance to pursue his own ends. The reverse it true. The merit lies in the assumption imposed upon every man that he shall serve

his fellow-men.... The concentration of every man on
his own interests has been the danger and not the safety
of democracy, for democracy contemplates that every
man shall think first of the state and next of himself....
Democracy assumes perfection in human nature.[146]

Given his moral orientation, Croly's antipathy toward
Jeffersonism makes all the more sense. Politically, Jefferson was
more of a democrat than were many other prominent Americans
of his time. But his genuine commitment to individual rights
reflected an implicit commitment to a morality of self-interest. A
political system dedicated to securing the rights of individuals is
grounded in a morality that considers it right for men to place a
higher priority upon their own interests than upon the interests of
society. "The triumph of Jefferson and the defeat of Hamilton
enabled the natural individualism of the American people free
play. The democratic political system was considered tantamount
in practice to a species of vigorous, licensed, and purified self-
interest." [147] Croly's hostility toward the Jeffersonian brand of
democracy was fundamentally a hostility toward its underlying
morality.

An obvious objection to Croly's conception of democracy is
that men do not naturally tend to grant preference to the interests
of others ahead of their own. Democracy, as Chapman said,
"assumes perfection in human nature." [148] Croly's answer was that
men must be taught to prefer the national interest above their
own. The democratic nation must "not accept human nature as it
is, but ... move in the direction of its improvement." [149]

The nation, like the individual, must go to school; and
the national school is not a lecture hall or a library. Its
schooling consists chiefly in experimental collective

action aimed at the realization of the collective purpose. [150]

Democracy, said Croly, echoing John Dewey, is an experiment. We cannot know ahead of time what will "work" and what will not work in the way of social, political, and economic arrangements. What is important is that the citizenry be willing to try new arrangements. Even those experiments that fail can be instructive if they are undertaken with sincerity. The lessons learned from these experiments will then constitute the schooling of the populace in the ways of democracy.

Croly emphasized that national democratic education consists not in a persuasive appeal to the rational faculties of the individuals who constitute the populace, but in collective *action*. (Again, echoes of John Dewey.) The unspoken premise was that these "experiments" must take place whether the citizenry agree to them or not.

> *The whole round of superficial machinery—books, subsidies, lectures, congresses—may be of the highest value if they are used to digest and popularize the results of a genuine individual and national educational experience, but when they are used, as so often at present, merely as a substitute for well-purposed individual and national action, they are precisely equivalent to an attempt to fly in a vacuum.* [151]

Whatever persuasion is to take place should occur after the fact as a way of rationalizing whatever "experiment" has already occurred. To attempt to persuade people ahead of time of the need to alter their political or economic institutions, in order to get them to undertake the alterations voluntarily, would be "superficial."

In the name of democracy, Croly would diminish individual liberty for the sake of increasing their collective "liberty."

A people to whom was denied the ultimate responsibility for its welfare, would not have obtained the prime condition of genuine liberty. Individual freedom is important, but more important still is the freedom of a whole people to dispose of its own destiny. [152]

To dispose of the destiny of a people is to dispose of the destinies of individuals. Croly did not call for the abolition of individual rights, *per se*, but he might as well have. Whatever rights individuals are to retain depends upon what the public good demands at any given time, as determined by "the people" and, especially, by their ruling elite. "The arduous and responsible political task which a nation in its collective capacity must seek to perform is that of selecting among the prevailing ways of exercising individual rights those which contribute to national perpetuity and integrity." [153]

To sum up, Croly's nationalist democracy was typically Progressive in its advocacy of a higher degree of democracy and of a diminished degree of security for property rights for the United States. His democratism had little to do with empowering "the people" and much to do with empowering government. He used the ideology of democracy as a battering ram against the bulwark of individual rights, and especially property rights, which stood in the way of the activist, interventionist government that he envisioned. Though an outspoken proponent of democracy on the order of Thomas Jefferson, he favored government by a powerful elite to an extent that exceeded anything Alexander Hamilton would have sanctioned.

In order for individuals to be free to order their lives as they see fit, the use of force, by private individuals and by government alike, must be prohibited except as a retaliatory measure. When the use of force is limited to retaliation, then those who govern have no choice but to rely on rational persuasion in their relations with those whom they govern. Therefore, in a constitutional republic in which individual rights are paramount, governance must of necessity be a matter of persuasion and rational consensus.

But in a polity in which the collective is supreme, whether it goes by the name "the people" or the state, things are otherwise. Any reliance upon rational persuasion presumes that the individual on the receiving end reserves the right to remain unpersuaded and to act accordingly. But in order for "collective supremacy" to have any meaning, the individual must forfeit the option to remain unpersuaded and to act accordingly. Therefore, in a polity in which the collective, and not the individual, is supreme, governance *can not* be a matter of persuasion and rational consensus. Government *must* rule by force. Croly is therefore entirely true to the logic of his democratic nationalism when he classifies the machinery of rational persuasion as "superficial."

Even in 2015, it is jarring to read lines like "every man shall think first of the state and next of himself." We are tempted to dismiss Croly as an irrelevant extremist. But, in one form or another, America has already adopted, or is in the process of adopting, virtually all of Croly's most important ideas. America is more democratic today, at least in form, than she was in 1909. If nothing else, American governments at all levels are far more able to override private rights, and especially property rights, in the name of "the people" than they were in Croly's time. Those same governments have assumed, in principle, the responsibility for overseeing the distribution of wealth among America's citizens. The chief instruments for this purpose currently at their disposal are the graduated income tax and the inheritance tax. To what extent these governments in fact exercise this power depends upon the political vicissitudes of the moment.

Her professions of democratism notwithstanding, America is far more governed today by powerful, centralized elites than she was in 1909. Perhaps there is no more glaring illustration of this than what has happened to America's public schools. Except in the big cities, parents in Croly's time probably knew most of the members of their local school board first hand. Those parents exercised genuine influence over the deliberations of those school boards. But with the consolidation of school districts and the centralizing of school funding, what used to be a strictly local responsibility has now become the shared responsibility a giant school district, the state, and the federal government. For all practical purposes, most parents today are powerless against the ruling elite of educational professionals and "experts" who run their schools.

It is not literally true that a great many Americans today are prepared, as Croly advocated they should be, to place the interests of the nation ahead of their own, especially if they have a say in the matter. On the other hand, with the advent of "national service" programs and school and university curricula that award course credit, and even require, "community service" from their students, younger Americans are now being taught that they do indeed have an obligation to serve "society." Of all of Croly's fundamental ideas for America, increased democracy, rule by centralized elites, government control of the economy, and politically institutionalized selflessness, this last is the least well developed today.[154] But our elites are working overtime to see this idea join the others as an integral part of the "promise of modern American life."

As we shall see in the next chapter, all of these elements of Crolyism, which are also elements of Progressivism, would play a role in the rise of the conservation movement.

Notes

1. Aristotle, *Metaphysics*, IV, 7, 1011b, *Basic Works of Aristotle*, ed. Richard McKeon (New York, NY: Random House), 1941.

2. Dewey quoting Peirce, *The Essential Dewey*, ed. Larry A. Hickman and Thomas M. Alexander (Bloomington, IN: Indiana University Press, 1998), 1, 4.

3. *Collected Papers of Charles Sanders Pierce*, ed. Charles Hartshorne and Paul Weiss (Cambridge, MA: Harvard University Press, 1931-35), 5, 276-277. Quoted in W.T. Jones, *A History of Western Philosophy* (New York, NY: Harcourt Brace Jovanovich, Inc., 1952), 4, 267.

4. Edward C. Moore, *American Pragmatism: Peirce, James, & Dewey* (New York, NY: Columbia University Press, 1961), 157.

5. Ibid.

6. Ibid., 158.

7. William James, "Pragmatism," *The Works of William James*, ed. Frederick Burkhardt and Fredson Bowers (Cambridge, MA: Harvard University Press, 1975), 143.

8. Peirce repudiated parts of James's brand of pragmatism.

9. James, "Pragmatism," *Works*, 103.

10. Charles Pierce, *Collected Papers of Charles Sanders Pierce*, ed. Paul Weiss and Charles Hartshorne (Cambridge, MA: Belknap Press, 1931), 5, 394-5. Quoted in John Dewey, *Logic: The Theory of Inquiry* (New York, NY: Henry Holt, 1938), 345n. Emphasis added.

11. John Dewey, "The Problem of Truth," *Essential Dewey*, 2, 114.

12. Ibid., 110.

13. John Dewey, "Democracy and Educational Administration," *School and Society* (April 3, 1937), reprinted in J. Ratner, ed., *Intelligence in the Modern World* (New York, NY: Modern Library, 1939), 400-04.

14. James, "Pragmatism," *Works*, 52-53.

15. Eric Goldman, *Rendezvous with Destiny* (New York: Alfred A. Knopf, 1952), chap. 5.

16. Oliver Wendell Holmes, *The Common Law* (Boston, MA: Little, Brown and Company, 1923), 1.

17. Ibid., 2.

28. Ibid.

29. "[W]hile the terminology of morals is still retained, and while the law does still and always will, in a certain sense, measure legal liability by moral standards, it never the less, by the very necessity of its nature, is continually transmuting those moral standards into external or objective ones, from which the actual guilt of the party concerned is wholly eliminated." Ibid., 38.

20. Ibid., 214.

21. Ibid., 43.

22. Samuel J. Konefsky, *The Legacy of Holmes and Brandeis* (New York, NY: DaCapo Press, 1974), 34. Laski quote from *Mr. Justice Holmes and the Supreme Court*, ed. Felix Frankfurter (Cambridge, MA: Harvard University Press, 1938), 144. Wrote Holmes, "Judges should be slow to read into [the Constitution] a *nolumus mutare* [We must not change.] as against the law-making power." The police power "extends to all great public needs." "It [the police power] may be put forth in aid of what is sanctioned by usage, or held by the prevailing morality or strong or preponderant opinion to be greatly and immediately necessary to the public welfare." *Noble State Bank v. Haskell*, 219 U.S. 104, 110 (1911). Quoted in Konefsky, *Legacy*, 34.

23. Holmes, *Common Law*, 41. Emphasis added.

24. *Missouri v. Holland*.

25. Testtimony before United States Commission on Industrial Relations. Senate Document # 415, 64th Congress, First Session, Vol. VIII, p. 7659. Quoted in Konefsky, *Legacy*, 72.

26. This quotation ends, "[a]bove all rights rises duty to the community." *Duplex Printing Press Co. v. Deering*, 254 U.S. 443, 488. Quoted in Konefsky, *Legacy*, 127.

27. *Truax v. Corrigan*, 257 U.S. 312, 376 (1921). Quoted in Konefsky, *Legacy*, 136.

28. Quoted in *Adams v. Tanner*, 244 U.S. 590, 595 (1917) from *Booth v. Illinois*, 184 U.S. 425, 425 (1902).

29. Louis Brandeis, "The Constitution and the Minimum Wage," Argument before the Supreme Court of the United States in *Stettler v. O'Hara*, reprinted in Osmond K. Fraenkel, *The Curse of Bigness* (New York, NY: Viking Press, 1934), 68-69.

30. Konefsky, *Legacy*, 84.

31. Ibid., 91.

32. Alpheus Thomas Mason, *Brandeis: A Free Man's Life* (New York, NY: The Viking Press, 1946), 543.

33. Testimony before United States Commission on Industrial Relations. Senate Document # 415, 64th Congress, First Session, Vol. VIII, p. 7659. Quoted in Konefsky, *Legacy*, 75.

34. Brandeis, testifying before the United States Commission on Industrial Realtions, 1915. Fraenkel, *Curse*, 73-74.

35. Roscoe Pound, *The Spirit of the Common Law* (Francestown, NH: Marshall Jones Company, 1921), 102. Pound was dean of Harvard Law School from 1916 to 1936.

36. Ibid., 99.

37. Ibid.

38. Ibid., 91.

39. Ibid., 92.

40. See Goldman, *Rendezvous*, 90-91.

41. Pound, *Spirit*, 196.

42. Ibid., 7.

43. Ibid., 20.

44. Ibid., 28-9.

45. Ibid., 213.

46. Ibid., 185-89. The eighth and final change that Pound cites is also worth noting: "Courts no longer make the natural rights of parents with respect to children the chief basis of their decisions. The individual interest of the parents which used to be the one thing regarded has come to be almost the last thing regarded as compared with the interest of the child *and the interest of society*." Ibid., 189. Emphasis added.

47. John Ruskin, *Sesame and Lilies, Unto This Last, and The Political Economy of Art* (London, Melbourne: Cassell and Company, Ltd., 1907), 162-3.

48. Ibid., 113-14.

49. Ibid., 114.

50. Ibid., 123.

51. Ibid., 125.

52. Ibid., 125.

53. Ibid., 119.

54. Ibid., 161-2.

55. Ibid., 176.

56. Goldman, *Rendezvous*, 106

57. Ibid., 117.

58. Edward Alsworth Ross, *Sin and Society* (Boston, MA: Houghton Mifflin Company, 1907), 146.

59. Ibid., 83-84.

60. Ibid., 151.

61. Ibid., 17.

62. Goldman, *Rendezvous*, 106.

63. Walter Rauschenbusch, *Christianizing the Social Order* (New York, NY: The MacMillan Company, 1912), 41.

64. Ibid., 1. "Capitalism has generated a spirit of its own which is antagonistic to the spirit of Christianity; a spirit of hardness and cruelty that neutralizes the Christian spirit of love; a spirit that sets material goods above spiritual possessions." Ibid., 315.

65. Ibid., 221.

66. Ibid.
67. Ibid., 233.
68. Ibid., 14.
69. Ibid., 14-15. In 1911 the Northern Baptist Convention advocated "The control of the natural resources of the earth in the interests of all the people." The resolve also advocates: "the gaining of wealth by Christian methods and principles, and the holding of wealth as a social trust; the discouragement of the immoderate desire for wealth, and the exaltation of man as the end and standard of industrial activity." Ibid., 16.
70. The exception is the prohibition against child labor; children do have needs that justify certain statutory prohibitions. Whether government ought to prohibit child employment altogether is a different matter.
71. Ibid., 60-61.
72. Ibid., 362.
73. Ibid., 163.
74. Ibid., 183.
75. "A subject working class, without property rights in the instruments of their labor, without a voice in the management of the shops in which they work, without jurisdiction over the output of their production is a contradiction of American ideals and a menace to American institutions." Ibid., 198.
76. Contrary to Rauschenbusch's wishes, the extent of American workers' *direct* control of the means of production would always be limited to what control the unions would eventually exercise over the workplace. But workers, or, more precisely, the people at large, would come to exercise a great deal of *indirect* control over the means of production through their democratic influence on the legislative process. (Perhaps it is more correct to say that the government would come to exercise a great deal of control over the means of production *in the name of* the people.)

77. Ibid., 379. "The aim of Christian legislation at the time [of the medieval guilds] was to secure to every business man a moderate circle of customers and a decent living, and to shackle those who would try to secure inordinate wealth by snatching the bread of their fellows. When the capitalistic method gathered force and headway, it swept away these protective laws which hampered free competition and the massing of capital and labor." (Ibid., 167.) "In the handicraft system which preceded capitalism, the aim was to fix a price that would be fair and reasonable for producer and consumer, and to curb any hankering for excess profit." Ibid., 222.

78. Ibid., 3

79. Richard T. Ely, *The Past and Present of Political Economy* (Baltimore, MD: N. Murray, Publications Agent, Johns Hopkins University, 1884), 10.

80. "As every individual, therefore, endeavors as much as he can both to employ his capital in the support of domestic industry, and so to direct that industry that its produce may be of the greatest value; every individual necessarily labours to render the annual revenue of the society as great as he can. He generally, indeed, neither intends to promote the public interest, nor knows how much he is promoting it. By preferring the support of domestic to that of foreign industry, he intends only his own security; and by directing that industry in such a manner as its produce may be of the greatest value, he intends only his own gain, and he is in this, as in many other cases, led by an invisible hand to promote an end which was no part of his intention. Nor is it always the worse for the society that it was no part of it. By pursuing his own interest he frequently promotes that of the society more effectually than when he really intends to promote it."

Adam Smith, *An Inquiry into the Nature and Causes of the Wealth of Nations*, ed. Edwin Cannan (London: Methuen & Co., Ltd., 1904), IV, 477.

81. Ely, *Past and Present*, 45.

82. Ely himself argued that the Americans should never have secured property rights against government infringement to the extent that they did.

83. Ibid., 48. Emphasis added.

84. Ibid., 145.

85. Richard T. Ely, *The Social Law of Service* (Cincinnati, OH: Curtis & Jennings, New York, NY: Eaton & Mains, 1896), 34. *Moral* principles were indeed, for Ely, immutable and timeless in a way that he denies to politico-economic principles.

86. Ibid., 133.

87. "How can I claim that I love my brother as myself when I see him need the very necessaries of life and expend money for that which contributes in no measures to my real well-being. For the whole tone of the Bible, from beginning to end, condemns in the strongest terms anything of the kind." Ibid., 230.

88. "It is possible to reconcile the different classes of society only by a higher moral development. The element of self-sacrifice must yet play a more important role in business transactions, or peace and good-will can never reign on earth." Ely, *Past and Present*, 28.

89. "It was an unbelieving age of materialism which asserted the all-sufficiency and even beneficence of unrestrained self-interest, and attempted to restrict economic inquiries to this one question: 'How produce the greatest amount of wealth?'" Ely, *Social Law*, 148.

90. Ely, *Past and Present*, 35.

91. Ely, *Social Law*, 221.

92. Ibid., 170.

93. Ibid., 169. "But what are religious laws? … laws designed to promote the good life. Factory acts, educational laws, laws for the establishment of parks and playgrounds for children, laws securing honest administration of justice, laws rendering the Courts

accessible to the poor as well as the rich – all these are religious laws in the truest sense of the word." Ibid., 173.

94. Ibid., 173-4.

95. Goldman, *Rendezvous*, 117.

96. Thorstein Veblen, *The Theory of the Leisure Class* (New York, NY: The Modern Library, 1934), 209.

97. Ibid., 7.

98. Ibid., 110.

99. Ibid., 26.

100. Veblen's application of Darwinism to the socio-economic sphere is explicit: "The life of man in society, just like the life of other species, is a struggle for existence, and therefore it is a process of selective adaptation. The evolution of social structure has been a process of natural selection of institutions." Ibid., 188.

101. Ibid., 190.

102. Ibid., 194.

103. Ibid., 191.

104. Ibid., 210.

105. Ibid., 209.

106. Goldman, *Rendezvous*, 150.

107. In what follows, I accept uncritically Beard's statements of fact such as to which Framers were owners of public securities. Because my purpose is to elucidate Beard's ideas as a manifestation of progression, it does not matter whether his factual assertions are correct.

108. Charles A. Beard, *An Economic Interpretation of the Constitution of the United States* (New York, NY: MacMillan, 1913), 149.

109. Ibid., 150.

110. Ibid., 149-150.

111. Ibid., 151.

112. Ibid.

113. Ibid., 291.

114. Ibid., 264.

115. Ibid., 267.
116. Ibid., 270.
117. Ibid., 274.
118. Ibid., 290.
119. Ibid., 324.
120. Ibid., 241.
121. Ibid., 325.
122. Ibid., 250.
123. Ibid., 325.
124. "They have in theory at least repudiated the eighteenth-century doctrine that the few have a right to thwart the will of the many. The majority has in such countries become the only recognized source of legitimate authority." James Allen Smith, *The Spirit of American Government* (Cambridge, MA: The Belknap Press of Harvard University Press, 1965), 301.
125. Ibid., 99-100.
126. Ibid., 305.
127. Ibid., 291.
128. Ibid., 293.
129. Ibid., 294.
130. Ibid., 310-11.
131. Ibid., 298-99.
132. Ibid., 165.
133. Ibid., 296.
134. One other democratizing tactic of Smith's deserves mention. None of the measures mentioned so far would have directly affected the Supreme Court, which Smith rightly saw as a major obstacle to democratic control of private property. Noting that there is nothing in the Constitution that specifies how many justices there must be on the Court, Smith conceived the idea of having the President "pack" the Court with new justices agreeable to the democratic viewpoint. It was the mere threat of such packing by FDR in 1937 that caused the Court abruptly to

abandon its longtime mission of protecting vested economic rights from government infringement.

135. Herbert Croly, *The Promise of American Life* (Cambridge, MA: The Belknap Press, 1965), 11.

136. Ibid., 275.

137. Croly was also a great admirer of Lincoln who, whether wittingly or not, laid the groundwork for eventual federal supremacy over the states.

138. Ibid., 43-4. In terms reminiscent of Oliver Wendell Holmes, Croly writes: "[G]overnment should in the end, and after necessarily prolonged deliberation, possess the power of taking any action, which, in the opinion of a decisive majority of the people, is demanded by the public welfare." Ibid., 35.

139. Ibid., 205.

140. Ibid., 23.

141. Ibid.

142. Echoes here of the medieval Christian conception of the "just price." Who is to say what constitutes "just" compensation for work performed?

143. Ibid., 203.

144. Ibid., 193.

145. Ibid., 381.

146. John Jay Chapman, *Causes in Consequences* (New York, NY: Charles Scribner's Sons, 1899), 121. Quoted in Croly, *Promise*, 418.

147. Ibid., 49. "With all their professions of Christianity their national idea remains thoroughly worldly." Ibid., 10.

148. Chapman and Croly assume that selflessness is the epitome of moral perfection.

149. Ibid., 413.

150. Ibid., 407.

151. Ibid., 407.

152. Ibid., 178.

153. Ibid., 190.
154. I should qualify this assertion by noting that such selflessness is implicit in the idea of the welfare state.

4: The Conservation Movement

PROGRESSIVISM AIMED TO expand the power of government in the United States. In the later decades of the nineteenth century there appeared a number of individuals and groups who prefigured Progressivism in this respect, who aimed specifically to extend the power of government to manage the nation's forests, waterways, wildlife, and even its scenic beauty. Around the turn of the century, these resource protection groups would coalesce into a larger movement that comprehended all their respective goals under the heading of "conservation." And, because all of these groups aimed, to one extent or another, to expand government power, they were easily absorbed into the larger Progressive movement. To understand the conservation movement, we must look to its roots in the decade of the American Civil War.

Marsh

George Perkins Marsh was born in 1801 in Woodstock, Vermont. His father was a lawyer, a district attorney, and a member of Congress. George graduated at the top of his class from nearby Dartmouth College and went on to careers in law, business, and politics. Although he never achieved great success in any one field, Marsh had a range of interests and achievements that is remarkable. Among other things, he "bred sheep, ran a woolen mill, built roads and bridges, sold lumber, edited a newspaper, developed a marble quarry, [and] speculated in real estate." [1] He also served as a Vermont legislator, a U. S. Congressman, and U. S. Minister to Turkey. Throughout his life, Marsh pursued myriad artistic and scholarly interests, which included architecture, Scandinavian and Gothic studies, social history, the English language, and the collecting of prints and engravings. Among his other accomplishments, he helped found the Smithsonian Institution, and he determined the final proportions of the Washington Monument. Of the three books that Marsh published in his lifetime he is best known today for *Man and Nature*, which appeared in 1864.

Man and Nature is a call, the very first, for the "conservation" of natural resources in the United States, although it would be another four decades before anyone would think to apply that term to what Marsh was advocating. As a harbinger of the conservation movement, and of environmentalism, Marsh's book offers a remarkably modern-sounding critique of man's treatment of nature. "It is, in general, true, that the intervention of man has hitherto seemed to ensure the final exhaustion, ruin, and desolation of every province of nature which he has reduced to his dominion," wrote Marsh. [2] He argued that man's alterations of nature tend to harm man's own interests when those alterations go so far as to upset the natural balance. This idea of balance, of a natural harmony among the elements of an environment, was the dominant theme in Marsh's understanding of nature.

Marsh's discussion of deforestation illustrates this principle of balance at work. He argued that the cutting down or burning of a forest can have unintended and far-reaching effects upon other elements of the environment and that these unintended consequences often constitute the greater part of the damage that human alteration of the land inflicts on the natural order. Among the harms, according to Marsh, that can follow from deforestation are: the forest floor loses heat to the atmosphere in cold seasons, and it suffers greater heat from an unobstructed sun in warm seasons; winds blow unobstructed across the ground, carrying away the snow that had insulated against winter freezes; these winds also tend to dry up the moisture in the soil. "The face of the earth is no longer a sponge but a dust heap."[3] In addition, water runs off more quickly than before, so upland soil erodes and gets carried away to settle in rivers and harbors, creating hazards to navigation; the rich soil of the forest, now deposited in aquatic environments, "promotes a luxuriance of aquatic vegetation that breeds fever, and more insidious forms of mortal disease, by its decay."[4] Also, the diminished absorptive capacity of the upland forest lands promotes flooding in lower regions. Eventually, "The earth, stripped of its vegetable glebe, grows less and less productive, and, consequently, less able to protect itself by weaving a new network of roots to bind its particles together.... Gradually it becomes altogether barren."[5]

Marsh did not oppose all human alternation of the landscape. He recognized that man lives by tailoring nature to suit his needs, and he understood that as man advances in knowledge and culture, his ability and his desire to alter nature advance commensurately. "[M]an ... cannot subsist and rise to the full development of their higher properties, unless brute and unconscious nature be effectively combated, and, in a great degree, vanquished by human art," he wrote.[6] Nevertheless, "in commencing the process of fitting [nature's arrangements] for permanent civilized occupation," Marsh wrote, "the transforming operations should be so conducted as not unnecessarily to derange and destroy what, in too many cases, it is beyond the power of man to rectify or restore."[7]

Marsh's counsel that man should refrain wherever possible from "unnecessarily" upsetting the balance of nature prefigures remarkably the nature preservationism of Rachel Carson and the environmental movement. And his belief that one alteration of an environment by man can trigger a long and intricate series of other, potentially devastating changes has a distinctly modern, ecological ring to it.[8] On the other hand, one does not find in *Man and Nature* the overt reverence for nature, the keenness to preserve it intact, or the hostility toward industrialism that would be so important a part of environmentalism.

As for the forests, Marsh wrote:

> It is a great misfortune to the American Union that the State Governments have so generally disposed of their original [forested] domain to private citizens. [9]

He would have preferred to see governments in the business of owning and managing forests. What about the enlarging of the role of government that this would entail? Wrote Marsh, "The apophthegm, 'the world is too much governed'…has done much mischief whenever it has been too unconditionally accepted as a political axiom." [10]

He mentioned with approval recent developments in France, where law regulated the clearing of forest land and promoted the restoration of private woods. Said Marsh:

> [T]he recent legislation of France, and of some other Continental countries, on this subject, looks to more distant as well as nobler ends, and these are among the public acts which most strongly encourage the hope that the rulers of Christendom are coming better to understand the true duties and interests of government.[11]

Marsh's favorable view of public ownership of forests implies a critique of private ownership as insufficiently conducive to proper long-range care and maintenance of forested lands, but he was ambiguous on this point, arguing that in America the very abundance of timber led to careless attitudes toward forests on private and public lands alike.[12] He never came right out, in *Man and Nature*, and criticized private ownership of natural resources as in itself destructive of those resources. For his part, Marsh seemed to appreciate the importance of the institution of private property. Indeed, he was quite aware, in property-minded America, of the need to base his campaign for better resource management upon rational persuasion and voluntary compliance, rather than on legislative compulsion. "For the prevention of the evils upon which I have so long dwelt, the American people must look to the diffusion of general intelligence on this subject, and to the enlightened self-interest, for which they are remarkable, not to the action of their local or general legislatures."[13]

Nevertheless, Marsh's idea that the United States would be better off if the government owned at least some of the forests anticipates Gifford Pinchot and his National Forests, which would become the flagship of the conservation movement. In addition, Marsh's idea implies that private owners of natural resources cannot be trusted to manage them responsibly, whereas disinterested government managers could be so trusted. One does not find in Marsh any overt antipathy toward capitalism, but his implicit questioning of the trustworthiness of private ownership of natural resources would be expanded upon by the conservationists and would, in the hands of environmentalists, become a full-blown assault on capitalism.

Forest Conservation

When *Man and Nature* appeared in 1864, George Perkins Marsh stood virtually alone in calling upon Americans to better manage their natural resources. Within a decade, though, others would begin to join in his call. In the beginning, these proto-conservationists focused their attention on the nation's forests. In the 1870s there were no professional foresters nor schools of forestry in the United States. American forests, especially those very extensive ones on government land, existed under a regimen of benign neglect. The idea of caring for and cultivating forest land the way one might care for and cultivate farmland, though well-established in Europe, was all but unknown in the United States. This would soon begin to change.

Franklin B. Hough was a physician, scientist, historian, and statistician.[14] He was also an amateur forester, and he had read Marsh. In 1873 Dr. Hough presented a paper to the American Association for the Advancement of Science titled "On the Duty of Governments in the Preservation of Forests." The topic of the paper is significant; expanding upon Marsh's idea that government should have *some* role in the management of natural resources, "conservationists" would soon come to envision a *central* role for government in this arena. In his paper, Dr. Hough praised the European, and especially the French, model of government-owned and managed forests, much as Marsh had praised them. He then laid out a seven-point plan by which federal and state governments could promote the planting and proper care of trees. Among his recommendations: governments could give tax breaks or bounties to owners of privately owned woodlots-turned-preserves, and owners of railroads and toll roads could be required to plant trees alongside their thoroughfares.

Although Dr. Hough was willing to advocate a certain degree of such government-directed resource management, he stopped short of advocating a system of state or national forests on the European model. In 1876, the U. S. Commissioner of Agriculture appointed Dr. Hough to study forests in the United States. Dr. Hough's subsequent report contributed to the creation, in 1881, of the Division of Forestry within the Department of Agriculture, with Dr. Hough as its first chief. Forestry had begun to forge its alliance with government.

Another early champion of forestry was John Aston Warder, a Cincinnati physician who turned to horticulture in mid-life and then to forestry. In 1873, Dr. Warder represented the United States at the World's Exposition in Vienna. While there, he learned a great deal about forest management from European foresters, and he wrote an official report on the subject. Two years later Warder founded the American Forestry Association in Chicago and became its first president. The association aimed to protect the country's forests from unnecessary waste. Before too long, the Association would become an influential advocate of the idea of government-owned and operated forests on the European model.

Still another champion of forestry in these early years was Carl Schurz, Secretary of the Interior from 1777 to 1781. Schurz was a German immigrant and was well-acquainted with the practices of professional forestry in his homeland. During his tenure at Interior, Schurz repeatedly called for stricter protection of the federal government's own, extensive forest lands. In urging the federal government to "preserve the forests still in its possession," Schurz was in effect advocating what would later become the system of National Forests.[15]

By the early 1880s, the nascent forestry movement was coalescing around two foci. One focus was silviculture, the science of forest care and maintenance; the other was protecting forests, especially those in the public domain, from unregulated commercial exploitation. The "public domain" refers, of course, to those lands "owned" by the state governments and, more importantly, by the federal government. At one time or another in the nineteenth and twentieth centuries, over 1.2 billion acres of land between the Appalachians and the Pacific passed into, or through, the hands of the federal government.[16] How the federal government ever got into the land business is itself an interesting question.

Congress had decided as early as 1780 that new territories outside the original thirteen states could be settled and eventually admitted to the Union as new states. When Virginia, Massachusetts, and Connecticut ceded their claims in the old Northwest Territory to the Confederation after the Revolution, but before these territories could be admitted as states, Congress found itself directly responsible for governing a vast, largely unsettled territory. Congress decided to sell this territory in parcels to private parties as a means of raising much-needed revenue for the federal government. Later, the Northwest Ordinance of 1787 stipulated that the federal government was to retain "ownership" of, along with the right to dispose of, all lands not yet in private hands, even after a new territory was admitted to statehood.[17] In this way, the federal government got into the business of managing and selling land within the states themselves.

By the 1880s, the federal government owned hundreds of millions of acres of land, much of it forested. But as a landowner, the government did not care for its land as private owners tend to do. This is understandable given that the primary mission of the federal government, as far as this land was concerned, was simply to dispose of it, at no special profit, to private parties. In the meantime, though, neither the government nor the many private persons who were making use of the public lands for commercial purposes had any special incentive to protect these lands from abuse.

An alternative to this policy of neglect would have been for the government, as soon as possible, simply to sell or give the land to private owners, who *would* have had an incentive to care for it. But the forestry advocates rejected this option. Instead, they argued that the federal government should declare the best of these forest lands to be permanent reservations, forever under the management of the federal government, and that the government should begin to care for its property as a private landowner might do. (Given the vast holdings of the federal government, the new forest guardians probably preferred the prospect of having to sell the idea of proper forest management to a single, public "owner" rather than to countless private ones.)

In 1882 the first American Forestry Congress met in Cincinnati. Among the organizers was Dr. Warder of the American Forestry Association. The focus of this first great convention of forest conservationists was silviculture, the science of forest care and maintenance. All future forestry congresses, and they would continue into the 1970s, would be sponsored by Dr. Warder's American Forestry Association. In attendance was Bernard Fernow, a German immigrant and America's first professional forester, who had occasion to lament that few owners of private forest land had chosen to attend the event. In Fernow, a product of the Prussian Forestry Department, the foresters had an explicit critic of private ownership of forests. As a result of the "unrestricted activity of private individual interests," said Fernow, a forest is "quickly exhausted, its restoration made difficult and sometimes impossible, its function as a material resource is destroyed." A nation must "exercise the providential functions of the state to counteract the destructive tendencies of private exploitation." [18]

In 1885, the State of New York established the Adirondack Forest Preserve. This was the first permanent, publicly owned forest preserve in the United States, and it was a harbinger of things to come in forest protection.[19] Its purpose was to forestall the exploitation of the region by commercial mining and timber companies. The driving force behind the establishing of this preserve was Franklin B. Hough, who had left the Division of Forestry in 1883.

Then in 1891 Congress passed the Forest Reserve Act, which authorized the President to set aside forest preserves on federal land. These were forests on land already held by the federal government that would be forever withdrawn from sale to private owners and would thenceforth be managed by the federal government. Among the chief proponents of this act were Bernard Fernow, who at the time was serving as the third chief of the Division of Forestry, and Dr. Warder's American Forestry Association. President Harrison quickly declared 13 million acres as reserves. President Cleveland soon added another 5 million acres.

The Forest Reserve Act did not specify exactly what the Federal government was supposed to do with these lands. But the Forest Management Act of 1897 took care of this, stipulating that the reserves should be used for timber production, watershed maintenance, wildlife preservation, livestock grazing, mining, and recreation, all to be managed by the government itself. (This was a remarkable undertaking for a nation that once was committed to strictly limited government.) In 1907 the government would rechristen these federal forest reserves the National Forests.

Back in 1864 George Perkins Marsh had implied that forest management was a proper task for government to take on. By 1907 the focus of the forest protection movement had become, by and large, the conservation—and the expansion—of *government* forests. As we shall see, the main focus of the broader conservation movement would also become the conservation of natural resources in the public realm. The first federal forest reserve was the Yellowstone Park Timber Reserve, established in 1891. Leading the effort to establish this reserve was George Bird Grinnell.

Grinnell

George Bird Grinnell (1849-1938) spent his childhood at Audubon Park in upper Manhattan. He attended Yale and occupied his summers during college hunting fossils and big game in the American west. Although intended to take over the family

Thomas McCaffrey

business, George closed it down after graduation and instead returned to Yale to study paleontology with O.C. Marsh. While working on his doctorate, he, along with his father, bought up enough shares of the sportsmen's weekly, *Forest and Stream*, to enable them to install George as editor in 1880. In addition to carrying the standard hunting and fishing fare, the magazine, under Grinnell's leadership, now took on many of the issues that would later coalesce under the name of conservation. One of those issues was forest protection.

Grinnell came to forest protection indirectly. As editor of *Forest and Stream* he cared most about fish and game. To protect fish and game, it was necessary to preserve forest habitat. "No woods, no game; no woods, no water; no water, no fish" he wrote.[20] In April of 1882, in the pages of *Forest and Stream*, Grinnell joined the growing chorus of voices promoting the twin goals of government control of the nation's forests, on the one hand, and of instituting a regimen of scientific silviculture on those forests, on the other.[21]

Grinnell had a specific forest in mind. It happened that in 1875 he had served as official naturalist on a reconnaissance expedition to the newly proclaimed Yellowstone National Park in Wyoming. On this expedition, he had come to fear that unregulated hunting would soon wipe out the big game of the Yellowstone region. By 1877, he was promoting the idea of making the national park a game refuge.[22] Owing largely to his concern to protect the big game of Yellowstone, Grinnell advocated extending the boundaries of the park to the east and the south. Although he failed ever to achieve this goal, he did eventually realize an approximation of it.

Partly as a result of Grinnell's efforts, Congress on March 3, 1891 authorized the President to establish federal forest reserves on the public domain, and the first such reserve that President Harrison created, on March 30, 1891, was the Yellowstone Park

Forest Reserve, a region adjacent to the park that included much of the area that Grinnell had wanted to include in the park by extending its boundaries. Thus did George Bird Grinnell help in the establishing of the first national forest.

But Grinnell's first love was big game. In 1885, with Theodore Roosevelt, he founded the Boone and Crockett Club. Early members included William T. Sherman, of Civil War fame, Henry L. Stimson, the future secretary of state, U.S. Senators Henry Cabot Lodge and Elihu Root, and Gifford Pinchot, who one day would become the greatest conservationist of them all. Boone and Crockett was a hunting club with a political agenda, to make wild game in the United States a *public* resource, on the one hand, and to have it managed in the interest of sport hunters, as opposed to commercial hunters, on the other. Just as the forest protectors advocated public ownership of the forests, so Boone and Crockett promoted public control of the nation's wild game. Their strategy was two-fold: to institute their brand of game management on the public lands, and to institute a similar degree of control, by means of local and state regulations, over wild game on private lands.

As to the public lands, so long as these remained in the hands of the government, the game upon them would be public property by default. How, and to whose benefit, this public resource should be managed would therefore be political questions, and their answers were yet to be determined in 1885. There was in the late 1800s a thriving industry in commercial hunting upon the public domain, akin to today's commercial fishing industry on the oceans. It is conceivable that the wild game on the public lands could have been cultivated by, say, a government agency and then harvested by private hunting companies, much the way the forests have been cultivated by the forest service and harvested by private logging companies. But Boone and Crockett and the champions of sport hunting won this political battle early on, and the commercial hunting companies ceased to exist. Wild game on the public lands would be a public resource, and it would be managed for the benefit of sport hunters.

There was still the question of the wild game that inhabited the private lands of the nation. It is conceivable that such game could, and should, have been the property of the owners of the land it inhabited. But the Boone and Crockett Club and others championed legislation, such as license requirements, hunting seasons, and bag limits, which, by depriving private landowners of the right to use or dispose of the wild game on their land, in effect transformed that game into public property. Today the Club claims credit for laying the foundation of hunting and game laws throughout the United States.

Just as he sought to protect big game from commercial hunters, so also did Grinnell set out to protect wild birds, primarily from market hunters. In the late 1800s, the commercial hunting of birds, especially shorebirds, for sale to restaurants and wholesalers was common. But an even greater threat to wild bird populations was the ladies fashion industry. Feathers were all the rage, and the feathers were supplied by hunters. Florida supplied more feathers than any other state, with Maine and Virginia following.[23] Several states had laws protecting wild birds at this time, but these laws were lightly enforced and largely ineffectual.[24]

In 1883 a group of scientists founded the American Ornithologists' Union to promote the scientific study of birds. Grinnell, who had written his doctoral thesis on the roadrunner of the desert southwest, joined the new organization and, with other members, urged the Union to involve itself in the protection of wild birds. The Union established a Committee on the Protection of North American Birds which, in 1886, issued its "Model Law" as a template for future state bird-protection legislation. The law prohibited the killing of all wild birds except such game birds as ducks, geese, shorebirds, turkeys, grouse, pheasant, and quail. The Model Law also prohibited removing the eggs or harming the nests of any birds (except the alien, English sparrow).[25] The AOU thus initiated the effort to do for wild birds what the Boone and Crockett Club would later do for wild game, make them, in effect, public property.

But Grinnell envisioned a broader campaign to protect wild birds, one that involved not just the passing of laws but also an effort to enlist private citizens in protecting wild birds. This was the primary focus of his early efforts at bird protection. On February 11, 1886, he wrote in *Forest and Stream*:

> *We propose the formation of an Association for the protection of wild birds and their eggs, which shall be called the Audubon Society. Its membership is to be free to everyone who is willing to lend a helping hand in forwarding the objects for which it is formed. These objects will be to prevent, so far as possible, (1) the killing of any wild birds not used for food; (2) the destruction of nests or eggs of any wild bird; and (3) the wearing of feathers as ornaments or trimming for dress.[26]*

Grinnell named his new society for John James Audubon (1785-1851), who himself had hunted wild birds on a grand scale, but who later in life had produced the 430 or so watercolors for *The Birds of America* for which he is famous. When George Grinnell was a child, his family had moved to a new subdivision for the well-to-do at Audubon Park, John James Audubon's estate in upper Manhattan. There, young George had attended a school run by Audubon's widow, Lucy, who played an important role in Grinnell's young life.[27] Here Grinnell was exposed to the paraphernalia of a life spent hunting, and later painting, wild birds.

This first Audubon Society was committed to protecting wild birds through:

> ...the education of our whole people to the usefulness of the birds and the folly of permitting their wholesale destruction. [28]

In other words, Grinnell planned to *persuade* Americans voluntarily to refrain from harming birds, rather than forcing them by means of legislation. One became a member of the Society simply by signing a pledge, provided by Grinnell, not to harm wild birds. By 1887 the Society had incorporated in New York and had 39,000 members. In that year, Grinnell launched the *Audubon Magazine* as the voice of the Society. But the workload proved daunting, and Grinnell grew discouraged by his failure to diminish the commercial hunting of wild birds. In December of 1888 he disbanded the Audubon Society. In his book on the founding of the Society, Frank Graham attributes Grinnell's ineffectiveness to his decision to base his campaign on voluntary cooperation rather than on legislative compulsion. [29]

The Audubon Society was reborn on February 10, 1896 at the home of Harriet Hemenway of Boston. This new effort brought together the more progressive society women of Boston with some of the more prominent men of the city—men who were also sportsmen. William Brewster, a member of the American Ornithologists' Union and the curator of Harvard's Museum of Comparative Zoology, and himself a sportsman, was elected president. The new society's by-laws stated its purpose: "to discourage the buying and wearing, for ornamental purposes, of the feathers of any wild birds except ducks and game birds, and to other wise further the protection of native birds." [30] Debate at subsequent meetings soon resulted in the elimination of the phrase "except ducks and game birds." [31]

The Massachusetts Audubon Society set out early to pursue the legislative approach to wild bird protection. The effort to transform wild birds into a public resource now began in earnest. George Mackay hunted waterfowl on islands off the Massachusetts coast. On his own he started to enforce an 1886 state law that specified a $10 fine for killing gulls or terns or for disturbing their nests. When the new Audubon Society invited Mackay to join their board of directors, they gained an advocate of bird protection enforcement and a State House lobbyist for stronger protective legislation. He advocated, for example, that it be made illegal not only to kill certain wild birds within Massachusetts but also that it be illegal to import such birds even when killed out-of-state. [32]

A second Audubon Society appeared in Pennsylvania later in 1896. Within a year, similar societies had sprung up in New York, New Hampshire, Maine, Wisconsin, New Jersey, Rhode Island, Connecticut, and Washington, D.C. In 1898 there followed societies in Ohio, Indiana, Tennessee, Minnesota, Texas, and California. The Audubon Society was here to stay. The next step in the Society's development was to achieve national cohesion.

William Dutcher was a self-made insurance man from New Jersey. He hunted birds, collected them, wrote scientific papers on them, and became an associate member of the American Ornithologists' Union, as well as a member of their Committee on the Protection of North American Birds. Dutcher promoted the Union's Model Law, and he had helped George Grinnell launch the first Audubon Society. With the rebirth of the Audubon idea, Dutcher became an active partner of several of the state societies. He set as his primary goal the defeat of the millinery industry, centered in New York City.

Dutcher's approach to defeating the milliners was two-pronged: short-range, inspired by the work of George Mackay in Massachusetts, he planned to build up a sanctuary-and-warden system of wild bird protection; over the longer range, he planned to work for the passage of national legislation outlawing commerce in feathers. For the short-range effort, Dutcher recruited a cohort of men to enforce existing state protective laws, using donated money to pay them. He also recruited private property owners to either protect the birds on their land themselves or to allow Dutcher's wardens to do the protecting for them.

As to the longer-range goal, Dutcher and his Audubon allies gained an important weapon in 1900 when Congress passed the Lacey Act, which used the Commerce Clause of the U.S. Constitution to prohibit the interstate shipment of birds killed in contravention of state laws. (This was an early use of the Commerce Clause for police-power type regulation that had never been contemplated by its authors.) Although by 1900 only five states had bird-protection laws inspired by the Model Law, the protectionists had made an important beginning toward transforming wild birds into a public resource on a national scale.

Audubon came a step closer toward nationalizing wild bird protection with the elevation to the U.S. Presidency in 1901 of Theodore Roosevelt. In 1897 Assistant Secretary of the Navy Roosevelt had become an honorary vice-president of the Washington, D.C. Audubon Society. Roosevelt was a wildlife conservationist of the sportsman variety, and he was eminently supportive of the Audubon Societies' campaign to outlaw the commercial hunting of wild birds, as his Boone and Crockett Club had done for big game. In 1903 President Roosevelt established by executive order the first federal bird sanctuary—at Pelican Island in the Indian River in Florida. Pelican Island was "set apart for the use of the Department of Agriculture as a preserve and breeding ground for native birds." [33] This was the first of 53 such federal sanctuaries that Roosevelt would establish as President.

To promote his twin goals of long-range legislation and short-range sanctuaries and wardens, Audubon President Dutcher envisioned a more cohesive national organization for the state Audubon Societies. In 1901 he succeeded in forming a loose federation called the National Committee of the Audubon Societies of America. Under this arrangement, "the several societies [would] retain their individuality." [34] A high degree of state autonomy has characterized the Society ever since. By the end of 1903 there were thirty-seven state societies, most of which belonged to the National Committee. Then, in 1905 the state societies incorporated as the National Association of Audubon Societies for the Protection of Wild Birds and Animals, with William Dutcher as the first national president. Wrote Dutcher of the new organization:

> *The object of this organization is to be a barrier between wild birds and animals and a very large unthinking class, and a smaller but more harmful class of selfish people. The unthinking, or, in plain English, the ignorant class, we hope to reach through educational channels, while the selfish people we shall control through the enforcement of wise laws, reservations or bird refuges, and the warden system.* [35]

The Audubon Society has, ever since, carried on George Bird Grinnell's original intention of enlisting the public in wild bird protection. But it has also succeeded eminently in transforming wild birds into public property, both on a local and on a national level. Americans today could hardly imagine a United States in which wild game and birds were not protected as public property. While the memberships of groups like the Audubon Society and the Boone and Crocket Club could hardly be characterized as wild-eyed socialists, their efforts helped to promote the idea that government possessed a special efficacy for protecting "natural resources," and the legislative fruit of those efforts helped to inure landowners to the idea that their private property rights could legitimately be infringed for the sake of protecting such resources.

National Parks

At the same time that Hough and Warder and Schurz and Fernow were developing the national forest idea, other early "conservationists" were inventing the national park. In contrast to the way the national forests came into being—the federal government committed itself to a system of forest reserves before any actual reserves existed—the first national parks were born one at a time, each the product of an *ad hoc* citizen effort. These parks differed from earlier parks in the United States in several ways: they were national, as opposed to city or town, parks; they were larger than the traditional municipal park; and, most importantly, they were composed of wildlands that were intended to be kept mostly wild.

Before national wildland parks came into being, parks were typically planned, cultivated, artificial versions of nature. Central Park in Manhattan represented an attempt to introduce an element of nature into the highly artificial landscape of the city, although it was a version of nature that was itself highly altered for purposes of human utility and esthetic appeal. This conception of the park as a version of nature rendered superior to its wild cousin by human artifice had predominated in Enlightenment-era Europe until the latter part of the eighteenth century. But, in 1761, Jean Jacques Rousseau introduced readers of his *La Nouvelle Heloise* to a new sort of park, one from which all sign of human artifice was absent.

Entering this so-called orchard, I was struck by an

> *agreeable sensation of freshness which the thick foliage,*
> *the animated and vivid greenness, the flowers scattered*
> *about on all sides, the murmuring of a running brook,*
> *and the singing of a thousand birds brought to my*
> *imagination at least as much as to my senses; but at the*
> *same time I thought I saw the wildest, the most solitary*
> *place in nature, and it seemed I was the first mortal*
> *who had ever penetrated into this desert island.*[36]

Rousseau did much more than introduce a new style of park; he helped to change the way Western man thought of nature and of man's relationship to it. But more about Rousseau later.

The following paean to wild nature from the early nineteenth century bespeaks the new, Rousseauean view:

> *Oh! She is fairest in her features wild,*
> *Where nothing polished dare pollute her path:*
> *To me by day or night she ever smiled*
> *Though I have marked her when none other hath*
> *And sought her more and more, and loved her best in*
> *wrath.*[37]

Praising nature was nothing new when Byron wrote these words between 1809 and 1812. Since the time of classical Greece, a certain strain of Western culture had idealized the virtues of country life and the pastoral landscape and had opposed these to the supposed corruption of city life and the urban landscape. But the ancient pastoralists, and their modern counterparts since the Renaissance, had contemplated nature "as a rule less for its own sake than as a background for human action," and when they did concern themselves with nature itself, "it [was] a nature that [had] been acted upon by man," the nature of the countryside, of pastures and orchards and planted fields, rather than the deep forests and desolate mountaintops of the romantic poets.[38] Historical pastoralism celebrated a *human* way of life; it celebrated the life lived close to nature as much as it celebrated nature itself.

Byron's exaltation of the truly wild, on the other hand, of nature in her "wrath," amounted to a glorification of nature apart from its relation to man, a glorification of nature, *per se*. Over the course of the nineteenth century, the romantic nature artists, led by giants like Coleridge, Blake, Wordsworth, Byron, Keats, and Shelley, would win mainstream respectability for this glorifying of the natural in contrast to the man-made.

One outgrowth of this cult of nature was the invention of the wildland park. The first such park was Yosemite Valley and the neighboring Mariposa Grove of Big Trees. In 1851 a company of militia from the gold camps entered Yosemite Valley while hunting for Indians during the Mariposa Indian War. Although there is evidence that a few white men had glimpsed the valley prior to 1851, no one had followed up on these earlier sightings because access to the valley was difficult, and because the valley was jealously guarded by its inhabitants, the Yosemite Indians.

By late 1853, though, the Indian threat was permanently ended. In the previous two years, members of the Mariposa Battalion had spread word of the valley's scenic wonders. Then in 1855, an entrepreneurial San Franciscan named James Hutchings made the arduous trip to the valley, recognized its potential as a tourist destination, and began publicizing it in his new *Hutching's California Magazine*. In 1856 the first toll road went in, though it accommodated only horses. (There would be no wagon road until 1874.) 1856 also saw the construction of the first permanent building in the valley. By 1864 there were two small hotels, and there were homes and orchards and animals grazing in the meadows, and hundreds of tourists visiting each year.[39]

In that year of 1864, one Israel Ward Raymond wrote to the junior U.S. Senator from California, John Conness, to propose that the federal government cede Yosemite Valley and the neighboring Mariposa Grove of Big Trees to the State of California as a park. Not much is known today about the events that lead Raymond to write his letter. Several other notable personalities who were in the neighborhood in the early 1860s could have been involved. There were John C. Fremont and his wife, Jessie Benton Fremont. They owned a 44,000 acre estate at Mariposa in the gold country foothills, about a day's ride southwest of Yosemite, where they entertained visitors from far and wide. When the Fremonts moved to San Francisco in 1863, their home at Fort Mason became a gathering place for the city's artists and intellectuals.

After the Fremonts left Mariposa, the new owners brought in Frederick Law Olmsted to manage the estate. Olmsted had co-designed Central Park in Manhattan, and he would serve on the first commission charged with managing Yosemite Valley and the Mariposa Grove for the State of California. There was also Carlton Watkins, who would help to publicize Yosemite through his photography. And Galen Clark, who ran a hostel on the way to Yosemite, and who would serve as the first superintendent of the state park. As for Israel Ward Raymond, he represented the Central American Steamship Company in California, who stood to gain if eastern tourists could be induced to book passage to California to see the wonders of Yosemite.

Finally, according to John Henneberger, there was Frederick Billings, who visited the valley in the early 1860s.[40] Billings was a successful lawyer and real estate developer in San Francisco, who would later head the Northern Pacific Railroad. He became an adherent of George Perkins Marsh's ideas after reading *Man and Nature*, which was published the same year, 1864, that Congress ceded Yosemite to the State of California. Billings would go on to distinguish himself as a protector of natural resources. Any or all of these persons could have had a hand in conceiving the idea of

America's first wildland park.

In the preserving of Yosemite, we find united two separate strains of "conservation" that derive from different, even contradictory, premises. As a conservationist in the mold of George Perkins Marsh, Billings might have supported the preserving of Yosemite for utilitarian reasons. Frederick Law Olmsted stated a utilitarian case for preserving Yosemite in the draft of a report that he wrote as a member of the park commission shortly after the State of California took control of the valley.

> *It is a scientific fact that the occasional contemplation of natural scenes of an impressive character, particularly if this contemplation occurs in connection with relief from ordinary cares, change of air and change of habits, is favorable to the health and vigor of men and especially to the health and vigor of their intellect beyond any other conditions which can be offered hem [sic], that it not only gives pleasure for the time being but increases the subsequent capacity for happiness and the means of securing happiness.*[41]

Here Olmsted gives a practical reason for preserving Yosemite: scenic beauty is good for people. In the same way that the utilitarian Marsh advocated government ownership of forests to ensure a future supply of timber, so Olmsted advocated government ownership of Yosemite to ensure a continuing supply of scenic beauty.

But there is also a sense in which the preserving of Yosemite represented a different, anti-utilitarian line of thinking. As Olmsted's concern with scenic beauty suggests, the preserving of Yosemite involved the preserving of an *esthetic* value, as opposed to the purely physical value represented by Marsh's timberlands.

And whereas the conserving of forests for timber entailed extensive human intervention in nature, in the form of cultivating the forests, the preserving of the scenic beauty of Yosemite required the *prohibiting* of human intervention to one extent or another. As Olmsted, again, put the matter:

> The first point to be kept in mind then is the preservation and maintenance as exactly as is possible of the natural scenery; the restriction, that is to say, within the narrowest limits consistent with the necessary accommodation of visitors, of all artificial constructions and the prevention of all constructions markedly inharmonious with the scenery or which would unnecessarily obscure, distort or detract from the dignity of the scenery.[42]

The desire to preserve the natural beauty of Yosemite owes as much to the 19th century's apotheosizing of nature as it does to the kind of practical considerations that motivated George Perkins Marsh. Indeed, the very idea that government should intervene to prevent the human alteration of a natural setting would have been unthinkable had not Rousseau and Wordsworth and Byron, et al, planted the idea in men's minds that such a setting can be in the most literal sense priceless. This, after all, is the meaning of authorizing government to withdraw a given tract of land from the marketplace of private economic transactions, which functions by translating human valuations into prices. The romantic view of nature, on the other hand, holds that the beauty of places like Yosemite Valley transcends the valuations of men. Again, we have Olmsted:

> It was...when the paintings of Bierstadt...had given to the people on the Atlantic some idea of the sublimity of

225

> the Yo Semite...that consideration was first given to
> the danger that such scenes might become private
> property and through the false taste, the caprice or the
> requirements of some industrial speculation of their
> holders; their value to posterity be injured.[43]

Conceptually speaking, there were two steps in the process
by which the romantic nature artists elevated the beauty of nature
to the highest tier of human values. First, they elevated the beauty
of nature to the level of the highest esthetic experience, on a par
with the greatest works of human art. Then they elevated esthetic
experience to the level of the highest of spiritual values. This last
is implicit in Keats' famous line,

> Beauty is truth, truth beauty,—
> that is all
> Ye know on earth, and all ye need to know.[44]

In this way the experience of nature came to represent a pseudo-
religious experience. A place like Yosemite would become a kind
of temple, and the preservation of such a place from commercial
development would be analogous to keeping the money-changers
out of the temple.[45] This is not to say that the romantic nature
artists convinced Europeans and Americans to subscribe to a
religion of nature. It is merely to say that they succeeded in
endowing places like Yosemite with an air of sacredness that even
the most traditional church-going Christian could relate to.

The protecting of a forest to provide men with an ongoing
supply of timber differs fundamentally from the preserving of an
esthetic natural value like Yosemite Valley. These two impulses,
the utilitarian conservation impulse, on the one hand, and the
anti-utilitarian preservation impulse, on the other, would enjoy an
uneasy alliance throughout the last half of the nineteenth century

and the first half of the twentieth, with the former dominating until about mid-twentieth century, when the latter would gain ascendancy and transform conservation into the environmental movement.

President Lincoln signed the bill ceding Yosemite Valley and the Mariposa Big Trees to the State of California on July 1, 1864, thus giving the United States her first wildland park. But Yosemite was a state park, not a national park, and it would remain a state park until 1890. The first national park was Yellowstone.

To a greater extent than even Yosemite Valley, the Yellowstone region remained largely unknown to the civilized world for a considerable time after white men had first glimpsed it. This was in part because it was the official policy of the U.S. Government well into the nineteenth century to discourage settlement of the Missouri River region from Iowa to the Rockies.[46] Zebulon Pike had first enunciated the "Great American Desert" thesis that this region was fit only for buffalo and Indians.[47] A decade later, in the early 1820s, a Major Long led an expedition to explore the headwaters of the Platte, Arkansas, and Red Rivers and opined in his official report that the region was best suited for use by the United States as a buffer on its western frontier.[48]

There is evidence that a white man visited the Yellowstone region as early as 1807-08. According to Aubrey L. Haines, John Colter had been a member of the Lewis and Clark Expedition; shortly thereafter, he journeyed alone into the Yellowstone to establish ties with the natives for the Missouri Fur Trading Company.[49] Other trappers followed in the teens and early 1820s. Then on September 27, 1827, there appeared in the *Philadelphia Gazette and Daily Advertiser* the first published description of the wonders of the Yellowstone. It was written by a young trapper, Robert T. Potts, who was working in the Yellowstone region for the concern of General William H. Ashley and Major Andrew

Henry.[50] Haines says that it "seems probable that the Yellowstone Plateau was visited by American trappers every year after 1826" until 1840, when fur trading ceased to be an organized business on the northern Rockies.[51] It was during this period that the most famous of the Ashley and Henry men, Jim Bridger and Jedediah Smith, also visited the Yellowstone country.

After 1840 the Yellowstone Plateau was left to the Indians, until the discovery of gold in Idaho in 1861-63.[52] Over the next two years, prospectors visited the Plateau "repeatedly," according to Haines.[53] But by 1869, although it had been over sixty years since the first white man had visited Yellowstone, and although countless others had been there since, the knowledge of the areas wonders was still limited to a small fraternity of mountain men and prospectors; among the general population, knowledge of those wonders had advanced little beyond the status of rumor and myth. As had been the case with Yosemite, the threat of Indian attack still kept all but the most intrepid from venturing into the high country of the Yellowstone.

In the late 1860s, Yellowstone would finally be "discovered" and its wonders made known to the world. Father Francis Kuppens, a Jesuit missionary, visited Yellowstone in 1865 with a small group of Piegan Indians. Later that same year, a party of travelers sought refuge from a snowstorm at the mission where Father Kuppens was stationed near present-day Great Falls, Montana. Among this group of visitors were the acting territorial governor, Thomas Francis Meagher, two territorial judges, and two deputy U.S. marshals.

This distinguished group got a first-hand account of the Yellowstone country from Father Kuppens and another visitor who also had been there. Typically, for the time, none of the group had ever heard of the place. In reaction to Father Kuppens' description of Yellowstone, "General Meagher said if things were as described the government ought to reserve the territory for a national park." [54]

There followed several exploratory expeditions, which would finally open the Yellowstone country once and for all. The third and final of these, led by Ferdinand V. Hayden, head of the U.S. Geological Survey of the Territories, was financed by Congress, and among the several dozen members of the party were scientists and academics, the photographer, W.H. Jackson, and the artist, Thomas Moran, the latter in the employ of Jay Cooke's Northern Pacific Railroad (who seems to have seen in Yellowstone a potential tourist attraction for passengers on his railroad). [55] The party departed from Fort Ellis, Montana on July 16, 1871 and returned on September 1, the expedition a great success.

When Dr. Hayden returned to Washington, he found on his desk the following letter on stationery of "Jay Cooke & Co., Bankers, Financial Agents, Northern Pacific Railroad Company" and signed by A.B. Nettleton:

> *Dear Doctor:*
> *Judge Kelley has made a suggestion which strikes me as being an excellent one, viz.: Let Congress pass a bill reserving the Great Geyser Basin as a public park forever—just as it has reserved that far inferior wonder the Yosemite valley and big trees. If you approve this would such a recommendation be appropriate in your official report?* [56]

A.B. Nettleton was Jay Cooke's Philadelphia office manager, and Judge Kelley was a Congressman from Pennsylvania, a railroad supporter, and an ally of Cooke.

As this letter suggests, the precedent set by Yosemite helped to make Yellowstone Park possible. The frequency with which the advocates of preserving Yellowstone invoked the precedent of

Yosemite suggests that the creating of Yellowstone National Park was, to an important extent, an elaboration of the idea pioneered by the preservers of Yosemite, that government could and should intervene to preserve certain expanses of wild nature from alteration by humans.

In 1871, for example, the Montana journalist, James Hamilton Mills, compared Yellowstone to Yosemite and then asked that "this, too, be set apart by Congress as a domain retained unto all mankind ... and let it be *esta perpetua*." [57] Later, in 1872, Dr. Hayden, writing in *Scribner's Monthly*, asked "Why will not Congress at once pass a law setting it [the Yellowstone] apart as a great public park for all time to come as has been done with that far inferior wonder, the Yosemite Valley?" [58] Then Henry L. Dawes, "the most influential man in the [U.S.] House at that time," according to Haines, in urging the passage of the Yellowstone Park bill, said, "This bill follows the analogy of the bill passed by Congress six or eight years ago, setting apart the Yosemite Valley ... with this difference: that bill granted to the State of California the jurisdiction over the land beyond the control of the United States." [59]

Although Yellowstone was the first national park, the preserving of Yosemite as a state park was an essential step toward its creation. Indeed, one reason that Yellowstone became a national instead of a state park was that Wyoming, within which most of the Yellowstone country lay, was not yet a state on March 1, 1872, when President Grant signed the bill creating Yellowstone Park.

As in the case of Yosemite, we do not know a great deal about the motivations of the men who promoted the preserving of Yellowstone as a park. Clearly, there was an element of commercial motivation on the part of Jay Cooke and the Northern Pacific Railroad, as there had been in the case of Israel Ward Raymond and his Central American Steamship Company. But this motivation would have had to remain in the background; it would have needed the cover of "higher" motivations that were more in keeping with the withdrawing of a vast region from commercial development, motivations such as Olmsted had enunciated in favor of preserving Yosemite. Indeed, Olmsted and Bierstadt and the romantic poets and painters had already made the "higher" case for preserving scenic natural beauty like Yosemite; all that was needed in the end to justify the preserving of Yellowstone was to invoke the Yosemite precedent.

In 1890, Yellowstone returned the favor to Yosemite; dissatisfied with California's management of Yosemite, John Muir and his Sierra Club successfully lead an effort to have Yosemite declared America's second national park. The federal government was now unequivocally in the business of preserving and managing scenic natural beauty.

Powell

Many persons deserve credit for starting the national forests and the national parks in the United States, but in the conservation of water resources one man's efforts dwarf all others. John Wesley Powell was born at Mt. Morris, New York in 1834 to a Methodist minister and his wife, who named their son after the founder of Methodism.[60] Young Powell grew up on the frontier, first in Ohio, then in Wisconsin, and then in Abe Lincoln's Illinois. From sixteen until twenty-five he educated himself at a series of schools and colleges, supporting himself by working off and on as a teacher. Five days after Lincoln called for volunteers to put down the Southern rebellion, Powell enlisted in the Union Army. He lost an arm at Shiloh, returned to duty, and mustered out as a major of artillery four months before Appomattox.

Although he never earned a college degree, Powell got himself a professorship after the war at Illinois Welseyan University teaching geology, his first love, and a range of other sciences. Two years later he left teaching to take on the curatorship of the museum of the Illinois Natural History Society. Intending to get out of the museum and into the field, Powell lead an excursion to the Rocky Mountains in 1867 in search of geological and other specimens. On this trip, he formed the idea of leading an expedition down the Colorado River and through the Grand Canyon, the last bit of unexplored territory between the Atlantic and the Pacific. Thus began Powell's life's work as a student of the American west, its topography, its geology, its hydrology, and its native peoples. He undertook his Grand Canyon expedition in 1869, and its success made him a national hero.

Powell followed up this expedition by winning federal funding to conduct a detailed survey of the Grand Canyon and its surrounding region. He occupied himself with this survey from 1870 until 1879, when the United States Geological Survey came into being and took over the effort. Powell then founded the Bureau of Ethnology, within the Smithsonian Institution, to pursue fulltime his study of the native peoples of the American west. But two years later he was back in the surveying business, this time as head of the USGS. He held this position until his retirement in 1894.

Through his surveying work, Powell became an expert on the "arid region," as he called it, of the United States. The lack of water in this region, Powell believed, rendered obsolete many of the laws and institutions relating to the ownership of land, as those laws and institutions had developed in the moist climates of northern Europe and the eastern United States. In Ohio a farming family could support themselves on a square parcel of land whose boundaries had been arbitrarily drawn on a map. They could trust that the water they would need for their crops and livestock would fall from the sky, so they could situate their patch of ground almost anywhere on the terrain, without regard to the availability of water from streams or rivers.

But in the west, where it was impossible to raise crops without access to a source of irrigation, the location of creeks and rivers dictated where homesteading was possible and where it was not. Powell believed this meant that a whole new approach to land ownership was needed. He believed the government should decide beforehand which portions of the land were irrigable—he estimated that throughout much of the west the total amount of irrigable land would amount to no more than twenty per cent of the all land. This should be the only land made available for homesteading.

Powell believed it necessary that the owners of these irrigated farms should control the water they used for irrigation all the way from where it originated in the high mountains down to their fields. This upper portion of the drainage basin he would close to settlement or to individual ownership of any sort. The farmers on the irrigable land below would own and control this upper watershed collectively and in the interest of protecting the sources of their water.[61] The model for Powell's ideas on communal land ownership in the west came from the old Spanish villages of New Mexico and from the Mormon settlements of Utah.[62]

At the height of his career, Powell got a brief opportunity to launch his grand design for controlled development of the west. Starting in 1887 the region suffered a severe drought that lasted into the 1890s. In March of 1888, in response to this drought, a group of western politicians in Washington, lead by Big Bill Stewart of Nevada, passed a vaguely worded Joint Resolution calling for the Secretary of the Interior to identify potential reservoir sites throughout the west that could be used for irrigation purposes.[63] What the Senator and his colleagues wanted was a quick study to get some badly needed irrigation projects under way. What they got was something else entirely.

233

Powell, whose USGS was within the Interior Department, had consulted with Stewart on the Joint Resolution with the intention that he, Powell, would lead the Irrigation Study. But the Senator and his colleagues had misread Powell, and they gave him a degree of power they would later regret. The Joint Resolution carried an amendment that withdrew from settlement all the lands that Powell's survey deemed irrigable. The purpose of the amendment was to allay public suspicions that land speculators might use advance knowledge of the findings of the survey to buy up choice irrigable sites ahead of genuine homesteaders. But, since no one yet knew which lands were irrigable—this depended on the findings of the survey—let alone how large a portion of the American west was subject to the survey, the effect of the amendment was to close the public domain to settlement all the way from the 100th meridian, a line running roughly from mid-North Dakota through mid-Texas, to the Pacific Ocean.

Worse still, the amendment gave the power to reopen this region to a single man.[64] Indeed, in combination with the Joint Resolution, it gave that one man, by virtue of his authority to designate where irrigation water should flow, the power to determine the entire pattern of future settlement throughout the American west. That man was John Wesley Powell, the geologist who had never graduated from college, whose highest position in government was as the appointed head of the United States Geological Survey. Powell intended to take full advantage of the opportunity.

Powell saw the Irrigation Survey as the chance of a lifetime to redirect the settlement of the west. As part of his new assignment, he intended to complete the topographical mapping of the west that was already underway, to complete a survey of the region's surface hydrology, which would include the inventorying of irrigable lands, and to conduct preliminary engineering studies regarding potential headworks and canals.[65]

He estimated that the entire project should take six to seven years and cost six to seven million dollars.[66] Powell aimed at nothing less than to replace the free settlement of the west by planned settlement directed from Washington.

This was too much for Stewart and his cohort; they shared neither Powell's desire for a comprehensive survey of the entire hydrology of the west nor his affinity for central planning. In August of 1890, a year and a half after the Irrigation Survey was first authorized, Stewart and his colleagues added a new amendment to the Joint Resolution that shut down the survey. Free enterprise had won out over central planning, for the time being. The closing of the Irrigation Survey amounted to the defeat of a lifetime for Powell, and the beginning of the end of his public career.

Powell's intention had been to manage whole drainage basins according to scientific principles, just as the government silviculturists would manage whole forests. And, although he advocated that each hydrologic system, from mountain springs to the edges of the farmers' fields, should be collectively owned and operated by the homesteaders themselves, he also envisioned a major role for the federal government in the management of the surface hydrology of the west, just as the foresters envisioned for their forests.

But an additional theme animated Powell's work that did not animate that of the early forest conservationists, democratism. Powell opposed free, capitalist development of the water resources of the west not because such development would fail to harness water resources, but, among other reasons, because he believed it would lead to water "monopolies" in the region. By controlling the water of a region, private companies could control the land as well. Powell's vision for the west stood diametrically opposed to land monopolies. He believed that water ought to be used for irrigation, that all land that could be irrigated ought to be irrigated, and that all irrigable land ought to be made available to homesteaders.

This vision reflected the old Jeffersonian idea that the ideal society is one of independent yeoman farmers. Powell would revolutionize land ownership laws and institutions in the west precisely to preserve Jefferson's democratic ideal in a region of sparse water resources. In his championing of the Jeffersonian ideal and in his opposition to monopolies, Powell resembled not so much his fellow conservationists, the early foresters, as he did a later conservationist, the greatest of them all, Gifford Pinchot, and he anticipated the larger movement within which Pinchot would lead conservation to full maturity, the Progressive movement.

Powell's plan for collective control of the headwater regions by the owners of the irrigated farms below never became a reality. Nor did has plan, presented to the Montana State Constitutional Convention, for drawing county lines in the arid states according to drainage basin boundaries. But many of his most important ideas have been implemented in one form or another. With the passage of the Reclamation Act in 1902, the federal government not only took on the responsibility of designating irrigable lands and reservoir sites, it went into the business of constructing whole irrigation systems. And with the formation of the Reclamation Service (later the Bureau of Reclamation) later the same year, the federal government in effect institutionalized Powell's Irrigation Survey.

Centralized planning of drainage basins and entire watersheds is today a reality, with local, state, and federal authorities all participating. Most of the high country headwater regions are now preserved, under the aegis of the U.S. Forest Service. And, with much of the west now irrigated, most of the rest has been closed to settlement, as Powell had urged, since the passage of the Taylor Grazing Act in 1934. Powell failed in his grand design only in that it came to fruition later than he had intended, and in somewhat altered form.

Powell was ahead of his time in another sense. On his first trip west he had solicited and won from Washington the right to draw military rations from posts along the way at reduced rates. From the time of his second trip west, down the Grand Canyon, to the end of his career, every scientific venture that he headed was funded in part or in full by government. Powell pioneered government science much as he pioneered government management of water resources.

It is worth quoting Wallace Stegner at length on this topic:

> *Alexander Agassiz had testily attacked government patronage of science five years before, and he had had a clear idea of what he was attacking. The concept of the welfare state edged into the American consciousness and into American institutions more through the scientific bureaus of government than by any other way, and more through the problems raised by the public domain than through any other problems, and more through the labors of John Wesley Powell than through any other man. In its origins it probably owes nothing to Marx, and it was certainly not the abominable invention of Franklin Delano Roosevelt and the Brain Trust. It began as public information and extended gradually into a degree of control and paternalism increased by every national crisis and every step of the increasing concentration of power in Washington. The welfare state was present in Joseph Henry's Weather Bureau in the eighteen-fifties. It moved a long step in the passage of what Henry Adams called America's "first modern act of legislation," when the King and Hayden Surveys were established in 1867. It had come much farther by ...*

1890, and it would assume almost its contemporary
look in the trust-busting and conservation activities of
Theodore Roosevelt at the dawn of the next century.[67]

All of these early forms of "conservation," forestry, bird and
wildlife protection, wildland parks, and water resource
management, had in common the idea that the management of
natural resources is, to an important extent, a responsibility of
government. Each of these early conservation efforts called,
therefore, for government control of a portion of the nation's
natural resources. This call for government control of these
resources brought these early conservationists into conflict with
advocates of the earlier idea that government had no business
owning or otherwise controlling natural resources, and that such
resources ought properly to be privately owned. Nevertheless,
the early conservationists succeeded, against stiff resistance, in
establishing a beachhead against the private property advocates in
the last decades of the nineteenth century. In the first two decades
of the twentieth century, the conservationists would find in the
Progressive movement a crucial ally in their attack on the private
property regime. Together they would eventually roll like a
juggernaut over the defenders of private ownership of natural
resources.

Pinchot

In 1864, George Perkins Marsh had called for Americans to better
manage their natural resources. By the end of 1890, Marsh's call
had swelled to a small chorus, but the members of that chorus had
yet to achieve very much of what they aimed at: there were, as
yet, no federal forest reserves, though there one state
reserve; there were two national parks, but no plans for more;
commercial game hunting on public lands had been extinguished,
but there was no Audubon Society, and wild bird protection was

virtually nil; and John Wesley Powell's grand design for government control of water resources in the west had just been shut down by Congress. And, although there had been some interaction among the various resource protection efforts, most notably in the person of George Bird Grinnell, there was, as yet, no unified effort to better manage natural resources in general.

But the next two decades would see a great surge in resource "conservation" in the United States. In 1891 the first federal forests would come into existence, with many more soon to follow; and in the early twentieth century the federal government would begin to manage its forests, as well as the rest of its land holdings, like a true property owner. During the same period, federal engineers and scientists, in league with western politicians, would begin to implement, bit by bit, much of John Wesley Powell's master plan for federal management of western water resources. And, in tandem with its efforts at water resource planning, the federal government would begin to practice land use planning throughout its immense holdings in the west.

In 1896 the Audubon Society would come back into being—to stay, and wild bird protection would become an ongoing legislative priority at the state and federal levels. In 1908 the Grand Canyon would be preserved as the first national monument, and the preserving of scenic natural beauty would become an established federal (and state) priority. And, above and beyond all of these specific achievements, there would come into being a (somewhat) unified conservation movement. By then, most Americans would have heard of "conservation," and many of them would agree with the idea.

In part, this explosion of conservation activity was a direct result of the long and dogged efforts of men like George Bird Grinnell and the other pioneers of resource protection and management. But it was also a result of the convergence of these efforts with a new force in American life—Progressivism.

Thomas McCaffrey

Progressivism, as we have seen, was a movement to expand the role of government into the economic realm. Progressivism viewed constitutionally protected property rights as a barrier to government intervention in and regulation of economic matters. It sought to weaken or to abolish those constitutional barriers and thereby to enable local, state, and federal governments to intervene in the economic affairs of the nation, legislating wages, working conditions, hours of employment, and countless other matters that earlier generations of Americans had considered to be private concerns that lay beyond the reach of the law.

As we have seen, in the years before Progressivism, the various resource-protection efforts also harbored a certain, at least implicit, distrust of property rights; they considered private ownership to be inadequate to the purpose of ensuring that certain natural resources—whether forests, or waterways, or wild game or birds, or scenic beauty—would be cared for and maintained as the citizen-guardians of those resources believed they should be cared for and maintained. In consequence, most of these guardians advocated, to one extent or another, government regulation or ownership of the resources in question.

When Progressivism arrived on the scene, about 1890 or so on the local level, and Progressives began to hammer away at the constitutional protection of property rights, their efforts helped to clear the way for just the sort of government intervention on behalf of resource protection that the various resource guardians had been promoting for years. Although many of the resource-protection efforts predated the advent of Progressivism, after 1890 many of the more prominent advocates of conservation were themselves Progressives. It was no coincidence that the most prominent conservationist of them all, Theodore Roosevelt, ended his political career as an arch-Progressive. By the end of Roosevelt's second term, conservation would come to be identified as an integral part of the broader Progressive movement.

Samuel P. Hays identifies the campaign to bring the federal government into the irrigation business in the west as the beginning of the conservation movement proper.[68] Two leaders of that effort were Frederick H. Newell and George H. Maxwell. Newell was an MIT-educated mining engineer who had worked with John Wesley Powell on his ill-fated Irrigation Survey of 1888-1890. He became chief hydrographer of the U.S. Geological Survey, and throughout the 1890s he carried on the surveying of western water resources. At the same time, Maxwell, a young California lawyer who specialized in water law, initiated a campaign to win federal funding for irrigation projects in the west. Working together, Newell and Maxwell helped move irrigation to the top of the nation's political agenda; in the national election of 1900 both political parties included a plank in their platforms calling for federal funding of western irrigation projects.

Meanwhile, Representative Francis G. Newlands of Nevada led the effort in Congress to get the federal government into the irrigation business. The resulting bill, passed on June 17, 1902, achieved Newland's purpose by authorizing the federal government to use the proceeds from the sale of federal lands to finance irrigation projects. President Roosevelt placed the administration of the Newlands Act in the hands of Newell, who was to head up a new Reclamation Service under the USGS. By 1910 the Reclamation Service (by then the Bureau of Reclamation) had completed or started over two dozen irrigation projects throughout the west.

The Newlands Act, or, as it was also known, the Reclamation Act, gave broad authority to the Secretary of the Interior to determine which irrigation projects should be undertaken and where. The purpose of this strategy was to keep Congress, and thus politics, out of the irrigation business. (The funding of specific irrigation projects was not dependent on Congressional approval; by getting their funding from the sale of federal lands, the Reclamation Service would not need to go running to Congress for approval every time they wanted to undertake a new project.) Science and expertise would guide the running of the irrigation business, not political machination. "This provision for considerable executive discretion in resource development and management"—as opposed to Congressional control—"became a central feature ... of the entire conservation movement," writes Hays.[69]

Thus did conservation become the incubator of technocracy in America. Note that rule by technocrats is a profoundly anti-democratic phenomenon, despite conservation's and Progressivism's claims to value democracy. The conservationists believed that private citizens were not equal to the task of irrigating the west properly; only federally sanctioned experts could do the job. And they could only do the job properly if they could operate by executive fiat, without the need to subject their decisions to the messiness of legislative, i.e., democratic, oversight and control. Federal irrigation was a paternalistic, anti-democratic enterprise, in a way that eventually would come to characterize the conservation movement as a whole.

The campaign for federal funding of irrigation marks the beginning of the transformation of "conservation" from an assortment of separate resource-protection efforts into a unified movement. Large-scale irrigation is dependent upon a steady, reliable supply of water from the headwater regions in the mountains. The steadiness and reliability of these headwaters depends partly, in turn, upon the health of the forests in those regions. Early on, the advocates of federal funding of irrigation recognized their dependence on the work of the forest preservationists, with whom they began to ally themselves.[70] The leader of the forest reserve effort, and the most prominent voice of "conservation" besides Roosevelt, was the President's close friend and advisor, Gifford Pinchot.

Pinchot was born in 1865 to a wealthy commercial family of New York City. His maternal grandfather was a successful real estate developer, and his father made a fortune in lumber before starting a business selling commercial and domestic furnishings. Upon graduating from Yale in 1889, Gifford followed his father's advice and embarked upon the unusual career of forestry. His father, James, had come to fear that lumber companies like his own might one day destroy the forests of the United States; he himself had served a stint as vice president of the American Forestry Association. So after graduating from Yale Gifford embarked for France, where he would spend a year at L'Ecole Nationale Forestiere at Nancy. Upon his return to America, he took a position as forester at George W. Vanderbilt's North Carolina estate, Biltmore. Several years later Pinchot moved back to New York City, where he opened an office as a consulting forester.[71]

Charles S. Sargent in his *Garden and Forest* magazine had suggested in 1889 that a Presidential commission be appointed to make recommendations to Congress regarding the care and administration of the forested lands on the federal domain. In 1894, at the instigation of Sargent, Pinchot, William A. Stiles, editor of *Garden and Forest*, and Robert Underwood Johnson of *Century Magazine* launched an effort to put Sargent's plan into effect. In February of 1896, Secretary of the Interior Hoke Smith spurred on by Pinchot, among others, wrote a letter to the National Academy of Sciences requesting that the Academy sponsor such a study of the public forested lands. Of the seven men who were appointed to the resulting National Forest Commission, Pinchot was the youngest. He had now begun what would become his primary life's work, the promotion of government-sponsored "conservation" of natural resources.

Since 1891, when Congress had authorized the President to designate forest reserves on the public domain, 17,500,000 acres had been so designated. But, just as Congress had earlier been slow to define the meaning of the term "national park," so no one had yet defined what exactly it meant for an area to be a federal forest reserve. In consequence, the Department of the Interior had simply closed these reserves until Congress spelled out just how, or whether, they were to be used at all. The closing of the reserves was causing hostility throughout the west toward the very idea of federal forest reserves. It was important, therefore, to the advocates of the forest reserve idea that the National Forest Commission take the lead in defining what a forest reserve should be and in recommending where new reserves should be established. The young forester, Gifford Pinchot, as Secretary of the National Forest Commission, had thus maneuvered himself into the very center of the forestry discussion in the United States.

Among the Commission's eventual recommendations was that Congress authorize the President to designate 13 new forest reserves totaling 21,000,000 acres in seven states, which reserves President Cleveland duly designated. The Commission failed to recommend a plan of administration for the reserves, but the Pettigrew Amendment to the Sundry Civil Act of June 4, 1897, for which Pinchot had lobbied hard, eventually provided the needed plan of administration. The Pettigrew Amendment put the Secretary of the Interior in charge of the forest reserves, opened them to use, and authorized the Secretary to "make such rules and regulations and establish such service as will insure the objects of such reservations, namely, to regulate their occupancy and use and to preserve the forests therein from destruction." [72] (This is the same strategy of executive-branch governance that would be used a few years later for federal irrigation projects under the Reclamation Act.)

Pinchot then went to work for the Interior Department as a

Confidential Agent for the Land Office. He set about researching and inventorying the new reserves, and he began devising a plan for a Forest Service to administer the reserves. After finishing his survey for the Land Office, Pinchot was chosen on July 1, 1898 to replace Bernard Fernow as the chief of the Forestry Division in the Department of Agriculture. He now had charge of federal forestry for the entire United States. But he had all of ten employees, a miniscule budget, and no forests—these were still under the Department of the Interior.

Pinchot's predecessor at the Forestry Division, Fernow, had conceived of his task as essentially one of preaching forestry, to the states and to private land-owners, rather than practicing it. Pinchot's idea of his own mission was altogether different. He aspired to nothing less than complete managerial control over the entirety of America's existing federal forest reserves, and to expanding dramatically the number and size of those reserves.

Pinchot made a huge step toward realizing the first of these goals when, after years of political maneuvering, he succeeded in having the existing forest reserves transferred from the Department of the Interior to his own Bureau of Forestry in the Department of Agriculture. This came on February 1, 1905, when President Roosevelt signed the bill effecting the transfer. On this day, Pinchot came into control of 86,000,000 acres of land, an area about the size of New York, New Jersey, and Pennsylvania combined. By this time, he had changed the name of his department from the Bureau of Forestry to the Forestry Service. Soon he would the re-christen the reserves the National Forests.

By the time Roosevelt left office in 1909, there would be 149 forest reserves covering 193,000,000 acres. By then, the federal government would no longer be just a temporary owner whose primary purpose was to transfer federal lands to private citizens; it would be a permanent owner and manager of a vast forest empire, and Pinchot would be instrumental in effecting the transformation. Pinchot once said that government needed to be run more like a business. This would have made little sense to the men of 1789 with their idea of limited government, but for a government determined to manage a nation's natural resources, it did made a certain sense, and in his management of the National Forests Pinchot showed how to do it.

In addition to building irrigation projects and managing forested lands, the federal government also got into the business of land-use planning on the vast public domain. In 1903 President Roosevelt appointed a Public Lands Commission "to report ... upon the condition, operation and effect of the present land laws, and on the use, condition, disposal, and settlement of the public lands." A common theme of the federal laws in question, among them the Homestead Act of 1862, the Mineral Land Act of 1866, the Desert Land Act of 1877, and the Timber and Stone Act of 1878, had been to limit federal land sales to individual smallholders and to prevent the aggregation of large holdings by speculators and corporations. The Homestead Act, for example, permitted the transfer of no more than 160 acres to an individual (at no cost, if he agreed to inhabit the land for five years).

But in the view of the conservationists, the restrictions in these laws were too easily evaded, and the aggregation of large holdings was all too common. The Public Lands commission was to advise the administration on how best to revise these laws. In the main, the conservationists' solution to this alleged problem was for the federal government to end the selling off its lands and to retain ownership of most of what remained in its possession. From this point onward, the federal government would be permanent landlord of a huge portion (one third, after the addition of Alaska in 1959) of the land in the United States.

While the Public Lands Commission was getting going, the U.S. Geological Survey continued its inventorying of the western lands, and the Roosevelt Administration used the information they gathered to classify lands according to how they believed those lands should be used. In addition, the Forest Service was busy classifying land within the forest reserves as timber producing, grazing, watershed, or wildlife areas. Then in December, 1908, the Department of the Interior organized a Land Classification Board in the USGS to conduct a systematic inventorying and classification of public lands, an effort that had begun less formally within the USGS in 1905. This was an effort to classify lands as agricultural or grazing, as water power sites, or as coal or oil bearing. The upshot of all these federal efforts at central planning and administration was that future decisions as to how much of the land in the west would be used would be made not by the free market or even by political office holders, but by government "experts." [73] We have seen this same phenomenon in the case of irrigation. Rule by technocrats would be an integral part of the Progressive plan for America.

To complement the comprehensive central planning of land use on the public domain, the conservationists conceived the idea of doing the same for waterways. The leaders of this effort were Pinchot, Newell, and WJ McGee, geologist, former assistant to John Wesley Powell, head of the Bureau of Ethnology until 1903 (Powell's old post), and a founder of the National Geographic Society and the American Anthropological Society. McGee, would become, along with Pinchot, a leading theorist and promoter of the conservation movement.

On March 14, 1907, at the instigation of McGee (although Pinchot suggested that he himself might have had the idea first), Roosevelt announced the formation of an Inland Waterways Commission to look into the possibility of the federal government's implementing comprehensive, multi-use development plans for the nation's river systems (not just in the west). It's guiding principles were two: first, that each river system should be treated as a unit, and second, that the development of each such unit should take account of the full range of uses incident to it, including flood control, irrigation, soil conservation, power generation, reclamation, and navigation. In addition, the Commission "insisted that local questions of waterway control should be treated as general questions of national concern." [74] (This was an important theme of Pinchot's, that natural resources were a distinctly *national* responsibility.) Marshall O. Leighton, Chief Hydrographer of the Geological Survey and advisory hydrographer to the Commission, drew up a comprehensive river-development plan for the Ohio River system that involved the construction of 100 reservoirs, which would be funded by selling hydro-electric power, much as federal land sales had been used to fund irrigation projects.

Senator Newlands sponsored a bill to implement the Commission's ideas for federal planning and control of waterway development in the U.S. Among the features of this bill was a plan to put the control and funding of individual projects in the executive branch, rather than in Congress, much as the Reclamation Act had done for irrigation projects in the west. Ultimately, Congress thwarted the Roosevelt conservationists' grand design for waterway development, though the idea would come to fruition two decades later under the auspices of the Franklin Roosevelt administration. As it was, between the National Forests' control of the headwater regions in the mountains and the Bureau of Reclamation's control of irrigation water, the federal government had already advanced a long way toward realizing John Wesley Powell's vision of unified management of entire watersheds.

The high water mark of the early conservation movement came in 1908 with the convening of the National Governors' Conference. The brainchild of Pinchot and McGee, the Conference was a publicity extravaganza for the conservation cause. Following an agenda put together by Pinchot and McGee, the governors formally acknowledged the importance of conserving resources, and they called for the creation of a National Conservation Commission to conduct an exhaustive inventory of all the natural resources of the United States. But Congress declined to fund the resulting commission, and it disbanded in 1909. With Theodore Roosevelt's leaving office in 1908, centrally organized conservation began to wane, although the idea of conservation, and a great many of the government initiatives and private organizations would prove lasting. The conservation movement continued to influence events right up until the 1950s and 60s, when it would be subsumed by a new movement called environmentalism.

Gifford Pinchot deserves special credit for shaping conservation in its early years. In addition to leading the effort to nationalize the nation's forests and waterways, Pinchot served an important role as theoretician of and spokesman for the larger movement to "conserve" America's natural resources in general. One day in February of 1905, while out for a ride on his horse in Washington's Rock Creek Park, Pinchot pondered the bewildering multitude of questions he and others were attempting to answer regarding the management of the nation's natural resources. Twenty-some government agencies were then engaged, in one way or another, in studying or managing America's forests, soil, minerals, waterways, and fish and wildlife. "Suddenly the idea flashed through my head that there was a unity in this complication … all these separate questions fitted into and made up the one great central problem of the use of the earth for the good of man." [75]

Thus did conservation become a self-conscious, centralized movement, and Pinchot became its chief spokesman. Through his books, articles, speeches, and publicity extravanganzas (like President Roosevelt's "whistle-stop" cruise down the Mississippi with the Inland Waterways Commission in 1907), Pinchot did more in these years than anyone else to make "conservation" a household word. He also did a great deal to maneuver conservation to the center of the nation's political attention through his various commissions, his lobbying, and his influence with President Roosevelt. No one contributed more during these years, with the possible exception of Roosevelt himself, to putting the federal government into the business of managing the natural resources of the United States.

But Gifford Pinchot was not just a conservationist; he was also a Progressive, and he merits singular credit for making conservation a fundamentally Progressive phenomenon. Pinchot helped to bring the various resource-protection efforts into the Progressive fold by recasting them as elements of a larger conservation movement, which he proceeded to define in distinctly Progressive terms. Progressivism, specifically that of the Croly-Roosevelt variety, was pointedly national, as opposed to local, in its orientation. Pinchot moved conservation toward the same nationalist orientation. He argued that national prosperity was a function of an abundance of natural resources, and that that prosperity was threatened because natural resources were being wasted and otherwise used up too rapidly.[76]

The reason they were being wasted and used up—and here Pinchot's Progressivism manifests itself clearly—was that permitting private ownership of resources resulted in their monopolization by the wealthy, who tended to over-exploit them for private profit. "The earth ... belongs of right to all its people and not to a minority, insignificant in numbers but tremendous in wealth and power. The public good must come first." [77] The only way to ensure national prosperity was for government, especially the federal government, to control and manage as much of America's natural resource wealth as possible. This idea that natural resources ought of right to belong to "the people" collectively implicitly amounts to an attack on the very foundation of capitalism, the private ownership of land. Neither Pinchot nor his fellow leaders of the conservation movement would pursue this idea to its logical conclusion, the abolition of private land ownership altogether. That would fall to a later generation of conservationists, who would go by the name "environmentalists."

In Pinchot's view, private ownership of natural resources threatened not only national economic strength but also individual liberty. Pinchot held to the Jeffersonian ideal of the yeoman farmer; because of his economic independence, the yeoman farmer was the backbone of liberty. The concentration of landed wealth in a few hands, especially landed wealth that included other natural resources, such as timber, water, or minerals posed a threat to liberty. Pinchot opposed concentrations of landed wealth or of other natural resources because they enabled their possessors to exercise undue political power, to the detriment of small landowners. Because conservation opposed private concentrations of natural resources, Pinchot promoted it as more "democratic" than any system that permitted private ownership of natural resources. Thus did Pinchot introduce democratism, the second of the twin pillars of Progressivism, after opposition to constitutional protection of property rights, into the ideology of conservation.

But liberty was of only secondary importance in Pinchot's view. Democrat that he was, he held "equality" to rank higher than Jeffersonian liberty in the pantheon of American political values. James Madison had believed that liberty requires the securing of private property rights, and that such security might well result in some citizens' becoming more wealthy than others. This was a price of liberty, and it was far preferable to the tyranny of government-enforced equality. Pinchot, on the other hand, would put government into the business of managing natural resources in order to prevent the accumulation of great wealth by individuals and thereby promote equality—of opportunity, as he put it. "Equality of opportunity is the real object of our laws and institutions." [78]

Such government-enforced equality, of course, turns Madison's Constitution on its head. Pinchot knew that what he proposed amounted to a fundamental change in the politico-economic system of the United States. "The question of the conservation of our natural resources ... requires ... thinking out along lines directed to the fundamental economic basis upon which this nation exists." [79] In his view, the United States had succeeded in securing "political freedom," and now it was time to unshackle the power of government to regulate property in the name of "industrial freedom," which had more to do with economic equality than with genuine liberty and the rights of individuals. Pinchot aimed not to secure the latter but, rather, "the greatest good of the greatest number," to which he added "for the longest time."

Under a system of private property and of individual rights, each person decides for himself how best to secure his own "greatest good." But under a system that aims at "the greatest good of the greatest number," someone other than private individuals must decide what best conduces to that end; in Pinchot's overtly paternalist view, scientifically trained government professionals would control the natural resources and do the deciding. Socialism advocates government ownership of the means of production: land, labor, and capital. Gifford Pinchot helped move the United States down the road toward permanent government ownership of a huge proportion of its land (and he helped undermine the moral foundation that underlay private ownership of the rest of the nation's land).

The character of conservation as a movement is nicely captured in a quote from WJ McGee. He described it as "a conscious and purposeful entering into control over nature, through the natural resources, for the direct benefit of man." [80] As this quote so clearly shows, the purpose of the conservation movement was entirely consistent with the idea, implicit in the work of the American Founders, that it is possible for man to command nature, and that it is right for him to do so. On the other hand, the conservationists' distrust of private ownership of natural resources, and their embrace of government ownership and control of such resources, placed them in direct opposition to the founders' commitment to private property rights.

Notes

1. George Perkins Marsh, *Man and Nature,* ed. David Lowenthal (Cambridge, MA: The Belknap Press of Harvard University Press, 1965.) The biographical information on Marsh comes from Mr. Lowenthal's introduction.

2. Ibid., 352.

3. Ibid., 187.

4. Ibid.

5. Ibid.

6. Ibid., 38.

7. Ibid., 35.

8. "[W]e can never know how wide a circle of disturbance we produce in the harmonies of nature when we throw the smallest pebble into the ocean of organic life." Ibid., pp. 91-92.

9. Ibid., 258.

10. Ibid., 51.

11. Ibid., 436.

12. Ibid., 258-59.

13. Ibid., 259. "This [halting the damaging of lands newly settled by European man] can be done only by the diffusion of knowledge on this subject among the classes that, in earlier days, subdued and tilled ground in which they had no vested rights, bur who, in our time, own their woods, their pastures, and their ploughlands as a

perpetual possession for them and theirs, and have, therefore, a strong interest in the protection of their domain against deterioration." Ibid., 46. Marsh did share with the conservationists of the Progressive era a hostility toward corporations, and even toward the "old legal superstition," grounded in the Dartmouth College case, that corporations merited special protection against government interference.
14. Dr. Hough headed the New York State census twice and the U. S. Census in 1870.
15. Carl Schurz, *Annual Report of the Secretary of the Interior on the Operations of the Department for the Fiscal Year Ended June 30, 1877* (Washington, D.C.: U.S. Government Printing Office, 1877), [iii], xv-xx.
16. "Organizing the U.S. Geological Survey," United States Geological Survey, last revised April 10, 2000, accessed December 17, 2014, http://pubs.usgs.gov/circ/c1050/organize.htm.
17. "The legislatures of those districts or new states, shall never interfere with the primary disposal of the soil by the United States in Congress assembled, nor with any regulations Congress may find necessary for securing title in such soil to the *bona fide* purchasers." Northwest Ordinance, Article 4.
18. Bernard Fernow, "The Providential Functions of Government with Special Reference to Natural Resources," *Science* (August 30, 1895): 262-264. Quoted in Char Miller, "Wooden Politics: Bernard Fernow and the Quest for a National Forest Policy, 1876-1898" in *The Origins of National Forests*, ed. Harold K. Steen (Durham, NC: Forest History Society, 1992), 294.
19. In 1878 the Wisconsin State Legislature had set aside 50,000 acres in Vilas and Iron Counties as a park, but these were sold off to private owners nineteen years later.
20. George Bird Grinnell, *Forest and Stream* xviii (April 13, 1882), 204. Quoted in John F. Reiger, "Wildlife, Conservation, and the

First Forest Reserve," Steen, ed., *Origins*, 107.

21. Ibid., 108.

22. Ibid., 110.

23. This account of the founding of the Audubon Society is based on that of Frank Graham, Jr., *The Audubon Ark* (New York, NY: Alfred A. Knopf, 1990).

24. Ibid., 6.

25. The Model Law exempted the scientific collection of birds.

26. Ibid., 9-10.

27. The biographical material on Grinnell is drawn from John Franklin Reiger, *George Bird Grinnell and the Development of American Conservation, 1870-1901* (unpublished doctoral thesis, Northwestern University, 1970).

28. Ibid., 12.

29. Graham, *Audubon Ark*, 13.

30. Ibid., 16.

31. Ibid.

32. In the tradition of George Bird Grinnell, the new society did a great deal more than focus on legislative protection. They also published pamphlets, sold Audubon calendars, and instituted an annual Bird Day in the schools. They produced a poster-size bird chart, modeled on one put out by the Society for the Protection of Birds in Germany. And they recruited "secretaries" around the state to sign up new members and to pass out pamphlets.

33. Ibid., 44.

34. Ibid., 41.

35. Ibid., 47.

36. Jean Jacques Rousseau, *La Nouvelle Heloise* (Paris: Garnier-Flammarian, 1967), translated by Judith H. McDowell (University Park: Pennsylvania State University Press, 1968), 305. In the Middle Ages, the park originated as a private parcel of land reserved for hunting. These were "wildland" parks, but their wildness had nothing to do with the kind of nature worship that

spawned the wildland parks in the nineteenth century.

37. Lord Byron, *Childe Harold*, Canto II, XXXVII.

38. Irving Babbitt, *Rousseau and Romanticism* (New Brunswick, NJ and London, England: Transaction Publishers, 1991), 270.

39. This sketch of the early years of white settlement in Yosemite is drawn from Carl P. Russell, *100 Years in Yosemite* (Berkeley, CA: University of California Press, 1931).

40. This account of the persons in the Yosemite vicinity in the early 1860s is drawn from John Henneberger, "State Park Beginnings," *George Wright Forum* 17, 3 (2000): 9-20.

41. Frederick Law Olmsted, "Draft of a Preliminary Report upon the Yosemite and Big Tree Grove," Library of Congress, accessed December 17, 2014, http://memory.loc.gov/cgi-bin/query/r?ammem/consrv:@field%28DOCID+@lit%28amrv mvm02div1%29%29, p. 3.

42. Ibid., 6.

43. Ibid., 1.

44. John Keats, "Ode on a Grecian Urn."

45. In the *Springfield* (Massachusetts) *Republican* Samuel Bowles compared the towers of Yosemite to the Gothic cathedrals of Europe. "[I]t is easy to imagine, in looking upon them, that you are under the ruins of an old Gothic cathedral, to which those of Cologne and Milan are but baby-houses." Samuel Bowles, *Across the Continent: A Summer's Journey to the Rocky Mountains, the Mormons and the Pacific States with Speaker Colfax* (Springfield, MA: Samuel Bowles and Co., 1865), 226-27. Quoted in Alfred Runte, *Yosemite: The Embattled Wilderness* (Lincoln, NB: University of Nebraska Press, 1990), 15.

46. *Scribner's Dictionary of American History*, vol. 2 (1940). Cited in Aubrey L. Haines, *The Yellowstone Story: A History of Our First National Park* (Boulder: Yellowstone Library and Museum Association in cooperation with Colorado Associated University

Press, 1977), 7.

47. Ibid.

48. Ibid.

49. Ibid., 34-37.

50. Haines, *Yellowstone*, 43.

51. Ibid., 52.

52. Ibid., 60-61.

53. Ibid., 62.

54. Father Francis X. Kuppens, "On the Origin of the Yellowstone National Park," reprinted from *The Woodstock Letters* (1897) in *The Jesuit Bulletin* 41, 4 (October, 1962): 10. See Haines, *Yellowstone*, 340.

55. A second, military party also embarked upon a parallel expedition, this one dispatched by General Phil Sheridan, who had been hugely impressed by the report he'd received from Lieutenant Gustavus C. Doane, who had escorted an expedition to Yellowstone the previous summer.

56. A.B. Nettleton, Letter in Record Group 57, National Archives, Washington, D.C., *Records of the Department of the Interior*, Geological Survey, Letters received by F.V. Hayden, 1871. Quoted in Haines, *Yellowstone*, 155.

57. "Geyser Land," *Deer Lodge* (Montana) *New North-West*, December 28, 1871. Quoted in Haines, *Yellowstone*, 169.

58. *Scribner's Monthly* (February 1872): 396. Quoted in Haines, *Yellowstone*, 171 (Brackets in Haines).

59. U.S., *Congressional Globe*, 42nd Congress, 2nd session, Feb. 27, 1872. Quoted in Haines, *Yellowstone*, 171.

60. The biographical material on Powell is drawn from Wallace Stegner, *Beyond the Hundredth Meridian* (Boston: Houghton Mifflin Company, 1953).

61. See John Wesley Powell, "Institutions for the Arid Lands," *Century Magazine* (May, 1890): 111-116.

62. Stegner, *Beyond,* 228.

63. This account of Powell's Irrigation Survey is drawn from Stegner, *Beyond*, chap 3.
64. A second amendment empowered the president to reopen portions of the lands that he might designate, a power that he never used.
65. Stegner, *Beyond*, 305.
66. Ibid., 305.
67. Ibid., 334.
68. Samuel P. Hays, *Conservation and the Gospel of Efficiency* (Cambridge, MA: Harvard University Press, 1959), 5, 26.
69. Ibid., 12. The Newlands Act also limited the benefits of federal irrigation projects to family farms, as opposed to corporate interests.
70. Samuel Hays goes so far as to say that Congress's primary purpose in creating the forest reserves was watershed protection. Ibid., 23.
71. This biographical material is from Char Miller, "Gifford Pinchot and the Conservation Spirit" by Char Miller in Gifford Pinchot, *Breaking New Ground* (Washington, D.C.: Island Press, 1998).
72. Pinchot, *Breaking*, 117.
73. Hays, *Conservation*, 69-71.
74. Ibid., 331.
75. Ibid., 322.
76. A country's prosperity is much more a function of its socioeconomic system than of its wealth of natural resources, as the Soviet Union illustrates.
77. Pinchot, *Breaking*, 509.
78. Gifford Pinchot, *The Fight for Conservation* (New York, NY: Doubleday, Page & Company, 1910), 24.
79. Ibid., 104.
80. *Proceedings, National Conservation Congress* (1911), 184. Quoted in Hays, *Conservation*, 124.

5: Nature Preservation

Rousseau

"NATURE, TO BE commanded, must be obeyed." To imply, as Bacon's dictum does, that it is both possible for man to command nature and right for him to do so, is to imply as well that it benefits man to command nature. This was the Enlightenment view. At the midpoint of the eighteenth century, this utilitarian view of nature reigned supreme. But about that time, a new view began to influence men's minds. Among its more important proponents was the Citizen of Geneva, Jean Jacques Rousseau. "[N]ature ... does everything for the best," wrote Rousseau.[1] This proposition implies that man cannot improve upon nature by altering it, and that he ought not try. Rousseau himself never developed this idea of the natural-as-the-good to its logical conclusion, that man should alter nature as little as possible. But he did advance several forms of the idea that would prove highly influential over the next two centuries. One of these concerns the nature of man.

Man is by nature good, taught Rousseau. He possesses an unerring guide to right and wrong in the form of his conscience.[2] As long as he remains "natural," i.e., guided by his conscience, he remains good. When he allows ideas to influence his actions, ideas of philosophy and science and the arts, he ceases to be natural, and he ceases to be good. Referring to the onset of corruption in ancient Rome, Rousseau wrote, "Till then, the Romans had been content to practice virtue; all was lost when they began to study it." [3] But of Sparta, "that republic of demigods rather than men," as he put it, he wrote with admiration, "While vices were being introduced into Athens under the guidance of the fine arts … you were expelling the arts and artists, the sciences and scholars, from your walls." [4] The more "natural" a man remains, the more virtuous he will tend to be. The most "natural" man is, of course, the uncivilized savage; the idea of the noble savage is a prominent feature of Rousseau's exaltation of the natural.

The savage is virtuous and happy because he has no knowledge of wealth or power, no knowledge of anything beyond what he can readily secure for himself through the most elementary kind of effort. As Rousseau put it, the desires of the savage do not exceed his powers. Civilized man, in contrast, continually encounters all manner of desirable things that must remain forever beyond his grasp. This makes him unhappy. "True happiness," wrote Rousseau, "consists in decreasing the difference between our desires and our powers, in establishing a perfect equilibrium between the power and the will." [5] He would resolve the disparity between the extent of one's desires and the extent of one's powers not by increasing those powers, but by diminishing the desires. "[T]he more nearly a man's condition approximates to this state of nature [the savage state], the less difference there is between his desires and his powers, and happiness is therefore less remote." [6] Rousseau therefore advocated the simple life, a life best exemplified by that of the savage.

The way man increases his "powers" is, most importantly, through technology. The whole purpose of the Baconian quest to increase man's command of nature through technological progress is to increase his power to satisfy his needs and desires. To counsel, as Rousseau did, the reducing of one's desires as a substitute for increasing one's powers is to counsel the abandonment of the Baconian project to command nature, which was an integral part of the Enlightenment. Rousseau's disdain for Baconian progress was explicit: "God makes all things good; man meddles with them and they become evil. He forces one soil to yield the products of another, one tree to bear another's fruit. He confuses time, place, and natural conditions. He destroys and defaces all things; he loves all that is deformed and monstrous; he will have nothing as nature made it." [7] Wrote Rousseau, "Take away our fatal progress, take away man's handiwork, and all is well." [8] But his opposition to technology was only a part of Rousseau's campaign against civilization.

Not content to undermine the idea of technological progress itself, Rousseau also sought to undermine the politico-economic infrastructure that would eventually make such progress possible on an industrial scale; I am speaking here of the division of labor, on the one hand, and of the institution of private property on the other. Under the division of labor, each member of a community specializes in a certain line of work. Rather than each man's doing everything for himself, each becomes expert at just one line of work; when the baker needs carpentry done, he calls the carpenter, and when the carpenter needs bread, he buys it from the baker.

Because each man specializes in one field, the level of expertise in any given line of work that becomes available to every member of the community is dramatically higher than the level of expertise that each man could ever develop for himself. Also, because the carpenter is an expert, he is able to complete a given amount of work much more quickly, and therefore less expensively, than a non-professional could if he were to do the work for himself. The division of labor is indispensable to the developing of any socio-economic group beyond the most primitive of levels. To oppose the division of labor is to advocate the shutting off of human progress at a level of development barely above that of the savages that Rousseau so admired. And oppose it Rousseau did.

To understand why Rousseau opposed the division of labor, we must understand his conception of freedom. "There is only one man who gets his own way—he who can get it single-handedly; therefore, freedom, not power is the greatest good. That man is truly free who desires what he is able to perform, and does what he desires. This is my fundamental axiom." [9] Freedom, for Rousseau, is independence, which he defines as the absence of dependence of any sort on other men. An expert carpenter in a society based on the division of labor is dependent on others for a great many of his needs; he is not independent, under Rousseau's definition, and therefore he is not free. The truly free man is the primitive savage, whose life, though short and brutal, is indeed characterized by freedom from dependence on others (at least in Rousseau's highly romanticized view of the life of the savage).

As for private property, John Locke, on the one hand, and the American founders, on the other, held that the fundamental purpose of government is to secure the property rights of individuals. Rousseau also agreed with this proposition, but he attached an entirely different significance to it than they did. [10] James Madison, and most of the founders, recognized that liberty requires secure property rights, and that secure property rights must mean unequal degrees of wealth among a citizenry; liberty entails economic inequality. Rousseau, in contrast, argued that liberty requires economic *equality*.

Rousseau's economic ideal was the propertyless condition of the savage. But, recognizing that there could be no returning to that condition for civilized Western man, he conceded that some degree of property ownership was necessary and desirable, although he would have limited each man's holdings to what was required to satisfy his own needs. [11] Needless to say, by prohibiting the accumulation of wealth by individuals, Rousseau would have prevented either capitalism or industrialism ever from happening. His advocacy of a sharply limited right of property, combined with his opposition to technological progress and the division of labor, if all taken to their logical conclusions, would have returned Western Europe to a level of socio-economic development little above that of the savages whom Rousseau so romanticized, and all in the name of doing justice to man's nature.

In his *Discourse on the Arts and Sciences*, Rousseau argued that philosophy, the arts, and the sciences tend to corrupt. Intellectual disciplines such as these tend to be products of the city, as opposed to the countryside. There is, in Western culture, a long history, stretching back to the ancient Greeks, of portraying life in the city as corrupting and unhealthy, and of portraying life in the agricultural countryside as virtuous and invigorating. Rousseau himself offered variations on this pastoralist theme in his writings. But he exceeded the pastoralists when he opposed to the corruption of city life not the virtue of the less refined but nevertheless civilized farmer, but that of the completely uncivilized savage. This idealizing of the truly "natural" man was new in Western history.

In a fashion similar to their idealizing of the inhabitants of the countryside, the pastoralists had also idealized the landscape of the countryside, as well. And just as he exceeded the pastoralists by his idealizing of the wild in man, so also did Rousseau idealize not just the settled countryside, but the uninhabited and truly wild landscapes of mountains and forests. "Nature flies from frequented places. It is on the summits of mountains, in the depths of forests, and on desert islands that it displays its most affecting charms." [12]

> [F]or in the high mountains where the air is pure and thin, one breathes more easily, his body is lighter and his mind more serene. Pleasures are less ardent here, the passions more moderate. Meditations take on an

indescribably grand and sublime character, in proportion to the grandeur of the surrounding objects, and an indefinable, tranquil voluptuousness which has nothing of the pungent and sensual. It seems that in being lifted above human society, one leaves behind all base and terrestrial sentiments, and that as he approaches the ethereal regions, his soul acquires something of their eternal purity. One is serious there but not melancholy, peaceful but not indolent, content to exist and to think.[13]

Rousseau believed that wild nature constituted a positive value to man.[14] This glorifying of wild nature was new in Western history. Until this time, enlightened Westerners had found wild nature "simply repellent."[15] Within a generation of Rousseau's death, Europe's greatest poets would be singing the praises of wild nature.

The Romantic Poets

In 1798, Samuel Taylor Coleridge and William Wordsworth published *Lyrical Ballads,* a collection of poems that introduced romanticism to England. *Lyrical Ballads* is noteworthy for its ideas about nature. Coleridge's *Rime of the Ancient Mariner* is the opening poem of the volume, and it develops a theme involving man's relationship to nature that is new in Western thought. Whereas men of the Enlightenment had tended to view nature as man's for the using, Coleridge castigates man for his propensity to violate the natural order.

On a voyage to the Antarctic, the mariner is dismayed by the apparent absence of life. Then one day an albatross joins the crew, and all are cheered by its arrival. The albatross returns each day, and the crew feed it and revel in its company. But one day the mariner takes his cross-bow and kills the albatross. By doing so, he offends nature, and for this he and the crew must be punished. The ship is becalmed, the crew run out of water, and all perish except the mariner. After a prolonged agony alone on the ship, the despairing mariner notices a throng of colorful sea snakes, whose motion causes the water about them to glow. This spectacle of color and light caused by fellow living creatures breathes new spirit into the mariner's demoralized soul.

> *O happy living things! No tongue*
> *Their beauty might declare:*
> *A spring of love gushed from my heart,*
> *And I blessed them unaware:*
> *Sure my kind saint took pity on me,*
> *And I blessed them unaware.*[16]

By this spontaneous exulting in the company of the lowly sea snakes, the mariner redeems himself with the natural order.

> *The selfsame moment I could pray;*
> *And from my neck so free*
> *The Albatross fell off, and sank*
> *Like lead into the sea.*

Coleridge concludes:

> *He prayeth well, who loveth well*
> *Both man and bird and beast.*
> *He prayeth best, who loveth best*
> *All things both great and small:*
> *For the dear God who loveth us,*
> *He made and loveth all.*

In this, the opening chapter of romanticism in England, Coleridge suggests that man's right to command nature is limited by more than just the laws of physics; it is limited as well by religio-moral considerations that would not have occurred to Francis Bacon or to his intellectual heirs of the Age of Enlightenment. Whereas Rousseau had opposed the Enlightenment conception of man, Coleridge implicitly opposed the Enlightenment conception of nature—as something that it is right for man to command.

William Wordsworth went further than Coleridge. In "Hart-Leap Well" an English knight on horseback has chased a deer for much of a day. After following the animal up a mountain, the knight eventually finds it lying dead, having leapt in three bounds from a precipice to its death on the bank of a spring. The knight is happy to have captured his quarry, and he decides to build "a Pleasure-house upon this spot" to commemorate the occasion, "A place of love for damsels that are coy." And such a place he builds.

In the poem, these events are described as having occurred in the distant past. The narrative then moves to the present, and the narrator of the poem has himself come upon the place, which lies in ruins and has fallen into decay. Little now grows where once the vegetation was lush.

> More doleful place did never eye survey;
> It seem'd as if the spring-time came not here,
> And Nature here were willing to decay.[17]

A shepherd who happens by explains of the place, "something ails it now: the spot is curs'd." "Oftentimes," he says, "This water doth send forth a dolorous groan.... The sun on drearier hollow never shone."

The poet himself then theorizes:

This beast not unobserved by Nature fell,
His death was mourned by sympathy divine.
That Being, that is in the clouds and air,
That is in the green leaves among the groves.
Maintains a deep and reverential care
For them the quiet creatures whom he loves.

Here Wordsworth has echoed Coleridge's idea that even the most humble creature deserves consideration from man simply because it is a being created and loved by God. But in this passage, Wordsworth suggests that God is present in all of nature, implying by this pantheist notion that all of nature might deserve special consideration from man. With this idea, which he had developed earlier in "Tintern Abbey," Wordsworth is suggesting that nature has value entirely apart from its relationship to man.

But to return to the poem, the narrator then says, referring to nature's treatment of the ruins of the pleasure palace,

She leaves these objects to a slow decay
That what we are, and have been, may be known;
But, at the coming of the milder day,
These monuments shall be overgrown.

"[W]hat we are, and have been," of course, is a creature who would cause the death of an innocent deer and then celebrate the death with a vulgar palace of pleasure.

One lesson, shepherd, let us two divide,
Taught both by what she shews, and what conceals,
Never to blend our pleasure or our pride
With sorrow of the meanest thing that feels.

In Coleridge's "Rime," the mariner reconciled with nature. But in "Hart-Leap Well" there is no reconciliation between man and nature, only the hope that in some "milder day" yet to come man will learn not to purchase his pleasure or indulge his pride at the expense of "the meanest thing that feels." Otherwise, though, Wordsworth's theme in "Hart-Leap Well" is similar to Coleridge's, that man's right to command nature is somehow limited by moral considerations—which derive not from the interests of man but from those of nature herself. With the enunciation of this idea Western man has left the Enlightenment and has crossed over into the Romantic era.

Wordsworth's most famous paean to nature is "Tintern Abbey." The poem is a tribute to the place itself, a bucolic and beautiful setting, and to the effect that the place has upon the poet. The setting is classically pastoral, but the thrust of the poem goes beyond praising the pastoral life. The scene inspires in Wordsworth a state of reverie that he suggests might be the source of profound wisdom:

> While with an eye made quiet by the power
> Of harmony, and the deep power of joy.
> We see into the life of things. [18]

This idea of knowledge achieved passively and extra-rationally stands in direct contradiction to the main current of Enlightenment's epistemology of reason, and it brings to mind Rousseau's antipathy for the Enlightenment.

Wordsworth developed this idea at greater length in "The Tables Turned."

> Books! 'tis a dull and endless strife;
> Come, hear the woodland linnet,
>
> How sweet his music! on my life,
> There's more of wisdom in it.

And hark! How blithe the thrustle sings!
He, too, is no mean preacher;
Come forth into the light of things,
Let Nature be your teacher. ...

One impulse from a vernal wood
May teach you more of man,
Of moral evil and of good,
Than all the sages can.

Sweet is the lore which Nature brings;
Our meddling intellect
Misshapes the beauteous form of things;
We murder to dissect.

Enough of Science and of Art;
Close up those barren leaves;
Come forth, and bring with you a heart
That watches and receives.[19]

There is an unmistakable echo of Rousseau in these lines. Rousseau wrote that as long as man listened to his "natural" guide, his conscience, he remained good. But as soon as he began to study virtue, i.e., to rely on his reason to guide his actions, he ceased to be good. But whereas Rousseau held that man should follow his own inner nature, Wordsworth counseled that man should take his guidance from outer nature. So in Wordsworth's idea, we have the natural as the *source* of the good. Clearly, this is not a nature to be paved over in the name of progress.

Wordsworth draws lasting emotional sustenance from his experience of nature at Tintern Abbey.

> *While here I stand, not only with the sense*
> *Of present pleasure, but with pleasing thoughts*
> *That in this moment there is life and food*
> *For future years.*[20]

Wordsworth also develops the theme of pantheism, the idea that God resides in nature:

> *And I have felt*
> *A presence that disturbs me with the joy*
> *Of elevated thoughts; as a sense sublime*
> *Of something far more deeply interfused,*
> *Whose dwelling is the light of the setting sun,*
> *And the round ocean and the living air,*
> *And the blue sky, and the mind of man;*
> *A motion and a spirit, that impels*
> *All thinking things, all objects of all thought,*
> *And rolls through all things.*

He goes so far as to suggest that nature is the fount of morality:

> *The anchor of my purest thoughts, the nurse,*
> *The guide, the guardian of my heart, and soul*
> *Of all my moral being.*

Nature becomes a source of emotional sustenance, a fount of morality, a possible source of profound wisdom, and, indeed, the dwelling-place of the divine. Nature, in Wordworth's "Tintern Abbey," is not a nature to be commanded in quite the way the Francis Bacon implied that it could or ought to be commanded; it certainly is not man's for the using without limit. Indeed, Wordsworth's desire to preserve nature from the Baconian command represented by the industrial era was much more than implied. In a manner more characteristic of the year 2014 than 1884, Wordsworth came out in opposition to the construction of a railway into his beloved Lake District.

> *Is then no nook of English ground secure*
> *From rash assault?*[21]

...he asked in his "Sonnet on the Projected Kendal and Windermere Railway."

And to William Wordsworth, who, with Samuel Taylor Coleridge invented English romanticism, must go the credit for one of the earliest enunciations of the national park idea. In the conclusion to his *Guide to the Lakes*, published in 1810, Wordsworth wrote:

> *In this wish [to maintain the simplicity and beauty of the Lake District] the author will be joined by persons of pure taste throughout the whole island who, by their visits (often repeated) to the Lakes of the North of England, testify that they deem the district a sort of national property, in which every man has a right and interest who has an eye to perceive and a heart to enjoy.*[22]

These new ideas about nature did not, of course, convince westerners suddenly to see God in every rock and tree. But they did cause many of them to view nature in a wholly new light, no longer as something whose value was a function of how effectively man could transform it for his own purposes, but now as something that man ought to appreciate in its own right.

Thoreau

Henry David Thoreau disdained the work ethic of his fellow New Englanders. He considered working to amass wealth to be a distraction from the real business of living.[23] (A friend once congratulated him for eschewing a life of *doing* in favor of a life of simply *being*.[24]) Thoreau was not opposed to work; he would readily have acknowledged that a certain amount of labor is necessary to sustain one's life. (He did work hard at his reading and writing.) But he opposed spending any part of his time on earth working to accumulate more than he needed to live.

His famous sojourn at Walden Pond from July, 1845 to September, 1847 was itself an experiment to prove that a man, starting from scratch, could support himself comfortably on only four hours of labor a day. During this time—and throughout his life—he labored just long enough each day to keep himself warm and dry and fed. To have worked more than this, to have spent his precious time on earth working to accumulate riches, would have been, he believed, worse than useless. "Most of the luxuries, and many of the so-called comforts of life, are not only not indispensable," he wrote, "but positive hindrances to the elevation of mankind. With respect to luxuries and comforts, the wisest have ever lived a more simple and meager life than the poor." [25]

Thoreau's attitude toward material wealth has much in common with the Christian view.[26] But Thoreau avoided any formal connection with traditional, organized religion. He preferred direct intercourse with "the universe" to intercourse mediated by ministers or prophets.[27] And in breaking with Christianity, Thoreau also broke with the otherworldliness of that tradition. For him, an "original relation with the universe" meant direct communion with nature here on earth. Thoreau found nature beautiful, and he found that it afforded him spiritual sustenance; nature was his temple. "I suppose that what in other men is religion is in me love of nature," he wrote.[28] "When I would recreate myself, I seek the darkest wood, the thickest and most interminable, and, to the citizen, most dismal swamp. I enter a swamp as a sacred place—a *sanctum sanctorum*." [29]

Thoreau's veneration of nature is nicely illustrated in a passage from *Walden*:

> *Flint's Pond! Such is the poverty of our nomenclature. What right had the unclean and stupid farmer, whose farm abutted on this sky water, whose shores he has ruthlessly laid bare, to give his name to it?... so it is not named for me. I go not there so see him nor to hear of him; who never saw it, who never bathed in it, who never loved it, who never protected it, who never spoke a good word for it, nor thanked God that he had made it. Rather let it be named from the fishes that swim in it, the wild fowl or quadrupeds which frequent it, the wild flowers which grow by its shores, or some wild man or child the thread of whose history is interwoven with its own; not from him who could show no title to it but the deed which a like-minded neighbor or legislature gave him ... whose presence perchance cursed all the shore; who exhausted the land around it, and would fain have exhausted the waters within it; who regretted only that it was not English hay or cranberry meadow,—there was nothing to redeem it, forsooth in his eyes,—and would have drained and sold it for the mud at its bottom. It did not turn his mill, and it was no privilege to him to behold it.... A model farm! Where the house stands like a fungus in a muck-heap, chambers for men, horses, oxen, and swine, cleansed and uncleansed, all contiguous to one another! Under a high state of cultivation, being manured with the hearts and brains*

of men! As if you were to raise your potatoes in the churchyard! Such is a model farm.[30]

Thoreau suggests here that man is morally obligated to "love" and "protect" nature and to "thank God" for it, that man makes use of nature not as a right but as a "privilege," and that he ought to revere nature (Flint's farm is compared to a church-yard) rather than regard it as merely a resource to serve his needs and wants.[31]

Although he spent his life in and around Concord, Massachusetts, a setting that lent itself to the idealization of the pastoral life—of which man is an essential part—Thoreau, like Rousseau, was enamored of wild nature, nature unaffected by the influence of man. "My spirits infallibly rise in proportion to the outward dreariness. Give me the Ocean, the desert, or the wilderness. In the desert a pure air and solitude compensate for want of moisture and fertility."[32] But Rousseau was more interested in the natural within man than he was in nature itself. For his part, Thoreau also was interested in "wild" man.[33] He considered civilization to be an enervating influence upon man.[34] In order to remain viable, a man, or a civilization, needed to keep alive a certain degree of "wildness." And the way to accomplish this was to avoid becoming so caught up in society and technology that one lost touch with nature.

But nature itself was more important to Thoreau than it had been to Rousseau. "I wish to speak a word for nature, for absolute Freedom and Wildness, as contrasted with a freedom and culture merely civil—to regard man as an inhabitant, or a part and parcel of Nature, rather than a member of society."[35] By emphasizing the "wild" within man, Thoreau sought to awaken in him a sense of kinship with nature and to diminish his tendency to view nature as something alien and apart, an object to be exploited for man's benefit. Thoreau hoped to counteract the idea, inherited from the

Enlightenment, that nature is something to be overcome (or to "command," as Bacon had put it.) Thoreau would turn the tables; rather than viewing nature as an obstacle or a threat to the designs of man, he sought to portray man as a threat to nature.

Man is, of course, "part and parcel" of nature, as Thoreau put it. He readily conceded that, as a part of nature, man is entitled to sustain his life as best he can, and to make use of the rest of nature to do so. It is "natural" for man to behave in this way; it is the way all of animal life behaves. But when man uses nature to produce more than he needs to subsist, when he produces *surplus* wealth, when he *profits* at the expense of nature, then he behaves *un*naturally.

Consider Farmer Flint, a "skin-flint, who loved better the reflecting surface of a dollar, or a bright cent, in which he could see his own brazen face … his fingers grown into crooked and horny talons from the long habit of grasping harpy-like." [36] Flint "thought only of [his pond's] money value." Wrote Thoreau:

> I respect not his labors, his farm where everything has its price, who would carry the landscape, who would carry his God, to market, if he could get anything for him; who goes to market for his god as it is; on whose farm nothing grows free, whose fields bear no crops, whose meadows no flowers, whose trees no fruits, but dollars; who loves not the beauty of his fruits, whose fruits are not ripe for him till they are turned to dollars. [37]

Wealth, of course, is the raw material of nature transformed in some way by human effort to make it useful to man. All material wealth-generation must come at the expense of nature. Even Thoreau's rude cabin at Walden Pond was a form of wealth. But it was a necessity. "Unneeded" wealth, on the other hand, is a violation of the natural order, in Thoreau's view. In an America that was turning increasingly to the pursuit of wealth, Thoreau was among the first anti-capitalists. "[T]rade curses everything that it handles," he wrote.[38]

But in Thoreau's own time, a new force was being loosed upon the land that would transform it in ways that would make Farmer Flint look benign. Just fifteen miles from Concord, at Lowell, Massachusetts, the industrial age was dawning in America. Thoreau was critical of the factory system: "[T]he principal object [of the textile factory system] is, not that mankind may be well and honestly clad, but, unquestionably, that the corporations may be enriched." [39] Of the division of labor, which is the foundation of the factory system, he complained, "Where will this ... end?" [40] And of the railroad he wrote, "It rides upon us." [41]

More fundamental than his complaints about industrialism, and more interesting because of its implications for the future of nature preservationism, were Thoreau's ideas about private ownership of land; he opposed it. "Enjoy the land," he wrote, "but own it not."[42] Without private land ownership there can be no property rights, and without property rights there can be no individual rights at all. Thoreau's love of nature carried him into implicit opposition to one of the most important of America's founding principles. But Lockean individual rights were not the only element of the original American idea that Thoreau would oppose.

Coleridge and Wordsworth had suggested that man ought to value nature as an end in itself and not merely in proportion to its utility for man.[43] In his explicit statements, Thoreau only hinted at such a conception.[44] But as his criticism of Farmer Flint shows, Thoreau clearly believed that man is not morally entitled to regard nature as entirely his for the using. He was offended by Flint's farm, "where everything has its price," implying that

276

certain things—in this case the landscape around Flint's Pond—ought to be considered price*less*.

But what does it mean to consider a thing priceless? It means that no amount of wealth could ever morally entitle a man to own it. Men are priceless in this sense; the idea of natural rights, in effect, prohibits men from being bought and sold as objects. (Thoreau was himself a staunch Abolitionist.) Thoreau did indeed believe that no man ought to be permitted to own a portion of nature, i.e., a parcel of land—thus implying that man ought to regard land as literally priceless.[45] To so regard nature would mean, in effect, that nature has "rights." [46] It would mean that just as a man is an end in himself and may not be considered a means to other men's ends, so nature is an end in itself and may not be treated as merely a means to human ends.

Up until the Romantic era, the idea that nature is man's for the using had been an unquestioned premise of Western civilization. Francis Bacon had assumed it when he said that "Nature, to be commanded, must be obeyed." Thoreau thus placed himself squarely in opposition to another idea that had been integral to the founding of the United States, the idea that it is right for man to command nature to serve his own ends. As we shall see, this opposition would become a defining characteristic of nature preservationism in the West.

Muir

John Muir was born in 1838 at Dunbar, Scotland, the third of eight children. His father, Daniel, was a religious man who practiced an austere form of Calvinism.[47] Daniel prohibited all music and pictorial art from the Muir household, and he employed corporal punishment on his children to the point of brutality. As an adult, John Muir would reject organized religion, but he would remain, in his own way, deeply religious throughout his life (much as Thoreau had done, on both counts). In Scotland

277

Daniel had married into a grain business, which he made profitable. In 1849 he moved the family to Wisconsin, where he became a lay preacher in a Cambellite sect. Young John was put to work on the farm, thirteen hours a day in winter and seventeen in summer. "We were all made slaves through the vice of over-industry," he would later write, sounding like Thoreau.

During this period, Muir developed an interest in mechanical devices. He began rising at 1:00 am to work on clocks and other contrivances that he designed himself. In 1860, at the age of 22, he set out from the family farm with three of his inventions, two clocks and a thermometer, to exhibit them at the Wisconsin State Fair. Here his inventions caught the attention of Jeanne Carr, the wife of Ezra Carr, professor of chemistry and natural history at the University of Wisconsin. Mrs. Carr would become an invaluable ally of Muir for years to come.

Muir entered the University of Wisconsin in 1861. There, with his untrimmed beard and his inventions, he would achieve a certain prominence as a campus eccentric, though a respected one. When wounded Union Troops were brought to Camp Randall on the campus for care, Muir ministered to their religious needs and tried to steer them away from the prostitutes who had congregated in the neighborhood. While at the university, Muir first encountered the writings of Emerson and Thoreau, which made a deep impression on him. He developed an interest in botany, and when he left the university in 1863 to avoid being drafted into the Union Army, he headed for Canada and the "University of the Wilderness."

After wandering about "botanizing" for a time, Muir went to work at a sawmill-furniture factory at Georgian Bay, where he proved himself a valuable mechanic. By 1866 he had developed several improvements in the plant machinery. When the plant burned down, Muir was offered a partnership to help rebuild it, but he decided that he could not in good conscience continue to work at a process that involved the destruction of trees, for which he had developed a love that would eventually define his life.[48]

Muir next settled in Indianapolis, where he found work as a mechanic in a carriage parts firm, and where he again proved valuable through his ideas for improving the plant's machinery. But an eye accident in March, 1867 that left him temporarily blinded precipitated a reevaluation of his direction in life. That summer, after a visit to the family farm in Wisconsin, he set out for the Gulf of Mexico with the ambition to become a sort of Alexander von Humboldt.[49] Muir had intended, once he reached the Gulf, to catch a boat for South America, there to follow in the footsteps of Humboldt. But he fell ill with malaria and was unable to get passage to South America. So, having come across a brochure for Yosemite Valley, Muir set sail for California. Arriving in San Francisco in 1868, he immediately set out for the Sierras. From this time until 1873 he lived at Yosemite or thereabouts and worked off and on to support his habit of "sauntering" about the mountains.

Muir's mode of sauntering was a remarkable achievement in privation and self-discipline. He would walk off into the mountains alone for days at a time without so much as a blanket. At night he would huddle beside a fire for warmth. For food he would carry nothing more than a couple of loaves of bread. And despite the presence of bears and mountain lions, he never carried a firearm. While off on his saunters, Muir lived about as close to nature as it is possible for a man to live (except that he eschewed hunting). It was during these years that Muir wrote his first magazine articles and that he decided to become a serious writer.

It is impossible to understand Muir's ideas about nature without seeing them as an outgrowth of his religious upbringing. Of organized religion, he wrote,

[God] is regarded as a civilized, law-abiding gentleman

in favor either of a republican form of government or of
a limited monarchy; believes in the literature and
language of England; is a warm supporter of the
English constitution and Sunday schools and missionary
societies; and is as purely a manufactured article as any
puppet of a half-penny theater. [50]

As an alternative, Muir sought and found God through his experience of nature, much as Thoreau had done. For Muir, nature was an expression of God's love.

Whatever we can read in all the world is contained in
that sentence of boundless meaning, "God is love." This
is the sum and substance of all that the sunshine utters,
and all that is spoken by the calms and storms of the
mountains, and by what we call terrible earthquakes
and furious torrents, and wild beating tone of the
ocean. [51]

Muir held an extraordinarily accepting view of nature. He was not oblivious to the harm that nature can do to man; he simply was not disturbed by it. In June, 1875, he wrote, "True, some goods were destroyed [in a Brownsville, California flood] and a few rats and people were drowned, and some took hold on the rooftops and died, but the total loss was less than the gain." [52] The "gain" achieved by the flood was not, of course, a gain for man. Muir did not view such phenomena as earthquakes and floods from an anthropocentric perspective. The gain was nature's. Man did not rank higher than nature in Muir's cosmology. "No dogma taught by the present civilization seems to form so insuperable an obstacle in the way of a right understanding of the relations which culture sustains to wildness as that which regards the world as made especially for the uses of man," he wrote.[53]

From a scientific perspective, we can agree with Muir that the universe certainly does not exist *for* man. But Muir took this to mean that man therefore has no right to use nature for any purpose he chooses. On the contrary, even though nature was not created for man, he could indeed have every right to use it as he chooses. Muir, however, deduced that since God did not create nature for man, then He must have created it for itself, and therefore man is obligated to respect nature for its own sake.

> Nevertheless, again and again, in season and out of season, the question comes up, "What are rattlesnakes good for?" As if nothing that does not obviously make for the benefit lf man has any right to exist; as if our ways were God's ways.... Anyhow, they are all, head and tail, good for themselves, and we need not begrudge them their share of life.[54]

Much as a man would be justified in killing another man only in self-defense, Muir believed that a man is morally prohibited from harming a rattlesnake as long as the snake does not threaten him. But if it did threaten him, then he would be justified in killing the snake, as Muir killed a rattler that found its way into his cabin. On the other hand, Muir fully conceded man's right to use animals for food.

> Plants, animals, and stars are all kept in place, bridled along appointed ways, with one another and through one another—killing and being killed, eating and being eaten, in harmonious proportions and quantities. And it is right that we should thus reciprocally make use of one another, rob, cook, and consume, to the utmost of our healthy abilities and desires.[55]

Man has the same right to consume animals that animals have to consume man. But man does not have a *superior* right to harm animals; just as animals kill men mainly for food or in self-defense, so man has a right to kill animals only for food or in self-defense.

For man to consume animals was acceptable to Muir because it is natural. Man is every bit as much a part of nature as are rattlesnakes and bears. Muir had no argument with "natural" man. But civilized man was another matter altogether.

> *I don't agree with you in saying that in all human minds there is poetry. Man, as he came from the hand of his maker was poetic in mind and body, but the gross heathenism of Civilization had generally destroyed nature, and poetry, in all that is spiritual. I am tempted at times to adopt the Calvinistic doctrine of total depravity.*[56]

Because nature is an expression of God's love, that which is natural is good. Inversely, that which is not natural is less good than that which is natural. Muir discoursed at length on the virtues of wild sheep, and on the profound shortcomings of domesticated sheep, as well as of other domesticated animals, and even of fruits and vegetables. During the decade that he worked as a—very successful—fruit farmer in Martinez, California, writes Thurman Wilkins, Muir "grew irritable, too; for he was beset by an intolerable conflict. The more he improved his [pear and other] trees and vines by grafting, the more he contradicted his dearest principle—that the natural was superior to the cultivated." [57] Man's "fall," as Muir referred to it in explicitly Biblical terms, was, in effect, his becoming civilized.[58] Civilized man was unfit for life in God's creation in the same way and for the same reasons that domesticated sheep were unfit.[59]

How was man to be redeemed? By returning to nature to

experience God's love.

> *Thousands of tired, nerve-shaken, over-civilized people*
> *are beginning to find out that going to the mountains is*
> *going home; that wildness is a necessity; and that*
> *mountain peaks and reservations are useful not only as*
> *fountains of timber and irrigating rivers, but as*
> *fountains of life.* [60]

By thus "saving" man, Muir would also save nature, because civilized man, i.e., "fallen" man, is a threat to nature. Industrial civilization entails continual transformation of the natural landscape. Muir was keenly aware, as Thoreau had been, of the threat posed to nature by an unrestrained profit motive. Upon discovering, for example, that a herd of domestic sheep had recently traversed a favorite mountain hideaway of his, Muir wrote that "all the gardens and meadow were destroyed by the horde of hooved locusts, as if swept by a fire. The money-changers were in the temple." [61]

So how would Muir save nature for man and from man? First, he would preach the gospel of nature to man. He would thereby bring man to nature, to experience God's love the way Muir had experienced God's love in the Sierras. In 1879 Muir wrote an article on Yosemite's glaciers for the New York Tribune. He would write other articles for that and other newspapers. He would also write for Harper's magazine, The Atlantic Monthly, Overland Monthly, and Century Magazine. And he wrote ten or so books. This was how he preached the gospel of nature to the people. By the time he was done, he would enjoy a national reputation as a nature writer.

The other way in which Muir would save nature for man and from man was to have the government take over and administer vast expanses of nature as preserves. Yosemite Valley had been

purchased by the State of California as a park in 1864. In 1889
Muir began a campaign, with Robert Underwood Johnson of
Century Magazine, to make the region surrounding Yosemite Valley
the country's second national park, after Yellowstone. Yosemite
National Park became a reality in 1890. In 1905 Muir succeeded
in having Yosemite Valley itself ceded from the State of California
to the Federal Government to become part of the surrounding
Yosemite National Park. He also played a role in winning national
park status for Sequoia in the southern Sierras, for Mt. Rainier,
the Petrified Forest, and the Grand Canyon. And, as important as
his role in the establishing of these parks, Muir lead early efforts
to define them specifically as nature preserves.

At the same time that Robert Underwood Johnson and Muir
conceived the idea of Yosemite National Park, in 1889, Johnson
had also urged Muir to start a society of private citizens dedicated
to preserving California's natural landscape. Thus was born the
Sierra Club. Muir was its president from its inception until his
death in 1914. The Club played an integral role with Muir in
getting Yosemite Valley ceded to the Federal Government, and in
defining the national parks as nature preserves. In 1891 Muir had
helped to convince Interior Secretary Noble under President
Harrison to include much of the Sierra headwater forests in the
new national forests. But Muir later lost the battle against Gifford
Pinchot to determine how the national forests would be used,
whether as nature preserves, as Muir preferred, or as utilitarian
sources of timber and hydroelectric power.

The same question would arise again, in the context of the
national parks, when the City of San Francisco launched an effort
in 1905 to dam the Hetch Hetchy Valley within Yosemite
National Park as a source of drinking water and hydroelectric
power for the city. Muir and the Sierra Club fought hard to
establish the principle that the national parks should function
essentially as nature preserves. (Wrote Muir, "[I]nstead of lifting

their eyes to the God of the mountains, [they] lift them to the Almighty Dollar. Dam Hetch Hetchy! As well dam for water-tanks the people's cathedrals and churches, for no holier temple has ever been consecrated by the heart of man." [62]) Muir and the Club lost this battle, but they won the war; from that time onward the purpose of the national parks has been essentially to function as nature preserves open to the public.

A generation before Muir, Henry Thoreau had opposed private ownership of land and the exploitation of nature for profit. Muir also opposed these things, but rather than condemning them in principle, as Thoreau had done, he sought to limit their application to areas outside of his "temples" of nature. Thoreau had advocated the idea of nature as an end in itself, an end that man is morally obligated to respect in the same way that he is morally obligated to treat other men as ends in themselves. Muir shared this view of nature and sought to give expression to it by setting aside preserves where the government would protect nature against man.

After Muir was through helping to fashion the national parks, they would function as capitalist-free zones in a capitalist, industrial country. Muir seemed to have shown that it was possible for a socio-economic system grounded in capitalism and the Baconian command of nature to coexist with a—delimited—natural landscape that both Thoreau and Muir could have loved. But he helped to establish the principle that a natural landscape can be a value worth preserving from profit-seeking, nature-transforming man, and that it is properly a responsibility of government to do this. A later generation of nature preservationists would apply this principle to an ever-widening expanse of the American landscape.

Leopold

Aldo Leopold was born in 1887 in Burlington, Iowa. After graduating from the forestry program that Gifford Pinchot had started at Yale, Leopold joined Pinchot's U.S. Forest Service in 1909 and spent fifteen years in New Mexico and Arizona. He left field work in 1924 to join the Service's Forest Products Lab at Madison, Wisconsin. In 1932 he left the Forest Service altogether to become professor of game management at the University of Wisconsin. The next year he published *Game Management*, a seminal book that would establish him as a founder of the field of wildlife ecology. In 1934 he co-founded the Wilderness Society with Bob Marshall.[63] By the time of his death in 1948, Leopold would be a nationally influential spokesman for the new field of thought known as ecology.

As a student of Pinchot, Leopold shared conservation's distrust of private land ownership. Conservationists had long argued that America's natural resources were the common property of the people as a whole—public property.[64] They did not want *all* land to be publicly owned, just certain categories of land that they considered important, such as headwater forests, wetlands, wildlife ranges, and other "sensitive" terrains. But, since all land is a "natural resource," all land could end up as public land if the conservationists' premise were carried to its logical conclusion. Indeed, since Leopold's time there has been a steady increase in government control of all land in the U.S.[65]

Leopold's own affinity for public ownership was not unqualified; he doubted that government could ever properly manage more than a fraction of the land in the United States, partly because of the limited effectiveness of government bureaucracy, and partly because only private ownership could engender the kind of affection for land that Leopold saw as an important ingredient of the man-land relationship.

But his critique of private land ownership was right out of the conservation handbook: "Land-use ethics are still governed wholly by economic self-interest," he wrote.[66] A farmer motivated by economic self-interest would have little reason to maintain habitat for native wildlife on his land. Maintaining such habitat is an integral part of land ownership, as Leopold saw it, but it is the kind of responsibility that might go unfulfilled in a society like America's, where a constant pressure to maximize the profit-generating potential of a farm tends to compel farmers to convert wildlife habitat to productive cropland. The maximizing of profit is a responsibility a farmer would owe himself, but the maintaining of wildlife habitat is one he would owe "the public," according to Leopold. "[E]very landowner is a custodian of two interests, not always identical, the public interest and his own."[67] In Leopold's view, only the latter would be accommodated sufficiently in a society like America's that tends to view land as an economic commodity.

The idea that every private landowner is a custodian of a public interest might not seem strange to modern-day American readers, so accustomed are we to thinking of land in just this way. When El Dorado County, California refuses a building permit to a developer of a housing tract unless he agrees to preserve a third of his land as greenbelts—a form of legalized extortion—no one protests; the County is simply looking after the public's (alleged) interest. We are, many of us today, quite Leopoldian in our view of land ownership.

But Leopold's contention that the public has an interest in every parcel of land represents a radical break with the American Founders' conception of land ownership. It undermines the very idea of private property, and of liberty as well, because it invites the public into the private sphere. By way of analogy, suppose a people were to adopt the idea that the public "has an interest" in each individual's choice of religious beliefs. For most of human history men believed precisely this. It was in America, for the first time, that a people formally rejected this idea. Since then, Americans have believed, rightly, that the coercive power of government has no role whatsoever to play in the sphere of religious belief. They have come to understand that *any* breach in the wall of separation between church and state could prove fatal to Americans' freedom of religion, and to other freedoms, as well.

But Americans have lost sight of the importance of keeping the public out of the private economic sphere. As I have described, the authors of the Constitution came close to erecting a wall of separation between the public and private economic spheres. But with the onset of the Progressive era, America began to abandon wholesale her constitutional protections of property rights. Leopold's idea that the public has an interest in every parcel of land, public or private, is simply a novel expression of Progressivism's hostility toward the idea of a private economic realm into which the public may not intrude. It represented a return to a pre-American way of thinking about the relationship of the individual to the state. Indeed, it was a step backwards toward medievalism—every bit as much as if Americans had decided to abandon the separation of church and state. (The defining characteristic of the feudal system was, after all, the absence of private ownership of land among the majority of the people).

The culmination of Leopold's work on the problem of land ownership came with the posthumous publication of "The Land Ethic," an essay that appeared in his book, *Round River* in 1953. "The Land Ethic" stands today as a manifesto of the nature preservation movement. Leopold opened "The Land Ethic" by describing how, when Homer's Odysseus returned from the Trojan War, he hanged a dozen of his slave girls whom he suspected of having misbehaved. Wrote Leopold:

> Concepts of right and wrong were not lacking from Odysseus' Greece,... [but] [t]he ethical structure of that day ... had not yet been extended to human chattels. During the three thousand years which have since elapsed, ethical criteria have been extended to many fields of conduct, with corresponding shrinkages in those judged by expediency only.... There is as yet no ethic dealing with man's relation to land and to the animals and plants which grow upon it. Land, like Odysseus' slave girls, is still property. The land relation is still strictly economic, entailing privileges but not obligations.[68]

Today "ethical criteria" prohibit the owning of slave girls, not to mention their summary hanging; girls today have *rights*. But land is still merely property, and it may, therefore, be treated entirely as its owner deems expedient. Leopold criticized this "strictly economic" relation. He implied that land, and the animals and plants upon it, should be accorded some sort of rights, just as Odysseus's slave girls would have rights today.

Despite using the slavery analogy in at least two places, Leopold was not advocating full-blown rights for nature like the rights that men possess, but for a kind of limited rights.[69] He did not intend the complete abolition of private land ownership. He advocated, rather, "[a]n ethic to supplement and guide the economic relation to land." His land ethic would not "prevent the alteration, management, and use" of nature. It would simply affirm the right to "continued existence" of animals and plants (much as we now protect "endangered species"), and it would affirm "their right ... at least in spots ... to continued existence in a natural state" (as we now set aside nature and wilderness preserves).[70] Leopold would do for wildlife, perhaps, what Muir did for his "temples" of nature.

But there is a problem with extending rights to nature; such rights must always come at the expense of man's rights. As I have argued throughout this book, the most fundamental of man's rights are property rights. In any settled, civilized society, all rights depend upon the ownership of land; without rights of land ownership no other rights are possible.[71] To the extent, therefore, that land (and the flora and fauna upon it) is accorded rights, man's rights of ownership must be diminished and his other rights be weakened. We have seen farmers, for example, deprived of their rights to plow their own land so that "endangered species" might be protected. Leopold's idea that land should have (limited) rights thus stands diametrically opposed to America's heritage of Lockean individual rights for human beings. To the extent that America adopts the one, it must cease to protect the other.

The idea of rights originated in a *human* context. Rights represent a legal recognition that a human being requires freedom from coercion in order to live a proper life. All men are better off if no men are allowed to initiate the use of physical force against other men, which is another way of saying that man lives better in a society of voluntary, i.e., non-coercive, relations rather than in one in which coercive relations are permitted, whether they be such anarchic forms of coercion as robbery, rape, and murder, or government-sanctioned forms like slavery or socialism. Man has given legal expression to the idea of individual rights because such rights fulfill a fundamental requirement of his nature, his need to be able to live according to his own judgment. The idea of such rights was integral to establishing, in the United States, the first society based in principle not on coercion but on voluntary cooperation, the first society of contract.

One can argue that fundamental rights are "God-given" in the sense that they are based on a fact of nature, man's need to be free. But reasonable men first adopted the idea of individual rights not because of some divine imperative, but because the idea so clearly benefits man—in this life here on earth. If we view morality as a set of rational principles to enable man to live a fulfilling life here on earth, then rights are a profoundly moral idea. But in declaring that nature has rights that man is morally obligated to respect, an idea that works to the detriment of man's life here on earth, Leopold severed morality from its rational purpose of promoting a person's long-range well-being in this life on earth. In so doing, he treated morality as a sort of arbitrary imperative that man is obligated to obey even to the detriment of his life here on earth. This is a distinctly Christian way of looking at morality.

But there is another Christian doctrine that Leopold's idea flatly contradicts. Christianity has long taught that the universe exists to serve the needs of man. By placing man, figuratively and literally, at the center of the universe, this *anthropocentric* doctrine provided a divine sanction for man's "commanding" of nature, to use Bacon's term. It was Galileo's challenging of the belief that man is situated literally at the center of the universe—which he effected on scientific grounds—that got him into trouble with the Catholic Inquisition. After Galileo, science would go on to destroy the belief that man is, either literally or figuratively, the center of the universe.

When Leopold argued that nature has rights that man is bound to respect, he implied that man is *not* the purpose for the existence of the universe, which science has indeed shown to be true. Leopold concluded that man is therefore not entitled to rule the earth in his own self-interest. But in order to be valid, the anthropocentric view does not require that man be literally at the center of the universe, nor that he be divinely intended as the purpose of the universe. All it requires is that man be morally entitled to behave *as though* the universe were his for the using. And he is so entitled—by any rational morality—because his health and well-being require it.

But if man adopts the idea that nature has rights that man is morally obligated to respect, then man is *not* entitled to consider the universe as his for the using, in which case he would have abandoned the anthropocentric view for what Leopold called the biocentric view. "Biocentrism" means "life-centered." In its purest form, it holds that all living things have an equal right to exist, and that no one species, especially man, is entitled to consider the rest of the earths' species as his for the using. "In its ultimate analysis the conservation movement may prove to be a denial of the anthropocentric philosophies," wrote Leopold.[72]

Aldo Leopold first began to think along biocentric lines during his days in the field for the Forest Service. He had begun his conservation career as a Progressive utilitarian, thoroughly committed to the idea that natural resources should be managed by government professionals for the "greatest good of the greatest number" of human beings. In the early years, one of his duties was to exterminate wolves in the national forests. But on a hunt one day, he experienced an epiphany that would have momentous consequences for the conservation movement.

> *We reached the old wolf in time to watch a fierce green fire dying in her eyes. I realized then, and have known ever since, that there was something new to me in those eyes—something known only to her and to the mountain. I was young then, and full of trigger-itch; I thought that because fewer wolves meant more deer, that no wolves would mean a hunters' paradise. But after seeing the green fire die, I sensed that neither the wolf nor the mountain agreed with such a view.*[73]

Learning to "think like a mountain," Leopold would eventually forsake Pinchot's man-centered, utilitarian brand of conservation for a nature preservationism along the lines of what Thoreau and

John Muir had practiced.

"Man must assume that the biota has a value in and of itself, separate from its value as human habitat." [74] Leopold argues here that man is morally obligated, to one extent or another, to treat other living things as a value, regardless of whether they are in fact a value to individual men. The ultimate meaning of this statement, then, is that, to one extent or another, man ought to treat living nature as the good. The corollary of this statement is that, to the extent that living nature is the good, human alteration of living nature must be bad. "A thing is right when it tends to preserve the integrity, stability, and beauty of the biotic community. It is wrong when it tends otherwise."**Error! Bookmark not defined.** [75]

The natural state of things thus becomes the standard by which to judge human actions; that which preserves the natural state is right, and that which alters the natural state is the wrong. As we shall see, this idea of the-natural-as-the-good is the defining characteristic of today's environmental movement. It is the reason why environmentalists seem forever to be moving toward the extreme. They do so not because they are taking a good idea too far, but because they are following a bad idea, that the natural is the good, to its logical conclusion.

If that which is natural is the good, and that which is man-made is, to one extent or another, the bad, then it follows that man's moral sanction to "command" nature is, that same extent, diminished. "The land must of course be modified," wrote Leopold, "but it should be modified as gently and as little as possible." [76] Leopold distrusted the Baconian enterprise to extend man's command of nature. Recall that it was Francis Bacon's recognition of the potential of technology to enhance the quality of man's life that led him call for a conscious effort to discover new technologies with which to extend his command of nature. Western man's success at realizing Bacon's dream of commanding nature—a success that must always come at the expense of nature—led Leopold, in turn, to question the desirability of pursuing scientific inquiry for the purpose of extending man's technological mastery of nature.

> *One of the facts hewn to by science is that every river needs more people, and all people need more inventions, and hence more science; the good life depends on the indefinite extension of this chain of logic. That the good life on any river may likewise depend on the perception of its music, and the preservation of some music to perceive, is a form of doubt not yet entertained by science.*[77]

It was not so much science's success at extending human domination of the landscape that Leopold objected to; it was its lack of success, as he saw it, at developing ways to mitigate or to compensate for the deleterious effects of man's domination of nature.[78] Indeed, despite man's apparent success at commanding nature, Leopold did not believe that man *could* command "the biota" in the long run. "The ordinary citizen today assumes that science knows what makes the [biotic] community tick; the scientist is equally sure that he does not. He knows that the biotic mechanism is so complex that its workings may never be fully understood."[79] This would become a defining theme of the environmental movement, the belief that man cannot successfully command nature, that his attempts to do so must inevitably disrupt nature, and that he is incapable of ever overcoming the harmful effects of those disruptions.

Partly because he believed that man could not command nature, especially on an industrial scale, Leopold argued that man should not try to do so.[80] "It is increasingly clear ... that there is a basic antagonism between the philosophy of the industrial age and the philosophy of the conservationist." [81] "Perhaps, [Leopold] mused, industrialism needed to be discarded entirely." [82] So, to an important extent, Leopold rejected Baconism. The great Baconian enterprise, which had given birth to the United States itself, which the Founders of the U.S. had implicitly embraced, and which the United States would incorporate into its very fabric, would slowly come to an end, if Leopold had his way. Not long after Leopold's death in 1948, a movement would come into being dedicated to realizing Leopold's dream of placing limits on the further growth and development of man's industrial command of nature.

Leopold began as a conservationist in the Pinchot mold, and he ended as a preservationist in the tradition of Thoreau and Muir. This change in his thinking presaged a similar transformation within the conservation movement; the progressive utilitarians, who were committed to government management of natural resources for the greatest good of humanity, would lose control of the movement to preservationists who believed that nature is good and that human alteration of nature should be avoided as much as possible. In the process, conservation would become environmentalism.

Notes

1. Jean Jacques Rousseau, *Emile*, trans. Barbara Foxley, (London: J.M. Dent and Sons, Ltd., 1911), 44.
2. "To learn [virtue's] laws, is it not enough for us to withdraw into ourselves and listen to the voice of our conscience while our passions are silent? That is true philosophy." Jean Jacques Rousseau, "Discourse on the Arts and Sciences," in *The Essential Rousseau*, ed. L. Bair (New York, NY: New American Library, 1975), 227.
3. Ibid., 214.
4. Ibid., 212.
5. Rousseau, *Emile*, 44.

6. Ibid., 45. "The other animals possess only such powers as are required for self-preservation; man alone has more. Is it not strange that this superfluity should make him miserable? In every land a man's labour yields more than a bare living. If he were wise enough to disregard the surplus he would always have enough, for he would never have too much. 'Great needs,' said Favorin, 'spring from great wealth; and often the best way of getting what we want is to get rid of what we have.' By striving to increase our happiness we change it into wretchedness. If a man were to live, he would live happy; and he would therefore be good, for what would he have to gain by vice?" Ibid.

7. Ibid., 5.

8. Ibid., 245.

9. Ibid., 48. "[T]he bands of servitude are formed only by men's mutual dependence and the reciprocal needs that unite them." Rousseau, "Discourse," 171.

10. "Such was, or probably was, the origin of society and laws, which gave new fetters to the weak and new strength to the rich, permanently destroyed natural freedom, established the law of property and inequality forever, turned adroit usurpation into an irrevocable right, and for the advantage of a few ambitious men, subjected all others to unending work, servitude, and poverty." Rousseau, "Discourse on Inequality," *Essential*, 186.

11. "[O]nly the express and unanimous consent of the whole human race could have entitled you to take more from the common resources [of mankind] than what you require for your own needs." Rousseau, "Inequality," *Essential*, 185.

12. Jean Jacques Rousseau, *La Nouvelle Héloise* (University Park: Pennsylvania State University Press, 1968), 312.

13. Ibid., 65.

14. There is one idea, though, that Rousseau advocated not for man's sake, but entirely for nature's. This is a limited conception of animal rights. "[B]ut since they share our nature to some extent

because of the sensitivity with which they are endowed, it follows that they must participate in natural right, and that man is bound by a certain kind of duty toward them. It appears, in fact, that if I am obliged to do no harm to my fellow man, it is less because he is an intelligent being than because he is a sentient creature, and since that quality is common to both men and animals, it should at least give the latter the right not to be needlessly mistreated by the former." Rousseau, "Inequality," *Essential*, 141.

15. Irving Babbitt, *Rousseau and Romanticism* (Boston, Houghton Mifflin 1919), 273.

16. Samuel Taylor Coleridge, "The Rime of the Ancient Mariner."

17. William Wordsworth, "Hart-Leap Well."

18. William Wordsworth, "Tintern Abbey."

19. William Wordsworth, "The Tables Turned."

20. Wordsworth, "Tintern."

21. William Wordsworth, "Sonnet on the Projected Kendal and Windermere Railway."

22. William Wordsworth, *A Guide through the District of the Lakes in the North of England* (Kendal, England: Hudson and Nicholson; London, Longman and Co., Moxon and Whittaker and Co., 1835.), 88.

23. "The order of things should be reversed; the seventh should be man's day of toil, wherein to earn his living by the sweat of his brow; and the other six his Sabbath of the affections and the soul." Thoreau, *The Writings of Henry David Thoreau: Familiar Letters*, ed., F.J. Sanborn (Boston, MA and New York, NY: Houghton Mifflin and Company, 1906), 6, 9.

24. "I honor you because you abstain from action, and open your soul that you may *be* somewhat. Amid a world of noisy, shallow actors it is noble to stand aside and say I will simply be." Letter from Harrison Gray Otis Blake to Thoreau, quoted in Robert D. Richardson, *Henry David Thoreau: A Life of the Mind* (Berkeley and Los Angeles, CA: University of California Press, 1983),187.

25. Thoreau, *Walden* (New York, NY:Holt, Rinehart, and Winston, 1961), 11. "Money is not necessary to buy one necessary of the soul." Ibid., 253. Thoreau placed a lower priority on worldly values than he did on spiritual ones. "Rather than love, than money, than fame, give me truth," he wrote. Ibid., 255.

26. "It is easier for a camel to go through the eye of a needle, than for a rich man to enter into the kingdom of God." Matthew, 19:24.

27. "The foregoing generations beheld God and nature face to face; we, through their eyes. Why should not we also enjoy an original relation to the universe? Why should not we have a poetry and philosophy of insight and not of tradition, and a religion of revelation to us, and not the history of theirs?" So wrote Thoreau's friend and mentor, Ralph Waldo Emerson in *Nature, Collected Works of Ralph Waldo Emerson*, eds., Robert E. Spiller and Alfred R. Ferguson (Cambridge, MA: Belknap Press of Harvard University, 1971), 1, 7.

28. *The Writings of Henry David Thoreau*, vol. 2, *Journal*, ed. Robert Sattelmeyer (Princeton, NJ: Princeton University Press, 1984), 55.

29. Thoreau, *Walking* in *The Writings of Henry David Thoreau*, vol. 5, *Excursions and Poems*, ed. Franklin Benjamin Sanborn (Boston and New York, Houghton Mifflin and Company, 1906), 228.

30. Thoreau, *Walden,* 152-153.

31. Of Walden Pond, Thoreau wrote, "[T]he villagers, who scarcely know where it lies, instead of going to the pond to bathe or drink, are thinking to bring its water, which should be as sacred as the Ganges at least, to the village in a pipe, to wash their dishes with! –to earn their Walden by the turning of a cock or the drawing of a plug!" Thoreau, *Walden*, 149.

32. Thoreau, "Walking" in *Writings*, ed. Sanborn, 228.

33. "Give me for my friends and neighbors wild men, not tame ones." Ibid., 234. "I would have every man so much a wild

antelope, so much a part and parcel of Nature, that his very person should thus sweetly advertise our senses of his presence." Ibid., 225. "I would not have every man nor every part of a man cultivated, any more than I would have every acre of earth cultivated; part will be tillage, but the greater part will be meadow and forest, not only serving an immediate use, but preparing against a distant future, by the annual decay of the vegetation which it supports." Ibid., 238.

35. Ibid., 205.

36. Thoreau, *Walden*, 152-153.

37. Ibid.

38. Ibid., 54. "Absolutely speaking, the more money, the less virtue; for money comes between a man and his objects, and obtains them for him; and it was certainly no great virtue to obtain it." Thoreau, *On the Duty of Civil Disobedience* in *Writings*, vol. 4, *Cape Cod and Miscellanies*, ed. Sanborn, 372.

39. Thoreau, *Walden,* 21.

40. Ibid., 36. As for his opinion of the factory worker, "... I think the fall from the farmer to the operative [employee] as great and memorable as that from the man to the farmer." Ibid., 50.

41. Ibid., 71.

42. Ibid., 161. "By avarice and selfishness, and by a groveling habit, from which none of us is free, of regarding the soil as property, or the means of acquiring property chiefly, the landscape is deformed, husbandry is degraded with us, and the farmer leads the meanest of lives. He knows Nature but as a robber." Ibid., 128. "At present, in this vicinity, the best part of the land is not private property; the landscape is not owned, and the walker enjoys comparative freedom. But possibly the day will come when it will be partitioned off into so-called pleasure grounds, in which a few will take a narrow and exclusive pleasure only,—when fences shall be multiplied, and man traps and other engines invented to confine men to the *public* road; and walking

over the surface of God's earth, shall be construed to mean trespassing on some gentleman's grounds ... Let us improve our opportunities then before the evil days come." Thoreau, *Walking* in *Writings*, ed. Sanborn, 216. The 1840's were, famously, a time of socialist utopian experimentation in the United States. Thoreau himself had been invited to join some of his fellow Transcendentalists at Brook Farm in West Roxbury, Massachusetts as a member of the commune immortalized in Hawthorne's *Blithedale Romance*. Never much of a joiner, Thoreau declined the invitation.

43. "This is my Lake Country," wrote Thoreau of several ponds in the vicinity of Concord—in homage to the Coleridge and Wordsworth. Thoreau, *Walden*, 153.

44. "These beans [which Thoreau had planted beside his cabin at Walden] have results which are not harvested by me. Do they not grow for woodchucks partly." Ibid., 129. "Shall I not rejoice also at the abundance of weeds [in the bean patch] whose seeds are the granary of the birds? It matters little comparatively whether the fields fill the farmer's barns." Ibid.

45. "By avarice and selfishness, and a groveling habit, form which none of us is free, of regarding the soil as property, or the means of acquiring property chiefly, the landscape is deformed, husbandry is degraded with us, and the farmer leads the meanest of lives. He knows Nature but as a robber." Ibid., 128.

46. The modern idea of "animal rights" is premised upon the idea that no man is morally entitled to own any of that portion of nature that we call "animals."

47. Thurman Wilkins, *John Muir: Apostle of Nature* (Norman: University of Oklahoma Press, 1995), 5-6.

48. Ibid., 42.

49. Humboldt was a German naturalist and explorer whom Muir had read. Muir's mother had urged him to pursue his interest in Humboldt. See Michael P. Cohen, *The Pathless Way: John Muir and*

American Wilderness (Madison: University of Wisconsin Press, 1984), p. 9.

50. John Muir, *A Thousand Mile Walk to the Gulf*, ed. William Frederic Badé (Boston, MA and New York, NY: Houghton Mifflin Company, 1916), 137.

51. John Muir to his sister, Maggie Lauder [Reid], March, 1873, Muir Manuscript Collection. Microform Edition, eds. Ronald H. Limbaugh and Kristen E. Lewis. (London:Chadwyck-Healey, Inc., 1986), 2:01248. Quoted in Dennis C. Williams, *John Muir, Christian Mysticism, and the Spiritual Value of Nature, 1866 to 1873* (Unpublished Masters Thesis, Texas Tech University, 1989). "We seem to imagine that since Herod beheaded John the Baptist there is no longer any voice crying in the wilderness. But no one in the Wilderness can possibly make such a mistake. No wilderness in the world is so desolate as to be without God's ministers. The love of God covers all the earth as the sky covers it and fills every pore. And this love has voices heard by all who have ears to hear. Everything breaks into songs of Divine Love just as banks of snow, cold and silent, burst forth in songful cascading water. Yosemite Creek is at once one of the most sublime and sweetest voiced evangels in the Wilderness of the Sierra." John Muir, *John of the Mountains: The Unpublished Journals of John Muir*, ed. Linnie Marsh Wolfe (Boston: Houghton Mifflin, 1979), 135-136.

52. Quoted in Wilkins, *John Muir*, 115. Wilkens writes that this quotation is drawn from an article titled "Flood-Storm in the Sierra" that Muir wrote for *Overland* Magazine.

53. John Muir, *Steep Trails* (San Francisco: Sierra Club Books, 1994), 7.

54. John Muir, *Our National Parks* (Boston: Houghton Mifflin, 1901), 57-58.

55. John Muir, *Steep Trails*, 8.

56. Letter to J.B. McChesney, 19 Sept., 1871, from the *John Muir Papers*, Stuart Library, University of the Pacific, Stockton,

California, Box 2. Quoted in Cohen, *Pathless*, pp. 187-88.

57. Wilkins, *John Muir*, 160.

58. Nature was "beautiful in the eyes of God" and "part of God's family, unfallen, undepraved, and cared for by the same species of tenderness and love as is bestowed on angels in heaven or saints on earth." John Muir, A *Thousand Mile Walk to the Gulf* (Boston and New York: Houghton Mifflin Company; Cambridge, The Riverside Press, 1916), 98-99.

59. "[P]ure wildness is the one great want, both of men and of sheep." Muir, "Wild Wool," *Steep Trails*, 11. Echoes, again, of Thoreau.

60. Muir, *National Parks*, 1.

61. John Muir, *The Mountains of California* (New York, NY: The Century Company, 1907), 116.

62. John Muir, *The Yosemite* (New York, NY: The Century Company, 1912), 262.

63. Marshall was an avowed socialist.

64. Conservation tended to present it as a settled fact that a nation's natural resources are the common heritage of the whole people.

65. Government control of private land in the name of preserving "wetlands" is one example among many. As I explained in chapter 2, when government takes it upon itself to regulate the use of privately owned land, it is, in effect, making itself part owner of the land, because determining how property should be used is, by definition, one of the characteristics of ownership.

66. Aldo Leopold, *A Sand County Almanac* (New York, NY: Ballantine Books, 1970), 245.

67. Aldo Leopold, "Some Thoughts on Recreational Planning." Quoted in Julianne Lutz Newton, *Aldo Leopold's Odyssey* (Washington, D.C.: Island Press/Shearwater Books 2006), 150.

68. Leopold, *Sand County*, 237-38. Emphasis added.

69. Referring to the United States' extirpation of slavery, Leopold

wrote of the 1860's "when thousands died to settle the question: Is the man-man community lightly to be dismembered? They settled it, but they did not see, nor do we yet see, that the same applies to the man-land community." Ibid., 16-17.

70. Ibid., 240.

71. I am speaking here of fundamental rights, such as life and liberty. I am excluding that class of procedural rights, such as the right to a jury trial or the right of *habeas corpus,* which do not depend on the existence of property rights.

72. Leopold, Untitled, Fragment, A. Leopold Papers, 10-6, 16. Quoted in Curt Meine, *Aldo Leopold: His Life and Work* (Madison, WI: University of Wisconsin Press, 2010), 297.

73. Leopold, *Sand,*. 138-39.

74. "Ecology and Economics in Land Use," unfinished, unpublished, ca. 1940's, Leopold Papers, 10-6, 17. Quoted in Newton, *Odyssey*, 420.

75. Leopold, *Sand*, 262.

76. Leopold, *Sand*, 262.

77. Ibid., 63.

78. Echoes of George Perkins Marsh.

79. Leopold, *Sand*, 240-41.

80. I say "partly" because I suspect that Leopold would not have wanted man to command nature even if man had the ability to do so.

81. "A Modus Vivendi for Conservationists" by Aldo Leopold, unfinished manuscript, no date (circa 1941?), p. 1, Leopold Papers, 10-6, 16. Quoted in Newton, *Odyssey*, 251.

82. Newton's words, citing "Ecology as an Ethical System," unfinished, circa 1940's, Leopold Papers, 10-6, 17. See Newton, *Odyssey*, 254.

Thomas McCaffrey

Part II: Environmentalism

6: Beginnings

IN 1956 PRINCETON University hosted a symposium titled "Man's Role in Changing the Face of the Earth." The conference was dedicated to the proto-conservationist, George Perkins Marsh. "As much as any event," wrote Donald Worster, "that Princeton gathering prepared the intellectual ground for the environmental movement." [1] Among the ideas discussed at the symposium, one would play a definitive role in the rise of environmentalism.

> Boulding pointed up what seemed to him the fundamental issue which had arisen in the symposium.... [T]here was a sharp division between the "biological-ecological" point of view and the "socioeconomic" point of view.
>
> The biological-ecological view is expressed in terms of equilibrium systems and of movements toward equilibrium.... Biologists and conservationists feel that man has to find his place within a cycle of this kind; if he violates it he is heading for trouble. [2]

Thirteen of the papers presented at the symposium endorsed the idea referred to here, that there exists a balance of nature that man should refrain from upsetting. Paul B. Sears, for example, a former president of the Ecological Society of America, wrote "There are many interesting approaches to the problem of man and his environment, and all, save perhaps the technological, seem to lead to the same conclusions—that humanity should strive toward a condition of equilibrium with its environment." [3]

The idea that man ought to refrain from upsetting the balance of nature is essentially a *preservationist* idea. Its effect is to inhibit men from altering the natural landscape. To the extent that man, in the name of protecting the balance of nature (or the web of life, or the health of ecosystems), must refrain from harvesting timber on mountainsides, or from erecting dams on rivers, or from filling in swamps, or from building on coastal dunes, or from drilling for oil off-shore or in "fragile" ecosystems, or from constructing housing tracts on farmland, or from a hundred other acts that would extend his command of nature, to that extent he has become a preserver rather than a user of nature. The idea that man is morally obligated to maintain the balance of nature would inspire, in environmentalism, a movement committed to transforming America from a country of land users into a country of land preservers.

In the early decades of the twentieth century, Gifford Pinchot's progressive utilitarianism had dominated what was then called the conservation movement. The preservationist element within that movement, exemplified by John Muir, was a minority, both in numbers and in influence. The ascendance of environmentalism in the 1960s represented the rise to dominance of the preservationist ideology as it overwhelmed the old utilitarian element within the conservation movement. Rachel Carson played an important role in bringing about that rise.

Carson

Rachel Carson was born the youngest of three children in 1907 at Springdale, Pennsylvania.[4] She was raised on 65 acres in the country with horses and cows and chickens. Her mother had briefly been a school teacher, and Rachel grew up a reader and a nature lover. At Pennsylvania College for Women (now Chatham College) she majored in zoology, graduating magna cum laude in 1928. She went on to earn an MA in zoology at Johns Hopkins.

During a five year period beginning in the mid 1930's, Carson wrote a series of articles on fisheries and related topics in the Baltimore Sun. She also taught part time at the University of Maryland and at Johns Hopkins. After her father died in 1935, she needed a full-time job to support herself and her mother, so she hired on with the U.S. Bureau of Fisheries in Washington, D.C. to write a series of radio broadcasts titled "Romance under the Waters." When her sister died in 1936 and left behind two school-aged girls for Carson to raise, she took the civil service exam and got a position as "junior aquatic biologist" within the Bureau of Fisheries.

In September of 1937 the *Atlantic Monthly* published her essay, "Undersea." This led to inquiries from publishers about writing a book, which came out in 1941 and was titled *Under the Sea Wind*. She continued her work at the Bureau (which had become the Fish and Wildlife Service in 1940) and eventually rose to the position of biologist and Chief Editor of the Service's publications, a position she held until her resignation in 1952. The reason for her resignation was that in July of 1951 Carson had published her second book, *The Sea Around Us*. This one made it onto the New York Times bestseller list, where it set a record for consecutive appearances. Carson subsequently won the National Book Award and the John Burroughs Medal for natural history writing. She was now famous. In 1962 Carson published *Silent Spring*, and the world has never been the same.

> Then a strange blight crept over the area and everything began to change. Some evil spell had settled on the community: mysterious maladies swept the flocks of chickens; the cattle and sheep sickened and died. Everywhere was a shadow of death. The farmers spoke of much illness among their families. In the town the doctors had become more and more puzzled by new kinds of sickness appearing among their patients. There had been several sudden and unexplained deaths, not only among adults but even among children, who would be stricken suddenly at play and die within a few hours.
>
> No witchcraft, no enemy action had silenced the rebirth of new life in this stricken world. The people had done it themselves.[5]

The blight that had crept over the area was from chemical pesticide. The passage gives Carson's impression of what lay ahead for American communities unless the use of these pesticides was stopped. No one could accuse Carson of having understated the seriousness of the threat these pesticides posed. "Along with the possibility of the extinction of mankind by nuclear war, the central problem of our age had therefore become the contamination of man's total environment with such substances."[6] *Silent Spring* would become a kind of *Uncle Tom's Cabin* for the twentieth century. Not only would it create a groundswell of support for banning the manufacture and use of certain chemical pesticides—most notably DDT—but it would spark the ideological conflagration we now call environmentalism.

Silent Spring does not appear to be an ideological book. There is nothing in it to compare with the philosophical musings of Thoreau or Muir or Aldo Leopold. There are no lengthy rants against capitalist greed or depredation of the landscape. The book addresses in concrete terms a single, specific issue, the alleged dangers inherent in the use of a handful of chemical pesticides and herbicides. Instead of arguing ideology, Carson let the "facts" speak for themselves: Clear Lake, California was sprayed with DDD (a derivative of DDT); hundreds of birds died; subsequent examination of the birds disclosed high concentrations of DDD in the birds' tissues.

Carson pointedly portrayed these pesticides as a threat to *human* well-being (as opposed to the well-being of wildlife), and she purported to base her critique of them on science.[7] These three characteristics of *Silent Spring*, the focus on specific environmental "problems" rather than on ideology, the focus on human well-being as opposed to the well-being of nature, and the apparent appeal to science, would become integral elements of the environmentalists' strategy for promoting their preservationist agenda.[8] (The casting of environmental issues as crises is a fourth element that Carson pioneered.)

But *Silent Spring* is very much informed by an ideology, however understated. Carson had the conservationist's characteristic antipathy for private property. She did not explicitly oppose property rights, but she consistently viewed elements of the free market as a problem to which increased government regulation was the solution. Regarding, for example, the question of insecticide residues on fruits and vegetables, she favored the instituting of a policy of zero tolerance by federal inspectors. And to enforce this she advocated a "vigilant and aggressive Food and Drug Administration, with a greatly increased force of inspectors."[9] More sweepingly, she wrote that "we have allowed these chemicals to be used with little or no advance investigation of their effects on soil, water, wildlife, and man himself."[10] Carson believed that any manufacturer of a new chemical pesticide should first be required to obtain government approval before he could bring the pesticide to market.

There is irony in Carson's advocating this vast expansion of government power as the solution to the pesticide "problem," since most of the serious cases of environmental damage she cited were cases in which governments were responsible for the use, or misuse, of the pesticides. At Clear Lake, California, for example, it was the County Mosquito Abatement District that administered the DDD. Judging from the cases cited in *Silent Spring*, the United Stated could have gone a long way toward solving the alleged pesticide problem simply by getting governments at all levels out of the insect extermination business or, at the very least, by requiring them to respect the property rights of private citizens.

Consider Carson's account of the use of DDT by the New York Department of Agriculture and Markets to eradicate the gypsy moth in 1957:

> *They sprayed truck gardens and dairy farms, fish ponds and salt marshes. They sprayed the quarter-acre lots of suburbia, drenching a housewife making a desperate effort to cover her garden before the roaring plane reached her, and showering insecticide over children at play and commuters at railway stations.* [11]

Carson wrote that the lawsuit brought by Long Island residents in response to this incident helped "focus public attention ... on the power and inclination of the [government] control agencies to disregard supposedly inviolate property rights of private citizens." [12] But rather than advocating that governments be made to respect private rights, Carson advocated, in effect, that government power to override private rights be legalized and expanded—as in her advocacy of government censorship of new man-made pesticides. A government with the power to suppress new technologies, even with the best of intentions, is capable of doing terrible harm. Consider the case of DDT.

Before the advent of DDT, malaria killed more than three million persons each year throughout the world. DDT was first used against malaria-carrying mosquitoes at the end of World War II. Within a few years of the end of WW II, it had virtually eradicated malaria from Europe and North America.[13] Within ten years of DDT's introduction into Ceylon, the incidence of malaria declined from 3,000,000 cases a year to 7,300.[14] In 1964 there were just 29. Before the introduction of DDT into India, 75,000,000 persons contracted malaria each year, and 800,000 of those died. By 1961, after the introduction of DDT, the number of cases annually declined to 50,000.[15]

Donald Roberts and Richard Tren point out that Carson failed to distinguish between the use of DDT in agriculture and its use for public health purposes.[16] In the latter case, DDT is sprayed on the interior walls of homes and acts as a spatial repellent, causing the mosquitoes to exit the home rather than killing them. Carson belittled this characteristic of DDT: "Some malaria mosquitoes have a habit that so reduces their exposure to DDT as to make them virtually immune. Irritated by the spray, they leave huts and survive outside."[17] She failed to acknowledge that this method of using DDT saved millions of human lives. And she failed to recognize that, precisely because this method did *not* kill the mosquito, it prevented them from building up a resistance to DDT. Mosquitoes' tendency eventually to evolve into DDT-resistant strains was one of Carson's criticisms of the pesticide.

In the culmination of a process that had begun with the publication of *Silent Spring*, the newly formed U.S. Environmental Protection Agency held hearings in 1971 and 1972 on the safety of DDT. In April 1972 Judge Edmund Sweeney, the examiner at the hearings, concluded that:

> *DDT is not a carcinogenic hazard to man.... The uses of DDT under the regulations involved here do not have*

a deleterious effect on freshwater fish, estuarine organisms, wild birds, or other wildlife.... The evidence in this proceeding supports the conclusion that there is a present need for the essential uses of DDT.[18]

But EPA head William Ruckleshaus, "who had never attended a single day's session in the seven months of EPA hearings, and who admitted he had not even read the transcript of the hearing," decided that DDT was a "potential human carcinogen" and he "banned it for virtually all uses."[19] Most of the world followed America's lead. The result? The World Health Organization reported that worldwide in 2006 there were 247 million cases of malaria and 881,000 deaths from it. This in just one year. And the vast majority of these deaths were of children.[20]

A silent spring is indeed upon us, but it is a silence caused not by the absence of birds, but by the absence of children, in Third World villages around the globe. And it has been caused not by the use of DDT, but by its prohibition. No private company could have caused such devastation on such a grand scale. Only governments are capable of this, governments with the power to censor technology.

Yet Carson favored extending government regulation of technology beyond the field of pesticides. "I remember that Barry Commoner pointed out, in a masterful address to the Air Pollution Conference in Washington last winter, that we seldom if ever evaluate the risks associated with a new technological program before it is put into effect." [21] Try to imagine an America with a technology czar responsible for approving or disapproving every major new technological innovation that comes along. Where could we be today if Alexander Graham Bell, Thomas Edison, and the Wright Brothers had needed to have their inventions approved by a government bureaucrat? Such a czar would rival Europe's Catholic Inquisitions for stifling scientific and economic development. And the impulse behind the idea of such a czar is identical to that behind the Inquisitions, the idea that, above all else, *someone must be in control*; if people were allowed to be free, then chaos would ensue.

Silent Spring is nothing if not alarming.

> [Exposures to dangerous chemicals] *have entered the environment of everyone—even of children as yet unborn. It is hardly surprising, therefore, that we are now aware of an alarming increase in malignant disease.... [M]alignant growth ... accounted for 15 per cent of the deaths in 1958 compared with 4 per cent in 1900.... This means that malignant disease will strike two out of three families.... A quarter century ago, cancer in children was considered a medical rarity. Today, more American school children die of cancer than from any other disease.*[22]

This is terrifying. It is also deeply misleading. In 1935, there were 6,260 childhood deaths per 100,000 from all causes, and 9.2 childhood cancer deaths per 100,000. In 1960, there were only 3,040 deaths per 100,000 from all causes, but 24.9 deaths from cancer.[23]

	1900	1935	1960
Child Deaths from All Causes	18,610	6,260	3,040
Child Deaths from Cancer	7.9	9.2	24.9
Percent Child Deaths from Cancer	.042	.147	.89

The number of childhood deaths attributed to cancer per 100,000 had indeed risen sharply between 1935 and 1960. But the cancer death rate reported for the general population had been rising steadily since 1900, long before the age of man-made pesticides, and it did not show a sharp increase after pesticide use began during WW II. Carson offered no reason why man-made pesticides should cause a sharp rise in the childhood cancer death rate but not in that of the general population. As for the relative increase in the number of cancer deaths compared to deaths from other causes, the incidence of death from all other causes had decreased dramatically, from 6,260 per 100,000 in 1935 to 3,040 in 1960. So even if the number of childhood cancer deaths per 100,000 had remained level from 1935 to 1960, cancer deaths would still have comprised a much larger percentage of the total number of deaths in 1960 than they had in 1935. At the very least, Carson was guilty of using statistics to create the false impression that man-made pesticides were killing children by the busload.

But Carson was guilty of much more than this. Even if the introduction of man-made pesticides in the U.S. between 1935 and 1960 had somehow caused an increase in the number of childhood cancer deaths from 9.2 to 24.9 per 100,000, it would have meant that 16 more children per 100,000 would have died of cancer in 1960 than in 1935, *but 3,196 fewer would have died of other causes.* The capitalist industrial system that had produced man-made pesticides had saved 203 times as many lives it had cost (*if* it had cost any lives at all). And, no doubt, pesticides themselves had saved some of those lives by making food more plentiful and less expensive. And yet, it was the system responsible for saving all those lives that Carson was indicting in *Silent Spring,* at least implicitly.

For centuries, Americans had believed that a man was morally entitled to use his land in whatever way he thought would best serve his own purposes. If a farmer wanted to fill in a swamp on his land, it was his business, and no one could stop him. The land itself and all the living things upon it were his property to do with as he judged would best serve his own private ends. When the conservationists came along, they argued that "natural resources" are the property of the people as a whole. This meant, in principle, that private land is not a man's to do with entirely as he pleases. But the conservationists concentrated primarily on adding more land to the public domain rather than on restricting the rights of private landowners. And, while they denied, in principle, the right of any one man to use his land as he pleased, they believed in the right of man, in the aggregate, to use land in any way that best served his own collective self-interest. This was the view of the majority of the early conservationists, and it was the view promulgated by the most prominent of those early conservationists, Gifford Pinchot.

Thoreau, John Muir, and Aldo Leopold held a different view of man's relationship to nature. At Walden Pond, Thoreau portrayed subsistence living as a moral ideal. Muir believed that some land, such as Yosemite Valley, is literally sacred, and that man is morally obligated to preserve such land in its natural state. Leopold believed that just as a man is morally prohibited from using other persons in any way he chooses, so is he morally prohibited from using the land in any way he chooses. All believed that man is not morally entitled to use land in whatever way best conduces to his own self-interest, whether collective or individual. This is the essence of the preservationist view of man's relationship to nature, and it is the view of nature that is implicit in *Silent Spring*.

Carson wrote in the preservationist tradition of Thoreau, Muir, and Leopold. She was a lover of nature, and she wrote as eloquently and lovingly of nature as those three ever did. And she certainly believed that there are limits to the degree to which man is entitled to use nature for his own purposes, limits of a moral kind. "I am glad to see Mrs. Harrison raise the question of how far man has a moral right to go in his domination of other life," she wrote.[24] Carson elaborates on these limits in the following passage from *Silent Spring*: "The earth's vegetation is part of a web of life in which there are intimate and essential relations between plants and the earth, etc. Sometimes we have no choice but to disturb these relationships, but we should do so thoughtfully."[25] We can now state the fundamental philosophical premise of *Silent Spring*: *As he goes about living his life on earth, man is morally obligated to refrain from upsetting the balance of nature any more than he has to.* This was the view of Thoreau and Muir and Leopold. And it is the premise of the popular movement to which *Silent Spring* helped give rise, environmentalism. Let us call it the "preservationist premise."

The idea that man ought to refrain from upsetting the balance of nature any more than he must involves a chain of reasoning that begins with an "is" and ends with an "ought." The first "is" in this logical chain is the contention that in any natural setting there exists a state of equilibrium of some sort, a balance of nature. Would the forests of the Sierra Nevada, for example, maintain themselves in a given state indefinitely if man did not intervene? This is a matter that falls within the purview of science. A second "is" in the logical sequence involves identifying the consequences of man's upsetting the alleged balance of nature. What would happen if man were to upset the alleged equilibrium within the forests of the Sierra Nevada? This also falls within the purview of science. Balance-of-nature advocates would spend a great deal of time and energy attempting to prove the existence of states of equilibrium in many different natural settings and in identifying what would be the consequences of man's upsetting those states of equilibrium.

From Rachel Carson's campaign against man-made pesticides to Al Gore's against global warming, environmentalists would consistently seek, with varying degrees of success, to back up their claims with science. So consistent would be the environmentalist appeal to science that one could be forgiven for forming the impression that environmentalism is not an ideology formed around an abstract principle, but, rather, an assortment of *ad hoc* efforts to solve specific problems, such as polluted waterways, shortages of fossil fuels, overflowing landfills, species in danger of extinction, and global warming. This emphasis on science at the apparent expense of ideology would become standard procedure for environmentalists. But environmentalism is not a series of efforts to solve specific problems with scientifically validated solutions. It is an ideology grounded in a specific moral view of what man's relationship to nature ought to be.

Accordingly, the third link in the chain of reasoning underlying the preservationist premise involves a major moral presumption. Having attempted to establish that a balance of nature exists, and having attempted to identify what would be the consequences of man's upsetting this alleged balance, the advocates of the preservationist premise then assert, implicitly, that it would be in man's *collective* best interest to refrain from upsetting this balance any more than he needs to. But who is competent to judge what is in man's best interest? "Man" is composed of individual men, and each man is the only competent judge of his own best interest.

Consider the global warming debate. Those who warn of a man-made warming crisis first presume that mankind in the aggregate would be better off if the alleged warming could be halted. They then advocate political measures, such as limits on the burning of fossil fuels, to force other men to act according to the "experts'" judgment.

The alternative would be to leave each person free to judge for himself, first, whether the scientific evidence for man-made warming is persuasive, and, then, how best to respond to that evidence. An individual might well conclude that man-made global warming is indeed occurring, but that it would be in his own best interest not to take any steps to avert it, because he himself stands to lose nothing if global temperatures rise a couple of degrees over the next century. But this outcome would be unacceptable to those who warn of a warming crisis, who presume to speak for, and who would force their judgment upon, all mankind. This presumption that man should act collectively (in effect, politically) to maintain the balance of nature is built into the preservationist premise as Rachel Carson conceived it and as the environmental movement has given effect to it.

Having thus made the moral assumption in the third link of the logical chain that man ought to deal collectively with the question of maintaining the balance of nature, the proponents of the preservationist premise then conclude with the moral imperative that man, both collectively and individually, ought to refrain from upsetting the balance of nature any more than necessary. Environmentalism, the ideology based on the preservationist premise, begins with a pair of scientific propositions: there exists a balance of nature, and if man upsets it bad things will happen. And it ends with a pair of moral imperatives: man ought to deal with questions involving the balance of nature collectively (i.e., politically), and, in the end, he ought to refrain from upsetting that balance any more than he needs to. At the Princeton Symposium in 1956, Paul Sears had enunciated the preservationist premise when he wrote that both science and ethics support the claim that "humanity should strive toward a condition of equilibrium with its environment."

As I have stated the premise it sounds almost innocuous. No doubt many readers will see it as not only unobjectionable, but as patently commonsensical. "Of course we should avoid disturbing the balance of nature if we possibly can." Yet the premise represents a radical departure not only from the view of nature regnant during the Enlightenment, but even from that held by Gifford Pinchot and the utilitarian conservatives. The latter saw nature as something that exists *for* man; Carson, and the environmental movement she helped set in motion, see it as

Thomas McCaffrey

something to be protected *from* man.

> *I am pessimistic about the human race because it is too*
> *ingenious for its own good. Our approach to nature is*
> *to beat it into submission. We would stand a better*
> *chance of survival if we accommodated ourselves to this*
> *planet and viewed it appreciatively instead of*
> *dictatorially.*[26]

So said E.B. White, in the epigraph with which Carson opened *Silent Spring*. It flatly contradicts Francis Bacon's "Nature, to be commanded, must be obeyed."

Whereas the Baconian view implied an ambitious, active orientation toward nature, Carson advocated a more passive stance. And whereas Bacon advocated technological innovation in order better to command nature, Carson advocated government censorship of technology in the interest of preserving the natural *status quo*. This anti-Baconian stance of Carson's is a different form of her preservationist premise. It holds, in effect, that man cannot command nature on an industrial scale without fatally upsetting the balance of nature, so he ought not try. The preservationist premise and its Baconian corollary place Carson—and environmentalists—firmly within the nature preservationist tradition of Thoreau, Muir, and Aldo Leopold, and they separate her and them both from the utilitarian conservationists of Gifford Pinchot's ilk and from those children of the Enlightenment, the American Founders.

White House Conference on Natural Beauty

One way that ideas like nature preservationism make their way into the public imagination is through events like the White House Conference on Natural Beauty, which was held in May of 1965. "We must not only protect the countryside and save it from destruction," President Lyndon Johnson told the conference, "we must restore what has been destroyed and salvage the beauty and charm of our cities." [27] Maurice K. Goddard was a conference participant and the Secretary of the Pennsylvania Department of Forests and Waters, a position that in an earlier age of conservation would have been filled by an advocate of *using* the forests and waters of Pennsylvania for the greatest good of the greatest number of human beings. But Mr. Goddard believed that man had become a "parasite" of the earth, and he told his fellow participants that "[t]echnological know-how must become the chief handmaiden for creating and *preserving* a *balanced*, healthy, and beautiful environment capable of supporting man and his fellow creatures indefinitely." [28]

Nature preservation means protecting land and waters from human development. It might involve the prohibiting of all almost human use, as in wilderness areas, the limiting of human use mostly to recreation, as in the national parks, or the prohibiting of more intensive human uses, such as the building of factories and homes, in favor of less intensive uses, such as farming. The White House Conference on Natural Beauty advocated a wide range of preservationist measures, including the preserving of land through zoning, tax breaks, and scenic easements. [29] They called for limitations on the development of steep slopes, and for the preserving of wetlands. [30] They also called for "stream renewal" from end to end on a scale comparable to urban renewal. [31] And they recommended "that the outstanding water areas and water courses in the United States, such as San Francisco Bay, Lake Tahoe, and the sources of major rivers be designated national scenic and recreational landmarks; and that all decisions affecting their development be reviewed by a Presidentially appointed board of citizens." [32] This is nature preservation on a grand scale, and most of what the conference advocated has become reality.

The premise behind the preserving of natural beauty is that a natural landscape is more beautiful than one that has been made over for human use. James Rouse, a participant in the conference, spoke of "commercial squalor" and referred to suburban sprawl as "disorder and squalor." It would be hard to find an environmentalist today who does not consider a mountain vista to be more beautiful than a city skyline. When land is withdrawn from productive use by government fiat in order to preserve its natural beauty, it means that there is less land available to those who would make productive use of it, thus raising the price of the remaining land, and thus raising the cost of living in general. Everyone is forced to purchase "scenic beauty" through a higher cost of living.

Now, some people consider an industrial plant to be a thing of beauty (because of what it represents—employment, productivity, prosperity). This is a perfectly rational viewpoint, and people who consider a factory to be more beautiful than a green valley are not morally or in any other way inferior to people who prefer the green valley. Even among people who value scenic beauty, some value it more highly than others. Some value it highly enough to pay to live in a place like Bar Harbor, Maine year round, while others are content to visit for two weeks a year.

Each of us has a unique hierarchy of values, and where natural beauty ranks in that hierarchy is a matter of individual taste. But the premise behind the effort to preserve natural beauty via government fiat is that natural beauty objectively ranks higher than other, competing values, and that it is right and proper for some people to force others to pay to preserve natural beauty through a higher cost of living (or through higher taxes, which are needed to purchase or to maintain the preserved land). This premise, that scenic beauty is objectively superior to a man-made landscape, would become an integral part of the environmentalist ideology.

National Environmental Policy Act

Events like the White House Conference on Natural Beauty are intended to influence legislation. One of the first important federal statutes of the environmental era was the National Environmental Policy Act of 1969 (NEPA). The first purpose stated in the act is "To declare a national policy which will encourage productive and enjoyable harmony between man and his environment." [33] If one wonders what is meant by "harmony between man and his environment," the First Annual Report of the Council of Environmental Quality provides an answer. The council, which reports to the President, was formed at the direction of NEPA to coordinate the efforts of federal agencies to abide by environmental statutes and regulations.

> *The first men upon this land, the American Indians, treated it with reverence, blended with it, used it, but left hardly a trace upon it. Those who followed have been less kind. They brought with them a different creed which called on man to conquer nature and harness it for his own use and profit.* [34]

Environmentalism is an ideology animated by a moral idea, and the American Indian, as this passage suggests, is its moral ideal. To say this is not to suggest that environmentalists expect everyone to live like an Indian, any more than Catholics expect everyone to live like Mother Theresa. But a moral ideal exerts a force nonetheless, and it reveals what an ideology aspires to. An ideology that holds the American Indian as its ideal clearly does not aspire to continue America's industrial transformation of the landscape. On the contrary, it aspires to limit that transformation, if not to end it altogether—and even to reverse it where possible.

A second purpose of NEPA is "to enrich the understanding of the ecological systems and natural resources important to the nation." [35] Here the act incorporates the balance of nature idea, and its authors endow their ideological enterprise with the aura of scientific legitimacy. The First Annual Report states that "The public has begun to realize the *interrelationship of all living things*—including man—with the environment." [36] It explains ecology as "the science of the intricate web of relationships between living organisms and their living and nonliving surroundings." [37] It states that "[i]n the future, the effect of man's actions on the complete ecosystems must be considered if environmental problems are to be solved." [38]

As to what all this concern with ecological relationships leads to, the report points out that "Man's understanding of these biological processes, particularly of the permanent damage that begins subtly, with piecemeal alterations of the land, is still limited. Yet his dependence on stability is enormous." [39] Small changes to a landscape can add up, in other words, and can threaten the ecological stability on which man's well-being depends; the less change, the better. This is nature preservationism, *par excellence*. "Planners and managers must begin to appreciate the enormous interrelated complexity of environmental systems, weigh the tradeoffs of potential harm against the benefits of construction,"—and, of course, decide *against* construction when "necessary."

Having embraced the balance of nature argument, the First Report follows up with its counterpart, the anti-Baconian idea that man is not capable of commanding nature on any grand scale, so he ought not try. President Richard Nixon, in a statement accompanying his transmittal of the First Report to Congress, set the tone. "In dealing with the environment we must learn not how to master nature but how to master ourselves, our institutions, and our technology." [40] Said the First Report,

The pace of technological innovation has exceeded our

> *scientific and regulatory ability to control its injurious side effects. The environmental problems of the future will increasingly spring from the wonders of 20th century technology. In the future, technology assessment must be used to understand the direct and secondary impacts of technological innovation.*[41]

Here the council is laying the groundwork for government censorship of technology, which Rachel Carson had advocated.

NEPA goes on to require that "proposals for legislation and other major Federal actions significantly affecting the quality of the human environment" include what would come to be called Environmental Impact Statements. Henceforth, before any federal action that might "significantly" affect the environment could go forward, it would first need to be scrutinized to identify what its effect on the environment would be. Based on the findings of the Environmental Impact Statement, a proposed project might be altered, or it might be denied approval altogether.

Following the passage of NEPA, California passed its own California Environmental Quality Act (CEQA) in 1970, which also required an Environmental Impact Report before any public project could be approved. In 1972 the California Supreme Court decided that (CEQA) applied to *private* projects as well as public ones.[42] This, of course, is nothing less than government censorship of land use, and it poses every bit as great a threat to freedom as any other form of censorship ever has. To argue, by the way, that California's application of the Environmental Impact Report requirement to private projects represents the action of environmental extremists would be a mistake. California environmentalists simply enjoy more power to enact the standard environmentalist agenda than do their fellow ideologues in other states. There are few environmentalists anywhere who would not relish having the power to censor private land-use decisions.

Thomas McCaffrey

Another purpose of NEPA is "to promote efforts which will *prevent or eliminate damage to the environment and biosphere* and stimulate the health and welfare of man." [43] Almost anything can qualify as damage to the environment: building a road, a bridge, a dam, a factory, or a tract of homes. Indeed, building anything whatsoever on undeveloped land, or in any way altering the natural landscape, can qualify as damaging the environment. If the purpose of NEPA is to prevent damage to the environment, then there is almost no productive endeavor that may not be prohibited in its name. Likewise, if NEPA's purpose is to "eliminate" environmental damage, there is virtually no existing human development that could not be removed in the name of eliminating some damage and restoring the environment to its natural state. An ideology that holds the American Indian as its moral ideal will implicitly hold the natural, unaltered landscape as its environmental ideal and will aspire to protect such landscapes from development where they still exist and to restore them elsewhere whenever possible.

Accordingly, the First Report is rife with nature preservationist ideas. Suburban sprawl is heavily criticized. "A jumble of developments sprawls over the landscape," it states. [44] "[I]nsufficient effort has been made to keep the most attractive rural lands near cities from being consumed in the massive conversion to urban life." [45] The automobile comes in for criticism for its role in promoting sprawl.

> As a consequence [of the auto], thousands of acres of undeveloped land fall prey each year to the bulldozer. More single-family, detached homes shoulder out the open spaces. Many of the developments are drab in design and wasteful of land. [46]

All this new development is presented as an unfortunate

occurrence that must be halted; sprawling development must be stopped, and undeveloped land must be preserved. But suburban sprawl is a consequence of economic growth, since economic growth fuels new development. Accordingly, the First Report suggests the possibility that government might need to restrain economic growth. "When the Governor's Conference on California's Changing Environment met last fall, it agreed that there was now a need 'to *deemphasize growth* as a social goal.'" [47] "Increasingly," the report explains, "concern has shifted [among Americans] from quantity to quality and from the desirability of growth to the desirability of stability." [48]

> *Many traditional assumptions are being questioned, and our attitudes toward population growth, economic expansion, and the use of natural resources are no exception. We have begun to ask what the optimum population of the nation or of a given metropolitan area should be. We are looking more critically at proposals for the construction of new power plants or factories and asking what the environmental effects of such projects will be.* [49]

The report thus anticipates American industrial decline—and welcomes it.

> *One result [of this decline] will be a reduction in the adverse environmental effects of increased national product since most such effects arise from manufacturing processes.* [50]

(F. Fraser Darling had raised the alarm about economic growth at the 1956 Symposium at Princeton: "A subsistence economy maintains itself for a very long time and engenders its own conservation ethos, but an export economy spells deterioration. With an increasing and expanding economy, we may, indeed without very great knowledge and without very great thought, be heading for disaster.")[51]

The automobile is further criticized for enabling the masses to gain access to as yet "unspoiled" areas. "The world wide boom in tourism [itself fueled by economic growth] teamed with rapid and cheap transportation [the airliner and the auto] threatens the very values upon which tourist attraction is based," the authors wrote.[52] "Unlimited access to wilderness areas may transform such areas into simply another extension of our urban, industrialized civilization."[53] *If we could just keep people out of nature.* One way to keep people out of wilderness areas is to be sure there are plenty of dangerous wild predators in them. "As a group, the large predators stand in great danger of extermination. The belief that most predators should be exterminated was central to the early days of ranching and wildlife management. In some areas this unfortunate belief endures."[54]

The report points up the need to preserve coastal wetlands. "Besides water pollution, the major adverse effect on the coastal lands and waters stems from physical alteration of submerged land and habitat—particularly shallow marshes and wetlands."[55] And the coastline itself must be protected, as is now being done in California. "Although some uses of coastal areas are undoubtedly necessary, many are not. Much industry, housing, and transportation could be sited elsewhere."[56] *What does it matter where people _want_ to live?*

Evoking memories of Thoreau's Farmer Flint, the report welcomes a questioning of the "commodification" of land. "Although it has been a long time coming, Americans are recognizing the need to examine carefully what government can do to assure that land is treated as a resource to be managed and not merely as a *commodity to be marketed*."[57] This, of course, is another way of saying that it's about time someone put an end to private land ownership.

President Richard Nixon made some remarkable statements on this subject in his transmittal statement of the First Report. "Traditionally, Americans have felt that what they do with their own land is their own business. This attitude has been a natural outgrowth of the pioneer spirit." [58] Who knew that property rights were nothing more than an American tradition, the mere outgrowth of an "attitude"? "The time has come," said President Nixon, "when we must accept the idea that none of us has a right to abuse the land, and that on the contrary society as a whole has a legitimate interest in proper land use." Imagine a U.S. president stating that "No one has a right to abuse the right of free speech," and that "society as a whole has a legitimate interest in the proper exercise of that right."

One other element of the First Report is worth mentioning: it endorsed the development of nuclear power generation.[59]

Notes

1. Donald Worster, *Nature's Economy: A History of Ecological Ideas* (Cambridge: Cambridge University Press, 1977), 352-353.

2. William L. Thomas quoting Dr. Kenneth Boulding, "Industrial Revolution and Urban Dominance," in *Man's Role in Changing the Face of the Earth*, ed. William L. Thomas, Jr. (Chicago, IL: The University of Chicago Press, 1956), 446-447.

3. Paul Sears, "The Process of Environmental Change by Man," in Thomas, *Man's Role*, 473.

4. The biographical material on Rachel Carson comes from Paul Brooks, *The House of Life* (Boston: Houghton Mifflin, 1972), 15-21.

5. Rachel Carson, *Silent Spring* (Boston: Houghton Mifflin, 1962), 2-3.

6. Ibid., 8.

7. "As you know, it has always been my intention to give principal emphasis to the menace to human health, even though setting this within the general framework of disturbances of the basic ecology of all things." Rachel Carson, letter to Paul Brooks, February 14, 1959 in Brooks, *House of Life*, 243-244.

8. Larry Katzenstein credits Carson in *Silent Spring* with originating the "Precautionary Principle." Larry Katzenstein, "The Precautionary Principle: *Silent Spring's* Toxic Legacy," in *Silent Spring at 50: The False Crises of Rachel Carson*, eds., Roger Meiners, Pierre Desrochers, and Andrew Morriss, (Washington, D.C.: Cato Institute, 2012), 245-270.

9. Carson, *Silent Spring*, 183.

10. Ibid., 13. Emphasis added.

11. Ibid., 158

12. Ibid., 159.

13. Richard Tren and Roger Bate, *When Politics Kills: Malaria & the DDT Story* (New Delhi, India: Liberty Institute, 2000), 15.

14. Gordon Harrison, *Mosquitoes, Malaria and Man: A History of the Hostilities Since 1880* (London: John Murray, 1978), 230. Cited in Tren and Bate, *When Politics Kills*, 16.

15. Ibid., p. 247. Tren and Bate point out that many of the early victories over malaria were not sustained, partly because the eradication efforts were not properly sustained. Ibid., 17.

16. Donald R. Roberts and Richard Tren, "Did Rachel Carson Understand the Importance of DDT in Global Public Health Programs," in Meiners, *Silent Spring at 50*, 190.

17. Carson, *Silent Spring*, 274.

18. Edmund M. Sweeney, EPA Hearing Examiner's recommendations and findings concerning DDT hearings, April 25, 1972, *Federal Register*, (40 CFR 164.32). Quoted in Adam J. Lieberman and Simona C. Kwon, *Facts Versus Fears: a Review of the Greatest Unfounded Health Scares of Recent Times* (New York, NY: American Council on Science and Health, 2004), 9.

19. Ibid.

20. There are limitations to the effectiveness of DDT. For one thing, mosquitoes develop immunity to it over time. But the world did not stop using DDT because it had ceased to be effective in eradicating malaria.

21. Rachel Carson, *Lost Woods: The Discovered Writing of Rachel Carson*, ed. Linda Lear (Boston, MA: Beacon Press, 1998), 232.

22. Carson, *Silent Spring*, 221. Emphasis in original.

23. Public Health Service, *Vital Statistics Rates in the United States, 1900-1940* (Washington, D.C.: Government Printing Office, 1947), 181, 250; Public Health Service, *Vital Statistics Rates in the United States,* 1960 (Washington, D.C.: Government Printing Office, 1963), 5-182, 5-192; and U.S. Bureau of Commerce, *Statistical Abstracts of the United States 1961* (Washington, D.C.: Government Printing Office, 1961), 29, 58. Cited in Meiners, *Silent Spring at 50*, 130.

24. Rachel Carson in preface to "Animal Machines," quoted in Lear, *Lost Woods*, 196. She continues, "Has he the right ... to reduce life to a bare existence that is scarcely life at all? Has he the further right to terminate these wretched lives by means that are wantonly cruel? My own answer is an unqualified no." As the rest of this quotation makes clear, Carson was speaking here specifically of man's treatment of animals.

25. Carson, *Silent Spring*, 64.

26. Carson quoting E.B. White in the epigraph to *Silent Spring*.

27. White House Conference on Natural Beauty, *Beauty for America: White House Conference on Natural Beauty* (Washington, D.C.: U.S. Government Printing Office, 1965), 1-2.

28. Ibid. Emphasis added.

29. Ibid., 641.

30. Ibid.

31. Ibid., 645

32. Ibid.

33. Sec 2 [42 USC § 4321].

34. "The First Annual Report of the Council on Environmental Quality", August, 1970, 165.

35. Sec 2 [42 USC § 4321]. Emphasis added.

36. "First Report," p. 6. Emphasis added.

37. Ibid.

38. Ibid., 18.

39. Ibid., 166.

40. Ibid., xii.

41. "First Report," 15. Emphasis added.

42. *Friends of Mammoth v. Board of Supervisors*, 8 Cal.3d 247 (Supreme Court of California 1972-09-21).

43. Sec 2 [42 USC § 4321]

44. "First Report," 10.

45. Ibid., 175.

46. Ibid., 15.

47. Ibid., 13-14. Emphasis added.

48. Ibid., 151.

49. Ibid., 164. Emphasis in the original.

50. Ibid., 156

51. Darling, "Man's Ecological Dominance," in Thomas, *Man's Role*, 401.

52. Ibid., 11.

53. Ibid., 10.

54. Ibid., 183.

55. Ibid., 177.

56. Ibid.

57. Ibid., 185. Emphasis added.

58. Ibid., xii.

59. Ibid., 160.

7: Pollution

IN 1965, THE Pollution Panel of the President's Science Advisory Committee identified pollution as a balance-of-nature problem.

> *Because living things are interdependent and interacting, they form a complex, dynamic system. Tampering with this system may be desirable and necessary, as in agriculture, which involves artificial manipulation of the balances of nature on a huge scale. But such tampering often produces unexpected results, or side effects, and these are sometimes very damaging. Many of the effects of pollution fall into this category.*[1]

The Panel paid special attention to water pollution, and they threw a wide net over the waters they intended to protect. "We recommend that interstate and Federal-state river basin development pacts and agreements cover related estuaries and marshes."[2] They then redefined the "the filling in of shallow waters essential in life cycles of fishes and shellfish" as pollution, and they recommended the prohibiting of all development on, and all except recreational use of, such waters.[3] They recommended, in other words, the *preservation* of a very wide swath of American waters, as a way of preventing pollution.[4] As we shall see, the environmentalist effort to clean up existing pollution, and the effort to prevent new pollution, would be predominantly a preservationist one.

Commoner

On Februay 2nd, 1970 *Time Magazine* ran a cover story titled "The Paul Revere of Ecology." The subject of the article, Barry Commoner, was a professor of plant physiology at Washington University in St. Louis, where he helped found the St. Louis Committee for Nuclear Information in 1958. One goal of the committee had been to raise public awareness of the dangers of radiation pollution arising from the testing of nuclear weapons. The Committee's signature effort in the anti-radiation campaign was the Baby Tooth Survey, in which they solicited the baby teeth of St. Louis area children and tested them for strontium 90 and other nuclear trace elements. They published their results in a series of articles in their newsletter on into 1963, the year the Nuclear Test Ban Treaty was passed. Commoner became an internationally known opponent of nuclear testing, and of nuclear weapons in general.

In 1967, having decided to take on a wider range of environmental problems, the Committee changed its name to the St. Louis Committee for Environmental Information. Their Magazine, *Environment*, is still published today. In 1966 Commoner founded the Center for the Study of Natural Systems to promote the study of ecology. He ran for President on the Citizens Party ticket in 1980.

The old-time conservation movement had concerned itself first with the conservation (i.e., government management) of natural resources and second with the preservation of wildlife and wild lands. In the decades after World War I, pollution had been gaining attention as a civic concern. The St. Louis Committee for Nuclear Information helped make pollution a major focus of the newly forming environmentalist movement. Commoner framed this issue in a manner that would become characteristic of the environmental movement, as a moral one.

Commoner published his environmental ideas in 1971 in a book titled *The Closing Circle*. Like Carson, Commoner presented his allegations in scientific terms. The central theme of the book was the need for man to restore the balance of nature that his pollution has upset. Commoner argued that where pollution is concerned natural systems are self-regulating and self-cleansing. A lake, for example, can absorb a certain amount of pollution without suffering lasting effects, but if that threshold is exceeded, the system will break down and the lake and everything in it will begin to die. Lake Erie, as a result of an excess of detergent phosphates and sewage, developed an oxygen deficiency and, by the 1960's, was in the process "dying." Also as Carson had done with pesticides, Commoner portrayed the pollution problem as a crisis. "My own judgment … is that the present course of environmental degradation … is so serious that, if continued, it will destroy the capability of the environment to support a reasonably civilized human society." [5]

The phosphate-laden detergents in Lake Erie exemplified a phenomenon that Commoner blamed for America's pollution problems, the substitution by American manufacturers after World War II of synthetic, chemically complex technologies for earlier, simpler, more natural ones. Another example is the high-compression automobile engine, which required that lead be added to gasoline, thereby increasing the lead content of the air in American cities. "The postwar technological transformation of the United States has produced not only the much-heralded 126% rise in GNP, but also, at a rate about ten times faster than the growth of GNP, the rising levels of environmental pollution."[6] Like other balance of nature advocates, Commoner was calling into question the Baconian idea that the only limitations on man's ability to command nature are the laws of physics and his own ingenuity.

Thomas McCaffrey

Commoner believed that new technologies needed to be designed to comply not only with the laws of physics, but also with the "laws" of ecology.

> *Clearly, if human activity on the earth—civilization—is to remain in harmony with the global system, and survive, it must accommodate to the demands of the natural sector, the ecosphere.*[7]

In Commoner's view, whereas modern technology was the cause of pollution, the ongoing pursuit of economic growth was the cause of modern technology. The pressure for ever higher profits fueled the search for ever more sophisticated technologies, which became ever more incompatible with the natural environment. Laundry soap was replaced by laundry detergent, which increased profits but polluted lakes and rivers. Wood and metal were replaced with plastics, which failed to decay back into the earth the way natural materials do but instead piled up in landfills. "Air pollution," Commoner wrote, "is a reminder that our most celebrated technological achievements—the automobile, the jet plane, the power plant, *industry in general*, and indeed the modern city itself—are, in the environment, failures." [8] Commoner argued (as Rachel Carson noted approvingly) for a system under which, before they could go on the market, new technologies would be subject to government censorship to make sure they would not harm the environment.

Ultimately, according to Commoner, "the productive system *must* reach a "no-growth" condition if man is ever to be prevented from destroying nature and himself.[9] A no-growth productive system is decidedly not a capitalist system. Commoner explicitly asserted in *The Closing Circle* that a politico-economic system based on private property is incompatible with the maintaining of a clean and healthy natural environment. In saying this, he was merely making explicit one of the fundamental logical implications of the environmentalist ideology, one that most environmentalists prefer to leave unsaid. (Commoner even went so far as to suggest that the Soviet system of central command and control would lend itself more readily to properly controlling environmental harm than does the American capitalist system.)

Commoner called for "altering the economic, social, and political priorities that govern the disposition of the nation's resources." [10] In calling for government censorship of new technologies, for a no-growth economy, and for the replacement of capitalism by socialism, Commoner was indeed radical. But it would be a mistake to dismiss him as unrepresentative of the true nature of environmentalism. Clearly, *Time Magazine* did not consider him unrepresentative.

Commoner's Third Law of Ecology states that "Nature knows best." It would be hard to devise a more eloquent rationale for wanting to preserve the landscape in its natural state. "[A]ny major man-made change to a natural system is likely to be *detrimental* to that system," wrote Commoner.[11] It is a perfectly logical application of this idea to conclude that man needs to limit the development of new technologies in order of preserve the natural balance. Since new technologies are a product of the freedom to pursue economic growth, and since wealth is the fuel that drives man's alteration of the natural landscape, it is also logical to want to limit economic freedom in the name of preserving the balance of nature. The United States has, since the dawn of the environmental era, been restricting the development of new technologies (nuclear power, pesticides—most notably DDT, and perhaps fracking), limiting economic growth, and substituting socialist-style control of the economy for private control. Barry Commoner was certainly a radical, but only in that he understood, and enunciated clearly and unequivocally, where the balance of nature premise would take the United States.

The Clean Water Act

Through most of the nineteenth century, the common law governed pollution. The common law is composed of previous court decisions. When a judge consults "the law" for guidance in deciding a common law case, he does not look to the text of a statute written by a legislature but instead to the legal precedents that have been set by courts in previous, similar cases. The first American colonists brought the common law with them from England. It continues in effect today, though in highly modified form.

The common law has long been the law used to settle disputes between private parties, especially disputes involving property or commercial matters. In the early years of the Industrial Revolution, if Smith was polluting a creek on Jones's property, Jones could sue Smith under the common law for an injunction or for monetary damages. Under this system, pollution is treated much the way trespassing is treated. Unlike police power regulation, the common law leaves it to the owner whose land is polluted to decide what to do, if anything, about the pollution. He even has the option to allow the polluter to continue if the polluter will agree to pay an acceptable price. The common law thus preserves the landowner's property rights intact. It also directs remuneration for damages to the injured party, rather than to the government. And it disperses the power to decide pollution questions among all the landowners of a polity (each acting individually), rather than concentrating it in a government entity, as police power regulation does. It thus makes impossible sudden sweeping changes in the law, such as those that have eviscerated American manufacturing since the 1970s.

When the United States came into being, however, the common law, which had been developed in England's static, agricultural economy, was not equal to the needs of a rapidly industrializing economy. It prohibited, for example, the damming of a waterway or the diverting of more than a small quantity of its water for private use without permission from the downstream owners.[12] Since 19th century mills used large quantities of water and needed the waterpower afforded by dams, it was necessary that the common law be changed. Change it did, though not entirely for the better.

At the time of the American Revolution, the common law was seen to embody timeless, universal principles of justice—the wisdom of the ages. But in the first generations after the Revolution, as the legitimacy of law came to be seen as deriving from the consent of the governed, law came to be seen as grounded in the will of the people, rather than in timeless principles.[13] The change is evident in the 1805 New York Supreme Court case of *Palmer v. Mulligan*, which involved a suit brought by a riparian owner against an upstream mill that had obstructed the flow of water. The court ruled that a newly arrived upstream user should be allowed to obstruct the flow of water, or else a downstream owner could use the law to block all further development upstream, and "the public, whose advantage is always to be consulted, would be deprived of the benefit which always attends competition and rivalry."[14]

Here is a just and desirable result—a river ought to be able to accommodate more than just the first mill built on its banks. But the result was justified by appeal to the "public advantage." This rationale is consistent with the idea that the law should be grounded in the "will of the people," but it is an unreliable foundation for securing individual rights of private property.

In the important 1827 case of *Tyler v. Wilkinson,* U.S. Supreme Court Justice Joseph Story reaffirmed the traditional eighteenth century prohibition against obstructing or significantly diminishing the flow of a river without all the effected owners' consent. But in his opinion, he stated that "There ... must be allowed of that, which is common to all, a reasonable use." This idea of reasonable use would become the basis of a new standard for determining riparian property rights.[15] By the time of the Civil War, according to Morton Horwitz, most courts employed a balancing test in which "reasonable use" became a function of how much harm the downstream owners suffered.[16]

Consider the 1856 Vermont case of *Snow v. Parsons*. A mill owner sued the owner of an upstream tannery for fouling the downstream mill's wheel with discarded bark. This should have been a clear-cut case of violating the downstream owner's rights, but the court ruled in the tannery's favor. Said the court:

> *Within reasonable limits, those who have a common interest in the use of air or running water must submit to small inconveniences ... The reasonableness of plaintiff submitting to this inconvenience must depend upon its extent, and the comparative benefit to the defendants, to be judged by the triers of the fact."*[17]

Here we have the pollution of a river justified as reasonable on the basis that the value to the defendant of discarding his refuse in the river exceeded, in the court's opinion, the harm to the downstream plaintiff.

We see this rationale echoed in the 1863 New Hampshire case of *Hayes v. Waldron,* in which a downstream riparian sued over shavings and sawdust that had been deposited on his land from an upstream sawmill. Again, the court ruled for the polluter. "Whether ... [the disposal of waste] may be rightfully done must depend upon the question whether ... it is or is not a reasonable use of the stream; and in determining that question, the extent of the benefit to the mill owner, and of the inconvenience or injury to others, may ... very properly be considered."[18] As these cases illustrate, the common law changed in the nineteenth century to accommodate the "public good" of allowing mills to proliferate, but it also changed in ways that undermined its effectiveness for protecting individual property owners against pollution.

The common law proved ineffectual against an especially noxious form of water pollution that became widespread in the second half of the nineteenth century. As running water in houses and other buildings became common, and as the proliferation of flush toilets followed this development, sewage from cities and towns began to pour into the waterways. Since most of this sewage came from publically—i.e., government—owned sewer systems, and since governments in the nineteenth century were immune from lawsuits under the common law, there was no way to regulate the flow of this pollution until legislatures stepped in and began to regulate it via statute law in the latter part of the century. In the twentieth century, it would be statute law that would finally get pollution under control.

There is no reason, in principle, though, why a common law approach to regulating pollution could not be made to work properly. A riparian owner ought to be able to go to court to stop an upstream manufacturer or a municipal sewer system from polluting a river. It is true that the common law of the nineteenth century proved unequal to the task. But no system of law that is grounded on the "will of the people" will long be able to protect the rights of individuals when those rights conflict with popular conceptions of the "public good." [19] In the nineteenth century, the will of the people dictated that the rights of individuals to keep their property free of pollution be sacrificed to the wants of a rapidly industrializing economy. Today the rights of individuals are again being sacrificed for the public good, this time in a process of de-industrialization undertaken in the name of cleaning up pollution.

While a system of property rights in air and water would diminish pollution, it would not necessarily eliminate it altogether, at least not quickly. A manufacturer wanting to dump waste into a river would be free to purchase from the owners of water rights downstream the right to pollute those waters, as was commonly done in the nineteenth century. If those owners agreed to sell the manufacturer a pollution easement, then the manufacturer would be free to pollute the waterway at will. If a government wanted to stop that pollution, it would have to negotiate with the manufacturer to purchase the easement from him.[20] This approach would enable manufacturers to better afford to develop or purchase pollution abatement technology than if he were simply forced by decree to stop polluting as is now done. In addition, manufacturers would be spared the injustice of having to shoulder the entire cost of society's decision to transform what had been riparian industrial zones into the functional equivalent of national parks.

But in today's world, the idea of paying a polluter to stop polluting would be anathema to the environmentalist mind. This is because, in the environmentalist view, to pollute a waterway or an airway is to commit an act this is absolutely, unmitigatedly evil. To pay a polluter to stop polluting would be the moral equivalent of paying a child molester to stop molesting.

Some harms, while bad for man, are an incidental byproduct of an activity that is beneficial. Auto accidents, for example, are a leading cause of accidental death in the United States, but to one extent or another they are an unavoidable concomitant of auto transportation. We could bring an end to auto accidents by abolishing the automobile. Instead, we judge the benefits of the automobile to outweigh the costs as measured in auto accidents, so we try to minimize the frequency of accidents while preserving the auto itself. In contrast to things that are categorically bad for man, auto accidents are an unavoidable concomitant of an automobile society.

Intellectual pollution is another example of an unavoidable concomitant of a beneficial activity. Pornography, "hate speech," and noxious ideologies such as Marxism are forms of intellectual pollution that we could eliminate, at least in their outward expressions, if we were willing to abolish free speech. But because we value free speech more than we would the eradicating of such offensive forms of expression, we choose to tolerate a certain level of intellectual pollution. (The current movement to transform "hate speech" from an unavoidable concomitant of life in a free society into a categorical evil threatens to bring an end to free speech.)

Gun crime presents an interesting case study in this regard. Gun crime is an unavoidable concomitant of the right to bear arms; it is part of the price one pays for living in a free society. Any effort to diminish the incidence of gun crime ought not threaten gun ownership itself. But for some, gun crime is a categorical evil; in their view, society ought to do whatever it takes to eradicate gun crime, even if it means abolishing private ownership of guns altogether.

Then there is pollution. It is a byproduct—to a certain extent unavoidable—of an industrial economy. It is a disvalue, in the sense that, if we could acquire the benefits of industrialism without pollution, we would prefer to do so. But because it is— given our present state of technological development—an unavoidable concomitant of life in an industrial economy, we should prefer to tolerate some level of pollution for the time being. Any effort to diminish pollution levels ought not endanger the existence of industrialism itself.

But for environmentalists pollution is a categorical evil. Consider the following description, from *Time Magazine*, of the 1969 Santa Barbara Oil Spill: "It looked like a massive, inflamed abscess bursting with reddish-brown pus. The huge bubble of oil and natural gas boiling up from beneath the surface ... coated an area of ... incomparable beach front with acrid, tarlike slime." [21] Words like "inflamed abscess" or "reddish-brown pus" are intended to induce anger and disgust—as one might feel toward some heinously evil act. We do not write about auto accidents in such terms, even though they kill tens of thousands in the U.S. each year. The Santa Barbara oil spill killed no one. And it was an accident, of course, not a deliberate act. Yet there is something about the spill that elicits profound outrage and revulsion in the environmentalist mind. That the oil wells were even there in the Santa Barbara Channel to begin with, that they were there for private gain, that they were there to fuel a prodigal consumerism—these all factor into the environmentalists' reaction to the oil spill.

They look upon it as a believer would look upon the intentional desecration of a sacred object, like Jesus upon the money-changers in the temple. For them, the pollution caused by the spill is not an unavoidable concomitant of a beneficial activity, oil production, but a categorical evil. In the environmentalist view, pollution like that caused by the Santa Barbara oil spill must never be accepted as an occupational hazard of life in an industrial economy. Society must do whatever it can to prevent such pollution, even if it means prohibiting new ocean drilling. Pollution is a categorical evil that must be eradicated whatever the cost—even if it means eradicating industrialism itself.

Treating pollution as a categorical evil entails treating polluters as evil, too. Again, *Time Magazine*, in a passage from 1970, illustrates the point nicely. "[Illinois Attorney General William] Scott treats blue-chip polluters as firmly as other prosecutors do the Mafia. His list of defendants facing court action reads like a Who's Who of big business: U.S. Steel, Republic Steel, Mobil Oil, American Zinc and Monsanto, to name a few." [22] Time Magazine approved of treating polluters such as U.S. Steel like the Mafia. Indeed, the criminalizing of pollution further attests to environmentalists' view of pollution as a categorical evil. Gun crime, by way of illustration, involves a deliberate act intended to physically harm another human being. Such acts should indeed be criminal. Pollution, on the other hand, is an incidental byproduct of an activity that is eminently beneficial, the producing of goods for human use. If pollution were properly handled, as a violation of property rights, it would be an offense on the level of trespassing. The criminalizing of pollution is a logical consequence of classifying it as categorically evil.

There once was a line which, just as the First Amendment prevents us from abolishing freedom of speech in our effort to eradicate intellectual pollution, would have prevented us from destroying our productivity and our prosperity in our efforts to eradicate environmental pollution. That line was our constitutional protections of property rights. Now that those are all but gone, our governments are free to choke off the growth of new industry and to destroy our existing industrial base, all in the name of regulating pollution.

The idea of pollution as a categorical evil is a preservationist one. Any activity that produces pollution, such as manufacturing, becomes morally suspect and is thus rendered more vulnerable to having regulatory limitations placed upon it, or even to being banned outright, as many communities have done to the more pollution-prone industries such as oil refineries and chemical plants. The ultimate effect is less and less manufacturing and more and more undeveloped land left in its natural condition. The idea of pollution as a categorical evil, and the preservationist impulse associated with it, are built into the Clean Water Act.

The Federal Water Pollution Control Act Amendments of 1972, also known as the Clean Water Act, are an early statutory expression of the environmentalist ideology. The Act's "Declaration of Goals and Policy" states "The objective of this Act is to restore and maintain the chemical, physical, and biological integrity of the Nation's waters." [23] This is a preservationist goal; its thrust is toward restoring and maintaining the balance of nature within the nation's waterways as opposed to—and, in fact, in opposition to—enabling Americans to make maximum *use* of those waterways for productive purposes. The act states that "it is the national goal that wherever attainable, an interim goal of water quality which provides for the protection and propagation of fish, shellfish, and wildlife and provides for recreation in and on the water (the "fishable and swimmable" standard) be achieved by July 1, 1983." [24]

To illustrate how preservationist these goals are, consider what it would mean if Congress set out to "restore and maintain the chemical, physical, and biological integrity" of Manhattan Island. Imagine if Congress arranged to provide for the "protection and propagation" of terrestrial wildlife—beavers, deer, wolves—in Manhattan. To mandate that the nation's navigable waterways be made "fishable and swimmable" is to require, in effect, that they be transformed from industrial zones, which is what many of them were in 1972, into the functional equivalent of national parks. [25]

Waterways were the birthplace of America's industrial might. Providence, Rhode Island was typical. The Industrial Revolution in the United States began on the Blackstone River in neighboring Pawtucket, where Samuel Slater started the first successful water-powered cotton mill in America in 1793. By 1900 Providence was one of the most productive manufacturing cities in the U.S., and its "Five Industrial Wonders of the World" were each among the largest producers of what they made in the world. [26] By the 1950s, about half of the city's work force was still employed in manufacturing.

But the Providence River was polluted, and it had been covered over by the "World's Widest Bridge." Today, thanks in part to the Clean Water Act, the river is clean, the bridge is gone, and in the evenings floating fires burn on its surface for the amusement of tourists. But the factories are gone too. Only a small fraction of the state's workers are any longer employed in manufacturing. As Rome in its decline relied on foreigners to man its legions, so Rhode Island, and much of the U.S., now rely on foreigners to produce their manufactured goods.

In 2012 the American Federation of Farm Bureaus filed suit against the EPA, accusing the agency of overreach in its proposed comprehensive pollution control plan for the Chesapeake Bay watershed. The plan seeks to limit the discharge of nitrogen, phosphorous, and sediment from farms, urban and suburban run-off, sewage treatment plants, and other sources over an area of 64,000 square miles in six states and the District of Columbia.

The charge of EPA (or state) "overreaching" has been repeated countless times as the EPA or the respective states have tightened limitations on the discharge of pollutants into waterways. Such tightenings have been numerous and will continue. Consider that the first enumerated goal of the Clean Water Act of 1972 is "that the discharge of pollutants into the navigable water be eliminated by 1985." [27] This goal was not achieved in 1985, and it still has not been achieved. And, as the EPA has developed and refined its elaborate system to enforce the Clean Water Act, it has never yet even attempted to achieve the "zero discharge" goal in practice. Indeed, it has yet to achieve the "fishable and swimmable" goal in all of the nation's waters, a goal which itself falls well short of eliminating *all* discharges of pollutants into waterways.

From the beginning the EPA has contented itself with reducing, but not directly eliminating, the discharge of pollutants from the sources it regulates. EPA's basic strategy has been to require polluters to employ EPA-prescribed technologies to limit discharges, in return for which polluters are issued permits to continue polluting. In waters where the level of pollution is so severe that even the best technology will not render them sufficiently clean, EPA can require that a determination be made as to the Total Maximum Daily Load (TMDL) of pollutants that the waterway could accommodate and still become fishable and swimmable, and then require that all dischargers reduce the amounts of their effluents accordingly. In other words, in any waters in which the standard technological approaches are unable to make the water sufficiently clean, the federal government has the authority to institute stricter effluent limitations. It is EPA's planned TMDL regimen for Chesapeake Bay, by far the most extensive ever proposed, that is causing the current controversy.

The plaintiffs in the Chesapeake Bay case are arguing that the EPA lacks the jurisdiction to override state governments' own pollution abatements in the watershed. As far as the preserving of liberty and property are concerned, the question of federal jurisdiction is not unimportant, since, at this point in our history, state regulation of pollution would be preferable to federal regulation.

The plaintiffs are also arguing that the costs of the EPA plan will outweigh the benefits, and that the EPA has not done an adequate job in their cost-benefit analysis. The question of cost-benefit is a hugely important one, from a practical standpoint. Since 1972, the EPA has generally moved toward more cost-benefit analysis in its efforts to limit pollution, rather than toward less. This is a good thing, as far as it goes. In any cases in which the costs, which can be estimated with some accuracy, clearly exceed the potential benefits (which cannot be estimated with any accuracy) by a wide margin, such cost-benefit analyses as are currently conducted can help to defeat an ill-conceived pollution-mitigation plan.

As is the case with all "public" goods, though, all efforts to estimate the value of, say, a pollution-free waterway, are futile. Valuations are performed by individuals, and individuals tend to differ from one another in the values they place on specific things. The most accurate measure of the value of any good or service yet discovered is the price that it commands in a free economy, because such prices reflect countless individual valuations in ways that no government-conducted cost-benefit analysis can ever hope to replicate.

If the water rights in a waterway were privately owned in a free market economy, then the owners would have an economic incentive to use the waterway in such a way as to maximize their profits, that is, in such a way as to command the highest prices from the market of potential users of that waterway (and, indirectly, from the potential users of any products produced by the users of the waterway, such as manufacturers). In a price system, the countless individuals who constitute the market for the waterway's services would "vote" with their dollars to determine how that waterway ought to be used. No more accurate method of cost-benefit analysis has yet been devised by the mind of man.

But environmentalists would strenuously oppose such a method of measuring costs and benefits, because they have already decided which use of the waterway is the "right" one—to preserve it in its natural state and use it primarily for recreation. They would have no desire to put the matter to a vote by the most genuinely democratic means yet devised, the free market.

As to whether, in the Chesapeake Bay case or in any of the pollution-limitation rulings it has handed down, the EPA has "overreached" its mandate in the sense of attempting to reduce pollution levels beyond what it is authorized to do under the Clean Water Act, it will be hard to argue with any limits the EPA places on a given waterway, however rigorous, until that waterway has at least achieved the fishable and swimmable standard. Even then, the goal of "zero discharge" of pollutants into waterways, which was literally the number one goal enunciated by the Clean Water Act in 1972, has the potential for ever more stringent pollution controls until every last mile of every last waterway is available for no wider a range of uses than is currently permitted in America's national parks. The Clean Water Act is a fundamentally preservationist statute.

Consider, for example, that as part of its strategy to restore the aquatic wildlife to America's waterways, the Clean Water Act redefines hot water as "thermal pollution." Many power plants use water from waterways for cooling purposes. When the water is discharged from the plant as hot water, it creates a zone in the waterway that is inhospitable to some fish and aquatic wildlife. Hot water is not pollution, though, and prohibiting its discharge into waterways is not cleaning up pollution. It is, rather, another example of the Clean Water Act's orientation toward diminishing the productive use of waterways and restoring them to their natural state, in this case, for the sake of fish and plant life.

The "anti-degradation policy" of the Clean Water Act is still another example of its preservationism. Under this policy any waterway whose water quality is cleaner than the "fishable and swimmable" standard may not be subjected to activities that would lower the water quality in any lasting way—even to a level that still left the waterway fishable and swimmable—except by special approval.[28] In other words, the preservation of clean waters in their current condition, and the exclusion, if necessary, of industrial use of those waterways is to be the rule, and the use of those waterways for industrial or other productive purposes, if they will cause even small amounts of pollution, is to be the exception.

Then there is the Clean Water Act's treatment of run-off from rainstorms and melting snow. Environmentalists contend that run-off is a significant enough source of waterway pollution that, unless it is regulated, the fishable and swimmable goal will not be attainable on all of America's waterways. Accordingly, in the 1987 Water Quality Act, Congress required the EPA to begin to regulate run-off from industrial sites, new construction sites of five acres or more, and the storm drains of municipalities of one hundred thousand or more persons. The EPA responded by instituting a permit system similar to that already in place to regulate industrial and municipal point sources.

According to environmentalists, modern urban and suburban development in the U.S. both causes and exacerbates the problem of run-off pollution. Of course, the very process of living and functioning in an American city or suburb causes all kinds of chemicals and other pollutants to end up on the ground. Environmentalist argue that by increasing the amount of run-off, through the construction of roadways, parking lots, and myriad other non-permeable artificial surfaces, modern cities and suburbs increase the amount of pollution finding its way into waterways. The general strategy of the environmentalists, therefore, is to minimize run-off as much as possible by minimizing the amount of undeveloped land that can be subjected to new growth and development. The strategy, in other words, is the *preserve* as much undeveloped land in its natural state as possible.

This strategy is nicely described in a pamphlet put out by the EPA titled "Using Smart Growth Techniques as Stormwater Best Management Practices." (Stormwater permits from the EPA are conditioned upon the adoption by applicants of so-called best management practices in dealing with run-off. By including smart growth techniques among "best management practices," EPA promotes their adoption by permit applicants.) The pamphlet adduces a number of "smart growth" principles. The first principle is "compact project and community design." The idea here is to squeeze more homes and businesses onto less land. In this way, some land that would have been devoted to new development can be left in its natural state. Just as environmentalists advocate the eventual abolition of the automobile in favor of mass transit, so the logic of "smart growth" would, in the name of controlling run-off, move us toward the eventual abolition of the single-family home in favor of "mass housing."

It is ironic that, in promoting mass housing, environmentalist central planners are aiming to counteract what an earlier generation of central planners helped bring into being, so-called "suburban sprawl." Before the advent of the Progressive practice of zoning, as the population of a city grew and the land at the center of the city increased in value, the density of development of that land tended also to increase. Manhattan Island, for example, which once accommodated only a few persons per square mile, by growing upward was made to accommodate over 70,000 persons per square mile.

But when zoning placed limits on the density of development in newer cities in the west, those cities were forced to grow outward rather than upward. A later generation of planners sought to halt sprawl by prohibiting new development in the suburbs, thus bringing into being the 1300 square foot, 3-bedroom, 2-bath house on land so valuable, in places like Palo Alto, California, that it commands a price of $1,000,000 or more. Now planners will "correct" their earlier mistakes by allowing some new, high-density development, but only in areas that are already developed. In the more enlightened regions of the west, people are, in effect, being herded onto reservations.

Another smart growth principle, the mixing of land uses, also reverses generations of central planning orthodoxy. By allowing people to live, work, and shop in the same neighborhood, smart growth communities require fewer run-off producing roads and parking lots. But, of course, mixing uses is precisely what cities did before central planners began to segregate homes, workplaces, and commercial land uses through zoning.

Still another principle calls for land-use planning on a regional or watershed-wide scale. Land use planning in California is already done on a county-wide scale, but efforts are well under way to coordinate some planning on a regional scale. The call for regional, and, especially, watershed-wide planning is important from a nature preservationist perspective because it enables the authorities to take into account the needs of entire "ecosystems" in their planning, something that preservationists have long aspired to. This level of central planning places the decision-making power over land use ever further from the individual landowner, just as the consolidation of public schools into giant, unified school districts has placed control of public schools further than ever from parents.

Other ideas include narrower streets, streets with sidewalks on only one side, and—an idea which is sure to appeal to homeowners already beset by ever-increasing utility bills—the discouraging of run-off from individual homes by charging homeowners for the privilege of allowing the rainwater to leave their property. The idea is to set up stormwater utility companies, who would charge a fee to their homeowner "customers," just as many municipalities now charge for sewage services. In general, the regulation of run-off vastly expands government's power to regulate land use, and to preserve land outright, all in the name of reducing water pollution.

John Rapanos was a developer who owned acreage in Midland, Michigan on which he planned to build a shopping mall. In 1987, he moved some dirt onto a portion of his land. The U.S. Army Corps of Engineers charged him with having filled in a federal wetland without the permit he was required to obtain by Section 404 of the Clean Water Act. Historically, federal jurisdiction of waterways had been limited to the "navigable waters" of the United States, which is the term used in Section 404. This was essentially the Corps of Engineers' own interpretation of the limits of their jurisdiction until the EPA forced them in federal court in 2001 to accept a much broader definition of "navigable waters," one which included "intermittent streams, mud flats, sand flats, wetlands, sloughs, prairie potholes, wet meadows, playa lakes, or natural ponds." [29]

It was this broader interpretation of "navigable waters" that Rapanos, whose land was over ten miles from a truly navigable waterway, challenged. His case reached the U.S. Supreme Court, who in a 4-1-4 plurality ruled against the EPA, although the Court's decision did not settle the question of what should constitute a wetland under the Clean Water Act. (Despite their "victory," the Rapanoses entered into a settlement with the EPA, under which they agreed to pay a $150,000 fine and spend $750,000 to restore the "wetland" they had disturbed.)

In the EPA's broad definition of the term, wetlands form, with lakes and streams and truly navigable waterways, an ecological whole. Wetlands mitigate flooding by absorbing flood waters; they enable water to seep into the ground, from whence it reemerges into stream beds during times of low flow; in places they work to stabilize shorelines against erosion; and they provide habitat for fish and other wildlife. Wetlands also filter pollution out of the water, and for this reason it is perfectly logical, from a clean water perspective, for the EPA to call for preserving them. But the practical effect of the EPA's broad definition of what constitutes a federal wetland would be to preserve millions upon millions of acres of swamp and marshland (and prairie potholes), from all human use except recreation. The EPA's ecologically-inspired attempt to expand its wetland purview is another expression of the essentially preservationist nature of the Clean Water Act.

A great many Americans today consider it entirely right and proper for the government to preserve wetlands, however defined. They could hardly imagine a world in which the government did otherwise. To suggest that an owner of private swampland should be free to fill it in would be morally equivalent, in their eyes, to advocating a return to slavery in the United States. That this moral view is so widely held today suggests that environmentalists have succeeded in getting their ideas enacted into law not only because a radical few have wielded disproportionate influence over the legislative and regulatory process, but also because environmentalists have succeeded in winning widespread support for their—radical—ideas. For their part, the defenders of private property have limited themselves to arguing about what constitutes a wetland (not that the definition of wetlands is unimportant), and they have completely conceded the larger, more radical premise—that the government ought to be in the business of preserving any wetlands at all, public or private, navigable or not.

The Clean Air Act

The Clean Air Act Amendments of 1970 were the first important federal statute to regulate air pollution in the environmental era. From the beginning, an important part of this statute has been the National Ambient Air Quality Standards (NAAQS). These dictate allowable levels of six air pollutants throughout the U.S., carbon monoxide, sulfur dioxide, nitrogen dioxide, lead, ground-level ozone, and particulate matter.

Section 109 of the Clean Air Act states that the NAAQS "shall be ambient air quality standards the attainment of which in the judgment of the Administrator [of the EPA] ... are requisite to protect the public health." [30] The Act also authorizes the establishing of "secondary" standards to protect the "public welfare." The EPA has long interpreted Section 109 to mean that in their setting of air quality standards their only object should be to protect the public health and welfare and that, therefore, the cost of Americans' compliance with those standards ought to be no concern of the EPA. In *Whitman v. American Trucking Associations*, which was decided in 2001, the U.S. Supreme Court ruled that, based on the language of the statute, it was indeed the intention of Congress that the EPA not take into account the costs of compliance when setting air quality standards. Congress has never acted to contradict this interpretation of the NAAQS.[31]

If the EPA's goal is to protect human health and welfare, and if they need not—indeed, may not—take into account the costs of complying with their NAAQS standards, then the EPA has every incentive to err on the side of caution and progressively to tighten standards until pollution is reduced to such a miniscule level that it can not possibly pose a threat to human health or the public welfare. The EPA is free to pursue this course, indeed is compelled to pursue it, even if it wipes out vast swaths of American industry and severely diminishes American economic growth. Such an approach is precisely what the EPA would take if it were treating air pollution as a categorical evil, rather than as an unavoidable concomitant of an industrial economy. The cost-is-no-object principle at the heart of the NAAQS standard setting is, in effect, the pollution-as-categorical-evil idea enacted into law.

Has the EPA, in fact, been tightening pollution standards so as to bring pollution levels down close to natural background levels, and have they, in the process, diminished American industrial growth? To answer the first question, consider the National Ambient Air Quality Standard for particulate matter. "Particulate matter" refers to small solid and liquid particles that float about in the air. These come from dust, soot, and water vapor, among other things. Much particulate matter, such as wind-blown dust and sea spray, is natural in origin. Among human-caused particulate matter, soot from fossil fuel burning is an important source.

As I have mentioned, the Clean Air Act authorizes the EPA to establish two sets of National Ambient Air Quality Standards. The primary standards are intended to protect human health, and the secondary ones to protect the "public welfare," which includes "protection against decreased visibility and damage to animals, crops, vegetation, and buildings." Note that if the EPA determines that any form of air pollution covered by the NAAQS poses a threat to nature, i.e., to animals and plants, then they are required to establish and enforce standards so as to bring that pollution down to "safe" levels.

There is no reason, in principle, why the health of plants and animals might not, in the estimation of the EPA, require more strict regulation of a certain pollutant than the health of human beings would require. In such a case, the secondary standard would be more strict than the primary one, as was indeed the case when the first particulate matter standards were established in 1971.[32] The Clean Air Act is designed to restrict human productive endeavor whenever necessary in order to protect nature (or crops, buildings, or visibility). This is the preservationist premise at work.[33]

In 1987 the EPA revised the particulate matter standard. From this time onward, the primary and secondary standards for the six NAAQS pollutants have been identical, with the exception of those for sulphur dioxide. This places the requirements of plants and animals *on an equal footing with* the requirements of human health.

In 1997 the EPA broke the particulate matter standard into two parts, PM_{10} to be applied to particles between 2.5 microns and 10 microns in diameter, and $PM_{2.5}$ to particles up to 2.5 microns in diameter. (A human hair is about 100 microns thick.) The annual average for $PM_{2.5}$ was set at 15 micrograms per cubic meter. In 2012, that standard was reduced to 12 micrograms per cubic meter.[34] For purposes of comparison, consider that at Point Reyes National Seashore north of San Francisco, where clean air blows in off the Pacific, one-hour average $PM_{2.5}$ concentrations commonly register above 20 micrograms per cubic meter.[35] Consider also that annual indoor concentrations of $PM_{2.5}$ can run

Thomas McCaffrey

well in excess of 15 micrograms per cubic meter. Such activities as cooking, showering, and cleaning can dramatically increase PM levels indoors. A study by Johns Hopkins that measured average annual $PM_{2.5}$ levels in children's bedrooms in Baltimore found that "Eighty-five percent of homes had [annual] $PM_{2.5}$ concentrations that exceeded [the EPA annual outdoor limit of 15 micrograms per cubic meter]," registering in many cases more than twice the EPA limit.[36]

Dr. Robert F. Phalen has written that, in epidemiological studies worldwide, associations have been found between very low particulate matter levels and human illness and death.[37] But "the associations linking particles to adverse effects imply relative risks that are small." In fact, the magnitude of the increase in illness and death associated with incremental increases in particulate matter levels was found to be so small as to call into question whether there existed any causal relationship between the one and the other.[38] "[P]erhaps only one in several hundred thousand individuals is made ill by [incremental increases] in [particulate] pollutant levels. With respect to the excess mortality produced by a single small episode [of $PM_{2.5}$ pollution], about one in a million might be a victim."[39]

The associations between increased levels of particulate air pollution and increased levels of sickness and death have been identified through the science of epidemiology, which studies patterns in large populations and can identify the possibility that a causal relationship exists between particulate pollution and sickness or death. The science of toxicology, on the other hand, *certifies* the existence of such a causal relationship. It can do so by conducting experiments to see if particulate air pollution induces illness.

"To date, toxicologists have not been able to replicate the low-dose PM findings seen in epidemiologic studies," writes Dr. Phalen.[40] They have not been able to prove that PM concentrations, in and of themselves, at the low dosages that the EPA has prohibited, are harmful to human health. "*Epidemiological associations are a starting point* in the process of identifying potentially harmful concentrations of agents.... *But the culprits that may be causing adverse health effects at very low concentrations in the air have not been identified.*"[41] Referring to the document on which the EPA based its tightening of particulate matter standards in 1997, Dr. Phalen wrote that the EPA's Clean Air Scientific Advisory Committee "had difficulty in achieving 'closure.'" The committee noted that the "*understanding of the health effects of PM is far from complete.*" [42]

"When scientific controversy surrounds a finding, or even a large set of findings, then the findings are probably not ready for automatic assimilation into public decision-making processes," writes Dr. Phalen.[43] So, scientists' understanding of the adverse health effects of $PM_{2.5}$ was not sufficient to justify the EPA's lowering of the $PM_{2.5}$ standard in 1997 to 15 micrograms per cubic (not to mention its further lowering, to 12 micrograms per cubic meter, in 2012). "The optimal levels of control and the timing of instituting additional controls of particulate air pollution in the interest of protecting human health are unknown." [44] Indeed, it is possible that setting the standard this low was *contrary* to the requirements of human health. "[T]he assumption that modern industry is harming public health and must therefore be forced to comply with ever more stringent regulation is subject to challenge. Modern industrial goods and services are, in fact, major factors in *protecting* public health and providing prosperity." [45]

In addition to the annual standard for $PM_{2.5}$ concentrations, the one that was lowered in 2012 from 15 to 12 micrograms per cubic meter, the EPA also enforces a 24-hour standard in order to police short-term elevations in particulate matter. The legal limit for 24-hour average $PM_{2.5}$ concentrations that is imposed on the United States is now 35 micrograms per cubic meter.[46] But, in fact, the EPA deems concentrations between 15 and 40 micrograms per cubic meter to be harmful only to "unusually sensitive" persons, and it deems concentrations between 40 and 65 micrograms per cubic meter to be harmful only to "sensitive" persons.[47] Not until concentrations exceed 65 micrograms per cubic meter does the EPA consider them harmful to the general population.

The amount that it costs Americans to limit 24-hour average $PM_{2.5}$ concentrations to 35 is enormous. (The EPA has estimated the total cost of Americans' complying with the Clean Air Act in 2010 to be $53 billion.[48] These costs do not include the crippling of U.S. industrial might, which I discuss in chapter 12.) In sounding the alarm about the dangers of indoor air pollution, the EPA has said that Americans spend 90 per cent of their time indoors. Surely, if protecting human health were the EPA's (and environmentalists') motivation in continuing to lower $PM_{2.5}$ concentrations, they could achieve far more good at far less cost and with far less collateral damage to industry by raising the outdoor limit from 35 to 65, and then supplying indoor air filtration systems to "sensitive" persons.

In a free society, the cleanliness of the air would be determined by the "democratic" workings of the market.[49] That is, it would be determined by the buying and selling (and litigating) choices of countless persons, acting individually and in their own best interest as they see it. It is not the business of the government to decide how clean the air should be. But if the government is going to take that decision upon itself, then it should not use the health of the weakest in society, of asthmatics and the elderly, as the standard. Imagine if the government passed a law requiring that all published material be written so as to be intelligible to the mentally deficient. Such a law would have a fatally debilitating effect on the intellectual vitality of society. To subject our industrial economy to an analogous standard for the sake of asthmatics and the elderly is to subject it to an equally debilitating standard, and the results, as I will show later, have indeed been economically debilitating.

The evidence does not support the contention that the annual $PM_{2.5}$ standard of 12 micrograms per cubic meter, with all the collateral damage it has caused to American industry, has been necessary to protect human health. Consider, then, more recent developments in $PM_{2.5}$ regulation.

"So if somebody wants to build a coal-powered plant, they can. It's just that it will bankrupt them because they're going to be charged a huge sum for all that greenhouse gas that's being emitted." These are the words of presidential candidate Barack Obama to the San Francisco Chronicle in January, 2008. The effort to render coal-fired electric power plants legally obsolete found expression in the American Clean Energy and Security Act of 2009. This bill narrowly won passage in the U.S. House of Representatives. When it became clear that "cap and trade" had no hope of passage in the Senate, President Obama said "cap and trade was just one way of skinning the cat." [50]

In 2011 the EPA instituted the Cross-State Air Pollution Rule. This rule regulates sulfur dioxide and nitrogen oxides emissions that, according to the EPA, cross state lines and contribute to $PM_{2.5}$ pollution.[51] Also in 2011 the EPA instituted the Mercury and Air Toxics for Power Plants rule (also known as the "Utility MACT"). Although nominally put into place to regulate mercury, the EPA's own Regulatory Impact Analysis shows that reductions in $PM_{2.5}$ concentrations will constitute 99% of the benefits from this regulation.[52] In other words, both measures were aimed at producers of $PM_{2.5}$, that is, coal-fired electric power plants. According to the Institute for Energy Research in a paper published in October of 2011 and updated in June of 2012, 34.705 gigawatts of coal-generated electric power generating capacity, or almost one tenth of all coal-powered generating capacity in the U.S., would have to be closed down because of these EPA rules.[53]

If the requirements of human health did not justify the EPA's tightening of the $PM_{2.5}$ standard to 15 in 1997, then they certainly did not justify these additional regulatory actions of 2011 and 2012. But they have served another purpose admirably well; they have helped to slow the rate of industrial and economic growth in the United States, and thereby to preserve America's remaining natural landscape from development.

The Clean Air Act is overtly preservationist in another important way. Section 101 (b) of the Act says that one of its purposes is "to protect and enhance the quality of the Nation's air resources so as to promote the public health and welfare and the productive capacity of its population." [54] In interpreting the meaning of this clause, the U.S. District Court of the District of Columbia stated "it is our judgment that the Clean Air Act of 1970 is based in important part on a policy of non-degradation of existing clean air." [55] In support of this interpretation, the court cited the Senate Report that accompanied the bill that became the Clean Air Act in 1970. "In areas where current air pollution levels are already equal to or better than the air quality goals, the Secretary shall not approve any implementation plan which does not provide, to the maximum extent practicable, for the continued maintenance of such ambient air quality." [56]

In other words, according to the district court, in a decision that was affirmed by the U.S. Supreme Court in *Fri v. Sierra Club*, the Clean Air Act prohibits "significant" new air pollution in areas where there is little or no air pollution, even if the added pollution would not increase the pollution level to a point in violation of the EPA standards. This principle of "non-degradation" was explicitly and unambiguously incorporated into the Clean Air Act in the form of the Prevention of Significant

Deterioration section of the New Source Review Program (a part of the NAAQS) in 1977.

Prevention of Significant Deterioration is essentially a preservationist policy because, given our current technology, some degree of air pollution is unavoidable in an industrial economy. To prevent "significant" increases in pollution in areas where the air is clean even if such increases would not violate EPA standards is, in effect, to severely restrict the development of new manufacturing in those areas. Note, also, that the purpose of Prevention of Significant Deterioration is *not* the protection of human health. If the EPA's primary pollution standards, such as the 24-hour $PM_{2.5}$ standard of 35 micrograms per cubic meter, represent levels that are "safe" for humans, then to allow 24-hour $PM_{2.5}$ concentrations to increase from, say, 20 to 30 micrograms per cubic meter would not, by the EPA's own estimation, pose a threat to human health. No, the purpose of Prevention of Significant Deterioration is explicitly to *preserve* clean air where it already exists—and to prevent the growth of new industry in such areas to whatever extent is necessary to that end.

The War on the Auto

"In announcing a sweeping new series of antipollution regulations last week, the Environmental Protection Agency outlined a fundamental, even traumatic change in an American culture that has grown deeply—and as the EPA believes, dangerously—dependent upon the automobile." [57] So said *Time Magazine* in June of 1973.

The automobile harms the earth, in the environmentalist view, in more ways that just about any other invention of the industrial era. The manufacture of automobiles requires huge quantities of metals and other natural resources; the mining and transport of these raw materials entails vast alterations of the natural landscape worldwide, and they generate a great deal of pollution, as does the manufacture of automobiles, especially as it was carried on before 1970. The automobile itself uses great quantities of fossil fuels, a precious resource that environmentalists consider to be in critically short supply. It also generates much pollution of its own, not to mention "greenhouse gasses."

Automobiles also require millions of acres of roadways and parking lots, which further derange the natural landscape and its systems. They enable people to live far from the cities in which they work, thus giving rise to suburban "sprawl," which increases the extent of the earth's natural landscape that is altered by human development. And when automobiles have outlived their usefulness and must be disposed of, they become another form of pollution. Finally—and this is not an exhaustive list of its alleged shortcomings—the automobile enables people to get out of the cities and into the backcountry, into places that environmentalists would prefer be left as free of human influence as possible.

A plethora of government policies have conspired, directly or indirectly, intentionally or unintentionally, to diminish the automobile and hurry it along toward extinction. These have included efforts to restrict the mining of the raw materials automobiles are made of, and efforts to limit drilling for the oil needed to fuel the auto. Both of these have been done, at various times and in various places, in the name of preventing pollution, conserving natural resources, or preserving land or wildlife. Anti-auto policies have included the government's requiring that auto manufacturers increase the gas mileage of their autos, again in the name of conserving natural resources, and more recently in the name of diminishing the production of green house gases. They have included efforts to restrict the building of new roadways, as well as related efforts to siphon off public funds that would have been devoted to the maintenance and construction of roads, to devote these instead to the construction of "mass transit."

Finally—and importantly—they have included the effort to eradicate the air pollution caused by automobile tailpipe emissions.

Before the passage of the Clean Air Act Amendments of 1970, Senator Gaylord Nelson had called for the abolition of the internal combustion engine by 1978 unless a way could be found to reduce its emissions to near zero. A more perfect expression of the pollution-as-categorical-evil idea could hardly be imagined. In its final form, the Clean Air Act of 1970 called for 90 percent reductions in automobile emissions of hydrocarbons and carbon monoxide in five years, and the same reduction in nitrogen oxides in six years. At the time this edict was issued, the technology did not exist that could reduce these emissions by anything close to the required amounts. In effect, the Clean Air Act had incorporated Senator Nelson's idea of "technology forcing" and, by implication, the idea of auto emissions as a categorical evil. The act implied that the problem of auto emissions was so severe that either the auto industry must invent a radical new technology (which no one was sure *could* be invented) or the industry would be extinguished.

"This bill," said Senator Edmund Muskie of the Clean Air Act Amendments of 1970, which he largely authored, "presents possibly the last chance to head off the disaster that air pollution could bring. Smog alerts could turn into death watches. A wave of public reaction could bring crisis legislation with *federal control over industry decisions—even nationalization*—things nobody wants." [58] In the event, following the passage of the Clean Air Act of 1970, the federal government granted several extensions to the auto manufacturers to meet the 90 percent reduction goal. It was not until 1981 that Detroit was finally producing new cars that met the EPA's emissions standards, and then only because the auto companies were finally persuaded to take the government's threat to put them out of business seriously.

Thomas McCaffrey

Despite the 90 percent reductions that had been achieved by 1981, the EPA continued to tighten tailpipe emission standards, in 1990 and again in 1999. In 2012, the California Air Resources Board adopted the "Advanced Clean Car" regulatory package. Part of the California program is called the LEV III Standards. Among other things, these would reduce tailpipe emissions of non-methane organic gasses and of nitrous oxides by 75 percent of what they were in 2012, and they would reduce emissions of particulate matter to .001 grams per mile, a level that was barely detectable with 2012 technology, by 2028. These LEV III standards were developed to be consistent with new national standards proposed by the EPA.

Another part of the California program requires that 1.4 million of the autos sold in California by the year 2025, or 15.4 percent of the total projected to be sold that year, be "zero emissions vehicles," i.e., cars powered otherwise than by a gasoline powered engine. Since a fully practical car of this nature had yet to be invented, this was another example of "technology forcing." The Board expects that 87 percent of vehicles sold in 2050 will need to be zero emissions vehicles in order to reach the state's goal of reducing greenhouse gas emissions to 80 percent below what they were in 1990.[59] California seems determined to realize both sides of the alternative Senator Nelson posed in the early '70s, to reduce emissions from the internal combustion engine to near zero *and* to phase it out (almost) completely.

The effect of all these anti-auto policies, including the EPA's emissions control policy, is a *preservationist* one, i.e., it tends to minimize human alteration and use of the natural landscape and its resources, in favor of preserving them. Through technology forcing and emissions regulation, the Clean Air Act has given the EPA the power to regulate the automobile as we know it out of existence. That the EPA has not yet done so is due to the ingenuity scientists and engineers acting in the free market, and to the continuing preference of Americans for private, gasoline-powered cars over mass transit.

But was it necessary to give the government this much power over the automobile? How would the auto emissions problem have been solved in a free market economy? As I have said, in a free economy pollution would be treated as a kind of property invasion akin to trespass. Under such a system, if a factory was spewing heavy smoke into my backyard, I could take the owner to court and force him to stop invading my property. All I would need to show would be that the smoke came from his factory, and that it entered my property in sufficient quantity either to threaten my health or to interfere with my use and enjoyment of my property.

But auto emissions would not be subject to court action in quite the same way that heavy smoke would be. Smoke is visible and, thus, immediately detectable. It constitutes a property invasion that a court could enjoin or for which it could order damages to be paid. But minor invasions, such as noise from children playing or from a power lawnmower, are invasions that courts overlook, as they should. Emissions from a passing auto would fall into this category. In most cases, such emissions would not even be detectable without electronic equipment. And no one motorist is likely to emit enough matter to harm the health of any given individual, even over a period of years. On the other hand, if heavy emissions were produced by the users of a privately owned roadway, then the owner of the roadway could be held liable in court. This is how a genuinely free market society would control auto emissions.

Governments that operate public roadways could, of course, similarly be held liable in court for auto emissions emanating from their roadways. But governments-as-owners tend to respond more to political forces than to market forces, so the long-range result of applying the common law approach to emissions from public roadways might be similar to the result that the police power regulatory approach is producing, which is the forced extinction of the automobile. (The closing of the national forests to productive use, which I describe in Chapter 10, exemplifies this tendency of government "owners" to respond more to political than to market forces.)

Even if we in the United States took no overt political action to curtail pollution from automobile emissions, the free market would in all likelihood eventually develop a technological solution to the problem. It was the free market that introduced the automobile as a replacement for the vastly more polluting horse. It would be unthinkable to try to run a city the size of New York on a system of horse transport—the volume animal waste would be overwhelming. It was the free market that replaced home heating via the wood stove and the coal furnace with vastly cleaner oil, gas, and electric-powered heat. Eventually, the free market will invent something better and cleaner than the gasoline-powered engine. In fact, the problem of auto emissions has already largely been solved, and it was not the government that produced that technology; it was the free minds of scientists and engineers employed by auto manufacturers that invented the technology to curtail auto emissions to a fraction of what they were in 1970. All government did was hurry the process along—at a cost.

Environmentalists will argue that if the government had not forced the auto manufacturers to develop new emissions technology, it would never have come into being. This thinking reflects the "tragedy of the commons" idea, that wherever there is a commons, such as the air above America's cities, into which polluters may dump their refuse—in this case, their auto emissions—they will continue to do so until they are forced to stop. But this is not necessarily true. How people treat the commons has much to do with attitudes about the commons.

It is highly probable that, had property rights in America's waterways been properly enforced throughout the nineteenth and twentieth centuries, they would not have become nearly as polluted as they did by the 1950s. But, as important as the non-enforcement of property rights was, it was also important that the vast majority of Americans simply did not care that their rivers were polluted. When people do care about the condition of their commons, they tend to take much better care of them than when they do not. As one drives the public roadways of the U.S., one will see a certain amount of litter alongside them. Some people litter. But the vast majority of Americans do not litter, and the reason they do not is not because they fear getting caught by the police. Most people refrain from littering because they care about keeping the roadsides clean and attractive. People will act voluntarily to care for the commons if the commons are important to them.

The reason why it took the threat of government force to cause auto manufacturers to invent the technology to control auto emissions in the 1970s was not that the manufacturers were greedy, but simply that the consumer demand for such technology did not yet exist. Most Americans did not care enough about automobile smog to do anything about it. (It was not just the producers of automobiles who resented and resisted the government's imposition of this technology upon them; consumers were equally resentful, both of the cost which the new technology added to the price of a car, and of the diminished gas mileage it caused.)

But in the more than four decades since the passage of the Clean Air Act, the environmentalist movement succeeded in creating a great number of environmentally-aware auto buyers who would have been willing to purchase a less-polluting vehicle if it could have been produced and sold at a price that was competitive with the older, polluting models. Manufacturers would have had every incentive to take advantage of this customer demand by developing pollution abatement technology. Such things happen all the time in a capitalist economy; it does not matter why the customers are demanding a new product, whether to reduce pollution or simply to keep up with some new fashion. It only matters that the demand exists in sufficient quantity and intensity to make it worthwhile for producers to invest in satisfying it. (Think Toyota Prius.)

If it had been left to the free market to develop the technology to reduce auto emissions, then it would have taken longer, perhaps significantly longer, than it did, to clear up the smog over America's cities. It is also possible, (though unlikely, considering the widespread support for cleaning up pollution that exists today), that the environmental movement would not have persuaded enough people of the importance of reducing auto emissions to make it worthwhile for the auto manufacturers to develop and market abatement technology. Had this been the case, then emissions would have continued unabated. But the question would have been decided in the most "democratic" way possible, by the actions of countless individuals voting with their dollars in the free market. Either outcome, though, a slow, gradual abatement or none at all, would have been unacceptable to environmentalists, because they view auto emissions as a categorical evil that needs to be eradicated as quickly and as thoroughly as possible, even if it would require the extinction of the automobile itself.

But what about the threat to human health posed by auto emissions? Is this not reason enough to want to eradicate them, categorical evil or not? Environmentalists say that the government must intervene to reduce auto emissions, which is the same as saying that either the government must force private individuals to spend their wealth on reducing auto emissions, or people will get sick or die. But how else would those private individuals have spent their wealth than on keeping themselves and their families healthy and alive, or on enhancing their lives in other ways? It simply is not possible to prove that "society" is safer or otherwise better off, on balance, because the government has dictated that individuals spend their wealth on reducing auto emissions than if those same individuals had been free to decide for themselves how best to spend their wealth.

Consider the car buyer who can purchase a new-style, low-emission car for an extra $500 or buy an old-style high-emission car and spend the $500 he would save to purchase air bags to protect himself and his family. He must weigh the risk that someone somewhere (perhaps, however unlikely, he himself or a member of his family) will become ill or die from exposure to his auto emissions—against the added safety for himself and his family that the air bags would provide. Is society "safer" if he chooses the low-emissions car instead of the air bags. Is it "right" for him to choose the low-emissions car? Of course not. The question involves a value judgment, the kind that each individual must make—and ought to be free to make—for himself.

But it is implicit in the environmentalist view that to choose the airbags would be to choose incorrectly. If the vast majority of car buyers in a free economy were to choose air bags over the low-emissions car, then we would have what environmentalists call a "tragedy of the commons," or a case of "market failure."

The environmentalist has science on his side, so he says—*and so even his opponents concede*. Science indicates that if the concentration of auto emissions were to reach a certain level, then a certain number of people would become sick or die. Clearly, therefore, science dictates that government must intervene in the private economy to reduce auto emissions. But what about the people who will be injured or die because, in order to pay for reducing auto emissions, they had to buy smaller, less safe cars, or they had to forego a home security system, or skip a few visits to the doctor for their children?

Yes, the Clean Air Act has reduced auto emissions to manageable levels; the benefit is plain for all to see. But what of the costs to private individuals? There simply is no way to identify all the safety and security they have had to give up in order to realize the environmentalists' dream of smog-free cities. Environmentalism has succeeded—magnificently—in framing as a purely scientific matter what is in fact a value judgment as to how private individuals should spend their wealth (and it treats as self-evident the related question of who gets to decide how individuals should spend their wealth, they themselves or their governments).[60]

Of course, if individuals had been free to spend their wealth as they chose, rather than being forced to spend it on reducing auto emissions, not everyone would have spent it on enhancing their safety and security; some would have bought TV sets, or beer. So the environmentalists can claim that the wealth was "better" spent reducing smog than if it had been left in the hands of the people.

But a great fallacy of the modern leftist view is that human safety trumps all; if a certain course of action, such as wearing a seat belt while driving, can be shown to enhance a person's safety, then morality requires that one wear a seat belt, and it becomes incumbent on government to force people to act accordingly. This applies *a fortiori* when the safety of children or the infirm is at stake. If one wants to remain a member of polite society, ones simply does not question the government's right to force people to ensure the safety of the children.[61]

In choosing the preserving of human health as their primary rationale for regulating auto emissions, the authors of the Clean Air Act have helped to promote, and in turn have benefited from, the new tyranny-in-the-name-of-safety. It is the new Puritanism, in which government forces individuals to save their bodies, just as under the old Puritanism government forced individuals to save their souls. In a free society, each individual would be free to spend his wealth as he chose, whether on reducing auto emissions for the public good, or on air bags for his family's good, or on beer for his own good. The new Puritanism will prove no less tyrannical than the old.

Notes

1. Restoring the Quality of Our Environment: Report of the Pollution Panel, President's Science Advisory Committee November, 1965, 5.
2. Ibid., 21.
3. Ibid., 17.
4. Ibid., 21,"We recommend that funds be made available to acquire title, either directly or through the states, to important coastal marshes, lagoons, and estuaries which could then serve incidentally, as national and state parks, wildlife refuges and public recreational areas."
5. Barry Commoner, *The Closing Circle* (Toronto, New York, London: Bantam, 1972), 215.
6. Ibid., 144.
7. Ibid., 119.
8. Ibid., 77. Emphasis added.
9. Commoner's recommendations for government control of the development and use of energy are pure socialism. "On ecological grounds it is obvious that we cannot afford unrestrained growth of power production. Its use must be closely governed by over-all social needs rather than by the private interests of the producers or users of power. This means that the allocation of power to a given productive activity, in turn, would need to be governed by a judgment of the expected social values to be derived" therefrom.
10. Ibid., 212.

11. Ibid., 37.
12. Morton Horwitz, *The Transformation of American Law*, 1780-1860 (Cambridge, MA and London, England: Harvard University Press, 1977), 39.
13. Ibid., 19.
14. Palmer v. Mulligan, 3 Cai. R. 307, 314 (1805). Quoted in Horwitz, Transformation, 37.
15. Horwitz, Transformation, 39.
16. Ibid., 40.
17. Snow v. Parsons, 28 Vermont, 459, 67 AM. Dec. 723 (1856, 725). Quoted in Juoni Paavolo, "Water Quality as Property: Industrial Water Pollution and Common Law in Nineteenth Century United States" in Environment and History 8, 3 (August, 2002): 304.
18. Hayes v. Waldron, 44 N.H. 580, 84 Am. Dec. 105 (1863, 108). Quoted in Paavolo, "Water Quality," 304-305.
19. Properly understood, the public good would be synonymous with securing the rights of individuals.
20. This is not to concede that government ought to be in the business of owning or controlling the use of waterways.
21. "Environment: Tragedy in Oil," Time Magazine (February 14, 1969): 23.
22. "Environment: Prosecuting Pollution," Time Magazine (January 5, 1970): 41.
23. 33 USC § 101 (a).
24. 33 USC § 101 (a)(2) The phrase "wherever attainable" refers to the deadline of July 1, 1983 and not to the ultimate feasibility of rendering all of the nation's navigable waterways "fishable and swimmable" by some date.
25. Many waterways in the U.S. still permit commercial traffic, whereas the national parks do not.
26. Brown and Sharpe (tools), Nicholson File, Corliss Steam Engine Company, American Screw, and Gorham (silver).

27. USC § 101 (a)(1).

28. "Water Quality Standards Handbook," U.S. Environmental Protection Agency, published online August 15, 1994, accessed October 10, 2014,

http://water.epa.gov/scitech/swguidance/standards/handbook/chapter04.cfm, chap. 4.

29. Solid Waste Agency of Northern Cook County (SWANCC) v. U.S. Army Corps of Engineers, 531 U.S. 159 (2001).

30. Sec 109(b)(1)(a).

31. The EPA does take costs of compliance into account in other sections of the Clean Air Act.

32. The primary 24-hour standard for total particulate matter was set at 260 micrograms per cubic meter, and the secondary standard was set at a much more strict 150 micrograms per cubic meter. As a result, the effective limit on 24-hour total PM concentrations was set at 150, even though, according to the EPA, humans could tolerate concentrations up to 260 micrograms per cubic meter without harm. (When the PM standards were originally established in 1971, the primary and secondary annual standards were set at the same level, 75 micrograms per cubic meter; it was only the 24-hour standards that differed from primary to secondary.)

33. Man has a moral obligation to refrain, as much as possible, from upsetting the balance of nature.

34. Averaged over three years. The secondary standard remained at 15 micrograms per cubic meter.

35. At Point Reyes, the 24-hour average concentration tends to run below 10 micrograms per cubic meter, as is true at many rural locations throughout the U.S..

36. Meredith C. McCormack, et al, "In-home Particle Concentrations and Childhood Asthma Morbidity," Environmental Health Perspectives 17 2, (February, 2009): 296.

37. Dr. Robert F. Phalen is a professor of toxicology at the

University of California, Irvine and a former member of the EPA's Clean Air Scientific Advisory Committee.

38. Robert F. Phalen, "The Particulate Air Pollution Controversey," Dose Response (October , 2004), 266.

39. Ibid., 267

40. Ibid., 276.

41. Ibid., 286. Emphasis added.

42. Ibid., 278. Emphasis added. Dr. Phalen cites Wolff, G.T., "Letter of Closure by the Clean Air Scientific Advisory Committee on the Draft Air Quality Criteria for Particulate Matter;" 1996b. EPA-SAB-CASAC-LTR-960005 and Wolff, G.T., the particulate matter NAAQS review, J Air Waste Manage, 1996c;46:926. The Clean Air Science Advisory Committee, which is composed of non-EPA scientists, performs independent reviews of EPA criteria on air pollutants.

43. Ibid., 285.

44. Ibid., 280.

45. Ibid., 281. Emphasis added. More recently (2012), Dr. James E. Enstrom, an epidemiologist at UCLA, has written: "Since 2000, overwhelming epidemiologic evidence that fine particulate matter is not killing Californians has been published by 26 accomplished doctoral level scientists (Ph.D. or M.D.), including myself." See James E. Enstrom ,"Misrepresentation and Exaggeration of Health Impacts in South Coast Air Quality Management District, Revised Draft 2012 Air Quality Management Plan, Appendix I Health Effects,", September 20th, 2012, accessed December 17, 2014, http://www.scientificintegrityinstitute.org/AQMP092012.pdf.

46. 98th percentile, averaged over 3 years.

47. "Guidelines for the Reporting of Daily Air Quality – the Air Quality Index (AQI)," "Pollutant-Specific Sub-indices and Health Effects Statements for Guidance on the Air Quality Index (AQI)," U.S. Environmental Protection Agency, published online May,

2006, accessed October 10, 2014, http://www.epa.gov/ttn/oarpg/t1/memoranda/rg701.pdf,   p. 11.

48. "Direct Costs Estimate for the Clean Air Act Second Section 812 Prospective Analysis," U.S. Environmental protection Agency, published online February, 2011, accessed October 10, 1014, http://www.epa.gov/cleanairactbenefits/feb11/costfullreport.pdf, pp. 1-11. $PM_{2.5}$ regulation is only one part of the NAAQS, which is only a part of the Clean Air Act.

49. This assumes that individuals could enforce their property rights in court against overt air pollution.

50. President Obama at a White House news conference on November 3rd, 2010.

51. A federal court in Washington DC declared this rule unconstitutional in August of 2012, but this does not diminish the rule's validity as an indicator of environmentalists' intentions.

52. "EPA Regulatory Impact Analysis for the Final Mercury and Air Toxics Standards," U.S. Environmental Protection Agency, published December, 2011), accessed October 10, 2014, http://www.epa.gov/mats/pdfs/20111221MATSfinalRIA.pdf, pp. ES6-ES7.

53. "Impact of EPA's Regulatory Assault on Power Plants: New Regulations to Take More than 72 GW of Electricity Generation Offline and the Plant Closing Announcements Keep Coming," Institute for Energy Research, published 2014, (accessed October 10, 2014, http://instituteforenergyresearch.org/wp-content/uploads/2014/10/Power-Plant-Updates-Final.pdf.

54. 42 U.S.C. § 1857(b) (1).

55. Sierra Club v. Ruckleshaus, 344 F. Supp. 253 (1972).

56. S. Rep. No. 1196, 91st Cong., 2d Sess., at 2 (1970).

57. "Environment: Life without Cars," Time Magazine (June 25, 1973): 54.

58. "Environment: Victory for Clean Air," Time Magazine (October 5, 1970): 44. Emphasis added.

59. Air Resources Board Staff Report "Initial Statement of Reasons for Proposed Rulemaking, Public Hearing to Consider the LEV III Amendments to the California Greenhouse Gas and Criteria Pollutant Exhaust and Evaporative Emission Standards and Test Procedures and the On-Board Diagnostic System Requirements for Cars, Light-Duty Trucks, and Medium-Duty Vehicles, and to the Evaporative Emission Requirements for Heavy-Duty Vehicles," California Environmental Protection Agency, published online (December 7, 2011), (accessed October 10, 2014),
http://www.arb.ca.gov/regact/2012/leviiighg2012/levisor.pdf, p. ES-3.

60. This criticism obviously applies to all "public goods," be they interstate highways or federal dams.

61. "The bill was passed unanimously after just two days on the floor. After the vote, Senator Eugene McCarthy commented to me, "Ed, you finally found an issue better than motherhood."" Senator Edmund Muskie, "NEPA to CERCLA The Clean Air Act: A Commitment to Public Health," The Clean Air Trust (accessed October 10, 2014),
http://www.cleanairtrust.org/nepa2cercla.html.
Originally published in The Environmental Forum (January/February, 1990).

8: The War on Energy

IN THEIR AUGUST, 1970 report, "Electric Power and the Environment," the Energy Policy Staff of the White House Science Advisory Committee wrote, "The relative costs and benefits of present policies as contrasted with a policy of *discouraging energy use* should be carefully evaluated. It may well be timely to re-examine all of the basic factors that shape the present rapid rate of energy growth in the light of our resource base and the impact of growth on the environment." [1] Energy is the lifeblood of an industrial economy. To slow the growth rate of energy usage would mean slowing the rate of economic growth, which would mean slowing the building of homes and businesses and factories and power plants and roadways, although it certainly would preserve "our resource base" and lessen "the impact of growth on the environment." Note that this idea was put forth as worthy of serious consideration by men at the highest levels of the federal government's scientific establishment. From early on in the environmental era, the United States would consistently discourage the development of new energy sources. Consider, first, nuclear generated electricity.

Nuclear Energy

"In fact, giving society cheap, abundant energy would be the equivalent of giving an idiot child a machine gun." [2] Paul Ehrlich

One nuclear fuel pellet, which weighs about a quarter of an ounce, produces the same amount of electricity as a ton of coal.[3] Although it can take ten tons or more of uranium ore to produce one ton of fuel pellets, the amount of ore that must be mined to fuel an electric power plant is still a small fraction of the amount of coal required to fuel such a plant. Therefore, providing fuel for a nuclear power plant involves a much smaller disruption of the environment by mining operations than does fueling an equivalent coal-powered plant. In addition, the mining operations themselves consume fewer resources, such as fossil fuels to power mining machinery, and they produce less pollution than equivalent coal mining operations. Likewise, the transporting of nuclear fuel consumes far fewer resources and pollutes far less than does the transporting of an equivalent quantity of coal. Finally, a nuclear power plant in operation produces virtually no pollution or "greenhouse gasses," very much in contrast to a coal-fired plant.

As for the safety of generating electricity by nuclear means, all fuels have their attendant dangers. In a 1998 study published in the journal *Risk Analysis*, Wolfram Krewitt and his colleagues found, in a study of European power generation, that the number of years of life lost to produce a terawatt of electricity by oil was 359, by lignite 167, by coal 138, by solar 58, by gas 42, and by nuclear 25 and a fraction. Only wind power, at 2.7 years of life, was lower than nuclear among the major methods of power generation. By almost every other measure used in that study, respiratory hospital admissions, cerebrovascular hospital admissions, restricted activity days, and six others, nuclear power generation was significantly safer than the other methods, except windpower.[4]

France produces 70 per cent of its electricity by nuclear means, and their injury and fatality rates have been no greater than by other methods of power generation. The problems with the Fukushima nuclear plant in Japan following the 2011 earthquake and tsunami, rather than proving the danger of nuclear power generation, in fact proved its safety. Few natural disasters will ever exceed in severity the 9.0 earthquake and devastating tsunami that Japan suffered, and yet no lives were lost due to the problems at the nuclear plant. Likewise, the incident at Three Mile Island in Pennsylvania in 1979 killed no one. As for the infamous meltdown at Chernobyl, it demonstrates the danger of Communist governance more than the danger of nuclear power generation. This is not to say that there will never be a serious nuclear disaster. There most certainly will be, just as there will be coal mine cave-ins and natural gas explosions. Accidents are part of the human condition. But to date, nuclear power generation has proven to be among the safest ways to produce electricity.

As for the infamous problem of what to do with nuclear waste, it remains a problem even for the French, though it is a political rather than a technological one. The amount of waste left from producing electricity for a family of four for twenty years would fit into a container the size of a cigarette lighter.[5] Storing the waste is not a problem, but *where* to store it is. In France, although the people are proud of their nuclear generated power and quite happy with it, no one wants to have the waste stored in his neighborhood. The United States is a much larger country, with vast, unpopulated regions, where storage would certainly not present a technological problem, and where it *should* not present a political problem. But, of course, it does.

Anyone who sincerely wanted to reduce pollution and minimize man's impact on the natural environment *without harming man's economic interests* would see nuclear power as by far the best way to generate electricity. But environmentalists halted the development of America's nuclear industry early on. Today only about twenty per cent of electricity in the United States is nuclear generated. Environmentalists have rejected nuclear power generation precisely because it would enable the American

Thomas McCaffrey

economy to grow, and a growing economy is one that would continue to transform the landscape from a natural into a man-made one. Nuclear power would be the "machine gun" in the hands of an "*idiot* child", as Paul Ehrlich put it. It is because environmentalism is essentially a *preservationist* ideology that it has rejected nuclear power in the United States.

Oil

Oil is an indispensable part of America's industrial economy. It comprised 36 percent of the energy consumed in the United States in 2011.[6] In 1970 the U.S. produced 69% of the oil it consumed; in 2011, it produced only 34%.[7] In 1970 a barrel of oil cost about $20 in inflation-adjusted dollars; in all but one year since then it has exceed that price, usually by a great deal. (In 2008, for example, it cost $99.)[8] In recent years Americans have had to spend far more of their wealth on oil and related products than they did in 1970, and much more of this wealth flowed out of the United States than in 1970, much of it to countries hostile to U.S. interests. If the U.S. were running out of oil in the ground, then there might be no alternative but to import it from other countries at high prices (at least until someone developed a substitute for oil—and assuming that the U.S. continued to eschew the development of nuclear power). U.S. production of oil peaked in 1970, and it has been declining almost steadily ever since. But is the U.S. in fact running out of oil?

From Maine to Florida, the outer continental shelf of the Atlantic coast contains an estimated 3.30 billion barrels of oil and an amount of natural gas equivalent to another 5.57 billion barrels of oil, for a total energy equivalent of 8.87 billion barrels of oil.[9] The outer continental shelf of the eastern Gulf of Mexico and off southern Florida contains, in oil and natural gas, the energy equivalent of an estimated 7.93 billion barrels of oil.[10] The Federal waters of the Pacific Coast, from northern Washington to southern California, contain the energy equivalent of an estimated 14.76 billion barrels of oil.[11] And the Federal waters off Alaska, not including the Beaufort Sea, Chukchi Sea, and Cook Inlet sectors, contain the energy equivalent of an estimated 6.45 billion barrels of oil.[12]

The total amount of oil and gas in all of these areas combined amounts to the equivalent of 38 billion barrels of oil. (The U.S. consumed 6.87 billion barrels of oil in 2011.) With the exception of a small portion of the eastern Gulf of Mexico, these areas have been off limits to oil drilling since the 1980s, and the primary reason they have been off limits has been to prevent pollution from oil spills.

The energy equivalent of 38 billion barrels of oil pales in comparison to the 159 billion barrels of oil equivalent in the central and western Gulf of Mexico, which has been America's primary source of oil in recent decades.[13] However, in 1996 the total undiscovered technically recoverable oil (not including gas) in the whole outer continental shelf of the United States was estimated to be only 45 billion barrels. By 2006 that number had increased to 85 billion, and almost all of the increase had come in the central and western Gulf of Mexico.[14] Technologically heroic efforts at discovering new deposits of oil and gas, and equally heroic efforts at extracting them from holes drilled 20,000 feet into soil and rock 10,000 feet underwater dramatically expanded the supply of known and "technically recoverable" oil and gas. In the Gulf, it was active exploration and drilling, combined with technological innovation, that increased the estimated supply of oil by almost 90 percent in ten years. One can only guess at the resources that might be found off the Atlantic and Pacific coasts, and off the coast of Alaska, if only oil companies were free to search for and develop them.

Since the start of the environmental era, the central and western Gulf of Mexico has been one of the few places in the

Thomas McCaffrey

outer continental shelf where oil companies are relatively free to explore and drill for oil. But in November of 2011 the Obama Administration reduced the number of oil lease sales in the outer continental shelf by half. They also raised the minimum bid for an oil lease from $37.50 per acre to $100 per acre, and they lowered the term of most leases from 10 years to 7. And they added new regulations, intended to prevent another blowout like BP's Deepwater Horizon blowout in 2010, which have diminished productivity in the Gulf.[15]

The Beaufort Sea lies off northeastern Alaska, and the Chukchi Sea off northwestern Alaska. The Beaufort contains an estimated 13 billion barrels of oil equivalent, and the Chukchi 28 billion barrels.[16] Officially, both areas are mostly open to oil drilling, but as of late 2012 there were no wells operating in federal waters in either one. Judging from the barrage of lawsuits that have obstructed Royal Dutch Shell's efforts to drill in the Beaufort and Chukchi, it appears that environmentalists are determined to ensure that oil drilling on the outer continental shelf never goes beyond the Gulf of Mexico (and Alaska's Cook Inlet), where it has a longstanding foothold. From 2005 until late 2012, Shell spent $4.5 billion in their efforts to begin production in the Beaufort and Chukchi Seas, and in that time they did not succeed in drilling one complete well.[17] Although most of that amount went toward purchasing leases from the federal government and toward satisfying federal regulatory requirements, considerable sums were also spent on battling lawsuits from indigenous peoples and from environmental groups.

Shell had planned to begin drilling in the summer of 2007, but a lawsuit filed by a coalition of environmentalist groups (one of numerous such lawsuits) succeeded in blocking drilling for two summer drilling seasons, costing Shell $200 million dollars.[18] The lawsuit questioned whether the needs of bowhead whales and the danger of oil spills had been adequately accounted for by the federal agencies who had issued Shell's permits. Another lawsuit forced Shell to forego the 2011 summer drilling season; it challenged the air pollution permits for ships at a drill site 70 miles out to sea. In early 2013, Shell decided to suspend their efforts to drill in the Beaufort and Chukchi Seas.

Critics of these kinds of lawsuits usually attribute them to "radical" environmentalists, and, indeed, the list of groups who have sued Shell over drilling in the Beaufort and Chukchi, a list that includes Greenpeace, the Sierra Club, and the Wilderness Society, reads like a hall of fame of what are commonly called radical environmental groups. But to attribute such lawsuits to environmentalist radicalism is to ignore that it is the official policy of the United States Government to encourage such lawsuits.

The Clean Air Act, the Clean Water Act, and most other major federal environmental statutes contains "citizen suit" clauses, which authorize private citizens to bring suit against other private parties or against government agencies in order to enforce statute law—a function that had in the past been reserved to attorneys general.[19] Implicit in this arrangement is that, much like an attorney general, a private party bringing suit under such a clause would not need to show that he personally had been harmed by the defendant's action, but only that the defendant had broken the law. Accordingly, virtually any citizen of the United States could sue any corporation for violating, say, the Clean Air Act, regardless of whether he himself had suffered direct harm as a result of the alleged polluter's actions.

Courts had been loathe to recognize the standing of complainants who had not suffered direct harm to bring suit under these citizen suit clauses, but beginning with U.S. Supreme Court's decision in *Friends of the Earth, Inc. v. Laidlaw Environmental Services, Inc.* in 2000, the courts have expanded the range of persons recognized as having the legal standing to bring suit against a private polluter or against a federal agency that allegedly has not been adequately enforcing a statute. The effect of the citizen suit clauses has been to encourage lawsuits against private companies and the government agencies who regulate them.

The playing field is further tilted in favor of environmentalists by the Equal Access to Justice Act, which President Carter signed into law in 1980. This statute enables private parties who successfully sue the federal government to have the government, under certain circumstances, reimburse them for their legal costs. So, whereas the "citizen suit" clauses make it possible for an environmental group to sue, say, the EPA to force a company to halt emissions, the Equal Access to Justice Act enables them to do so *at little or no cost to themselves.*

In addition, losing these suits can be so costly to private companies that they often find it less expensive to settle out of court. These settlements often require the companies to donate millions to environmental groups.[20] Through such legalized extortion, citizen suits, combined with the Equal Access to Justice Act, can actually make it profitable to sue private companies or the federal agencies that regulate them. In this way, organizations that are relatively small and poorly funded, compared with the corporations they sue, are enabled to inflict serious financial damage on much larger, much wealthier corporations, damage that is out of all proportion to the size and funding of the environmental groups.

Regardless of whether these lawsuits are being prosecuted by what are commonly called "radical" environmental groups, however, it is inaccurate to view the lawsuits and the havoc they wreak on our industrial economy as products of environmental radicals; they are, rather, products of official federal policy, and they have been so since the early 70s. They are as much in the mainstream of the environmental movement as are the Clean Air Act and the Clean Water Act.

At 30,135 square miles, the Alaska National Wildlife Refuge

is almost as large as South Carolina. It is located in the northeast corner of Alaska, bordered on the east by Canada and on the north by the Beaufort Sea. In 1998 the United States Geological Survey estimated there to be 10.36 billion barrels of oil and an amount of natural gas equal to .66 billion barrels of oil, for a total quantity of oil and gas equal to 11 billion barrels of oil.[21] Three quarters of this total is located in a small area on the coast called Section 1002, which comprises just 7.8% of the area of ANWAR.[22] Of the 1,500,000 acres that comprise Section 1002, only about 2000 would be affected were Section 1002 to be opened to oil and gas development.[23] But, because of fear of pollution, on the one hand, and fear for the well-being of wildlife, on the other, all of ANWAR, including the entirety of Section 1002, has been and continues to be off limits to any oil or gas development.

The National Petroleum Reserve-Alaska, in the northwestern part of the state, is an area about the size of Indiana. It was set aside in 1923 as a petroleum reserve for the U.S. Navy. In 2010 the USGS estimated there to be .896 billion barrels of oil and 53 trillion cubic feet of natural gas, for a total barrels-of-oil-equivalent of 10 billion.[24]

Except for some drilling by the Navy and the USGS in the 1940's, the oil and gas in the reserve is largely untouched. In 1998 the Secretary of the Interior opened about 17% of the reserve in the northeast to oil and gas leasing. Another 37% of the reserve, in the northwest, was opened to lease sales in 2004. The remaining 46% of the reserve is closed to oil and gas development. In December, 2012 there were 186 Federal oil and gas leases comprising some 3 million acres (about 13% of the whole reserve) in the northeast and northwest sectors, though there were no producing oil or gas wells.

A 2006 lawsuit brought by environmentalists had blocked the sale of oil or gas leases on some 600,000 acres in the northeast sector. Then in 2012 the Obama Administration introduced a plan to close about half of the reserve, including areas with existing oil and gas leases, to any further development. In addition to blocking access to oil and gas, this closure would, according to Alaska's Congressional delegation, block off the most economically feasible route for a new pipeline that is needed to transfer oil and gas from the Chukchi Sea and the North Slope to the Trans Alaska Pipeline System.[25]

Most oil comes from wells drilled into deposits in porous rock from which the oil is readily pumped to the surface. This is the easiest, and least expensive, way to extract oil from the earth. But as the demand for oil rises around the world, it has become profitable to develop other, less accessible types of deposits that are more expensive to exploit. Among these are deposits trapped in non-porous shale formations. Oil in these deposits can be made to flow by fracturing—or fracking—the rock with fluids pumped in at high pressure. Fracking technology has been around at least since the 1940's, but only in the last fifteen years has fracking been used to produce significant amounts of oil and gas.

On the North Slope of Alaska, from the western edge of the Alaska National Wildlife Refuge, across the Prudhoe Bay region, and then westward across and beyond the National Petroleum Reserve-Alaska, there lies a deposit of shale oil and gas that the USGS has estimated to contain the equivalent of 8.42 billion barrels of oil.[26] Most of this oil and gas is situated in the National Petroleum Reserve-Alaska, and a good deal of it will be placed off limits by the Obama Administration's decision to close about half of the NPR-A to development.[27] In the lower 48 states, though, oil and gas from shale are transforming the economic landscape.

The Shale Revolution

The Barnett Shale is a horizontal layer of rock that lies beneath 20 or so counties in north central Texas. When George Mitchell sank his first test well into the Barnett in 1981, he struck natural gas, but the flow rate was not sufficient to be profitable. Over the next quarter century, Mitchell continued to explore the Barnett and to experiment with different ways of getting the gas out of it. Back in 1947, Stanolind Oil and Gas Corporation, at the Hugoton Gas Field in Kansas, had tried to increase the flow rate of a gas well by injecting gelled gasoline under pressure into the well to induce fractures in the rock. A couple of years later, the Halliburton Oil Well Cementing Company made this process commercially viable.

But in the Barnett, Mitchell found the use of gel to be too expensive to be profitable. In the late 1990's, he tried injecting water combined with certain chemicals (slickwater), and this, combined with another innovation, horizontal drilling, enabled Mitchell to extract gas from the Barnett Shale at a profit. Horizontal drilling makes it possible to run as many as eight wells, each up to a mile and a half long, from one pad, thus minimizing the amount of disturbance to the surface landscape. In 1997 there were 410 wells operating in the Barnett; by mid-2012 there were almost 18,000. The USEIA has estimated that there are 43 trillion cubic feet of natural gas in the Barnett Shale.[28] (In 2011 the U.S. consumed about 24 trillion cubic feet of natural gas.)

Mitchell's success in the Barnett led Southwestern Energy to begin exploring and experimenting on the Fayetteville Shale in Arkansas in 2002. In 2011 the Fayetteville produced 943 billion cubic feet of natural gas. The EIA has estimated the Fayetteville to contain 32 trillion cubic feet of natural gas.[29] There followed the Woodford Shale in Oklahoma (22 trillion cubic feet), the Haynesville Shale in Texas and Louisiana (75 trillion cubic feet), the Eagle Ford in Texas (21 trillion cubic feet), the Barnett-Woodford in Texas (32 trillion cubic feet), and many others, including the Marcellus Shale, which extends from New York to Virginia (84 trillion cubic feet).

In all, there are estimated to be 424 trillion cubic feet of natural gas in shale formations in the lower 48 states.[30] (This is about 17 years' worth of natural gas at the 2011 rate of consumption.) In 2000, shale gas represented 1% of U.S. natural gas supplies; by 2012 that number had exceeded 30%.[31] In 2005 the U.S. produced 84% of the natural gas it consumed, and the price averaged $7.32 per thousand cubic feet. In 2012, the U.S. produced 99% of what it consumed, and the price averaged $2.66 per thousand cubic foot.[32]

Then there are the shale oil formations, the best known among them the Bakken, which have also proven capable of exploitation by horizontal fracking. In 2008 the USGS estimated there to be 3.65 billion barrels in the Bakken Formation, which straddles Montana and North Dakota. In the Eagle Ford Shale in Texas, there are an estimated 3 billion barrels, in the Avalon & Bone Spring in Texas and New Mexico another 2 billion, and in the Monterey/Santos Formation in California a whopping 15 billion barrels.[33]

The United States is now producing more natural gas than at any time since 1973, which is as far back as EIA records go.[34] As for oil, U.S. production rose by 760,000 barrels per day in 2012, the largest increase in annual production since commercial oil production began in 1859.[35] In fact, production has been on the rise since late 2008, the first sustained increase since 1986. Thanks to the Bakken Formation, between 2010 and 2011 North Dakota's gross state product grew by 7.6 percent.[36] Only six other states had growth rates greater than 2 percent. In November of 2012, North Dakota's unemployment rate was 3.1 percent, as compared with a national rate of 7.7 percent.[37]

One reason that fracking has so quickly become a significant source of oil and gas is that it has been largely unregulated by the federal government. Congress exempted fracking from EPA regulation under the Safe Drinking Water Act when it passed the Energy Policy Act of 2005. (Under that act, only fracking that used diesel fuel in the fluid injected into the earth could be regulated by the EPA. But the EPA did not produce guidelines for diesel fracking until 2012-2013, by which time little diesel was being used in fracking.)

Another reason that fracking went unregulated by the federal government is that it quickly became an important source of natural gas, which has long been regarded in environmentalist circles as a "clean" fuel. Still another reason for the rapid rise of fracking is that most of it is done on private or state land, as opposed to federal land. (Consider that from 2005 through 2011 it took an average of 307 days to get a drilling permit from the U.S. Bureau of Land Management, but in 2011 the State of Colorado reported average permit application times to be only 27 days, and the State of Ohio only 14 days.[38])

Gary D. Libecap points to a further reason, related to private land ownership, why the Shale Revolution has been succeeding in the U.S..[39] Because Americans own the subsurface rights to (non-government) land, they stand to benefit economically from the development oil and gas beneath the surface. In places like Germany, where the government owns the subsurface rights, even beneath private land, the surface owners, who would have to bear all the disadvantages of fracking in their neighborhoods but would enjoy no direct economic benefits, have every incentive to oppose fracking.

Fracking has been around since the late 1940's, and it has been used on over a million wells in the United States.[40] But since the onset of the Shale Revolution, three developments have occurred that have raised fracking's profile: the frequency of fracking has increased dramatically; fracking has begun to be employed in locales outside places like Texas and Oklahoma, locales where the residents are not used to oil and gas operations; and, perhaps, most importantly, fracking has increased oil and gas production levels to such a degree that a "re-industrialization" of the United States is beginning to look like a realistic possibility.[41] One consequence of fracking's increased prominence is that it has attracted the attention of environmentalists.

One response of environmentalists to the Shale Revolution has been to try to subject fracking to federal regulation. It is already regulated by the states in which it is practiced. But state regulators tend to be more sensitive to the economic effects of their regulations. They are therefore less likely than federal regulators to treat pollution as a categorical evil to be eradicated at all costs and more likely to treat it as an occupational hazard of industrialism that needs to be mitigated where possible, though decidedly *not* at all costs. Environmentalists tend to prefer federal to state regulation.

The Fracking Responsibility and Awareness of Chemicals (FRAC) Act would subject fracking to EPA regulation under the Safe Drinking Water Act. Among other things, it would require oil and gas companies to disclose the chemicals in the fluid they inject into their wells to fracture the shale. The FRAC Act was first introduced to both houses of Congress in 2009, but it has yet to win passage, which would likely subject oil and gas drillers to a whole new set of lawsuits.

Because of Congress's failure to pass the FRAC Act, federal agencies have begun to expand the regulation of fracking on their own. The EPA in 2011 initiated plans to regulate fracking under its authority under the Clean Water Act to regulate wastewater, of which fracking produces great quantities. EPA has also announced that, under the Toxic Substances Control Act, it will propose rules to regulate the manufacturers, processors, and distributors of the materials used in fracking fluids. In 2012 the EPA also announced new rules under the Clean Air Act that are intended to prevent the escape of methane and other volatile organic compounds into the air during the removal of fracking water from wells after the completion of fracking operations.

In addition to these initiatives by the EPA, the Bureau of Land Management in 2012 announced new regulations that will apply to federal lands. These will require frackers to disclose the chemicals used in fracking fluids, "strengthen" regulations governing the integrity of well bores, and regulate the disposal of waste water. This new wave of federal regulations by the EPA and the BLM will most certainly slow the pace of fracking, on the one hand, while increasing its costs, on the other. The result will be lower quantities of oil and gas produced and higher prices.

Environmentalists are especially eager to prove that fracking threatens to pollute underground aquifers. It is important to distinguish among the different types of pollution that might be associated with fracking. Groundwater aquifers tend to lie at shallow depths in the earth, while most shale oil and gas are found thousands of feet farther down. Under such conditions, when up to a mile of rock separates the oil and gas deposits from the groundwater, it has been difficult for opponents of fracking to prove that the injection of fracking fluids into the deep shale has resulted in the contamination of shallow aquifers by the fluids. There has yet to be a case in which such contamination has been scientifically proven.

Another kind of groundwater pollution might occur, though, when methane and other volatile organic compounds either leak directly from the frack zone into groundwater (in places where the frack zone is not far below the aquifer) or leak through openings in the vertical well-bore as they travel from the deep shale up through the groundwater to the surface.

In recent years, the environmental group Earthworks, along with the Sierra Club, the Natural Resources Defense Council, the Center for Biological Diversity, and other mainline environmental groups, along with a raft of *ad hoc* groups, have been working to halt the Shale Revolution.[42] Among such opponents of fracking, the search for a connection between fracking and groundwater pollution has become something of a Holy Grail. Their premise has been that if fracking can be shown to cause groundwater pollution, then governments at all levels will be more open to banning it.

"This bill will ensure that we do not inject chemicals into groundwater in a desperate pursuit for energy," said Governor Peter Shumlin as he signed the statute making Vermont the first state to ban fracking.[43] This view of fracking, of course, reflects the environmentalist idea of pollution as a categorical evil that needs to be extinguished whatever the cost. (That no proven instance of groundwater pollution has occurred despite the use of fracking over a million times speaks volumes about the flimsiness of environmentalists' charges and the speciousness of their motives.) This quest to find a connection between fracking and groundwater contamination has, in recent years, centered on three rural communities, Pavilion, Wyoming, Parker County, Texas, and Dimock, Pennsylvania.

In 2008 some residents of Pavilion, Wyoming complained to the EPA that gas drilling and fracking by Encana Corporation had polluted their well water. The EPA investigated in 2009, testing 39 wells. Based on the results of those tests, the EPA decided to drill two monitoring wells of its own in 2010 and commenced additional testing. Then, in a December, 2011 draft report, the EPA announced to the world that "ground water in the aquifer contains compounds likely associated with gas production practices, including hydraulic fracturing."[44] Here, apparently, was the smoking gun that environmentalists had been seeking.

Or was it? According to Thomas Doll of the Wyoming Oil and Gas Conservation Commission,

> *The Pavillion Draft Report was issued with incomplete data and technically inadequate conclusions. There was no opportunity to review and verify the data by*

Wyoming state agencies. The data was not verified by further testing or vetted through a peer review process. Based on a limited sampling and an inconclusive data set from Pavillion Wyoming ground water, EPA's conclusion is now national and international fodder for the hydraulic fracturing debate.[45]

In response to sustained criticism from Wyoming State authorities and from Encana Corporation, the EPA agreed to a further round of testing, this to be conducted by the USGS. While this testing did show contamination of some wells by methane (there had been methane in the water at Pavilion before any gas drilling or fracking had occurred), it failed to corroborate unequivocally the EPA's conclusion of contamination of the wells by chemicals from fracking fluid. It appears that, at best, the EPA had been too eager to conclude that fracking fluids had contaminated the drinking water at Pavilion.

This is not the only ground for questioning the EPA's involvement at Pavilion. The EPA intervened at Pavilion under powers granted to it under the Comprehensive Environmental Response, Compensation, and Liability Act of 1980 (CERCLA). CERCLA is the "Superfund" law of Love Canal fame. It grants the federal government authority to intervene when "there may be an imminent and substantial endangerment to the public health or welfare or the environment because of an actual or threatened release of a hazardous substance."[46] But at Pavilion, Encana had already reported themselves to the Wyoming Department of Environmental Quality for, and were working to clean up, some shallow groundwater contamination that had come from pits that had been there when Encana bought the property in 2004.

In addition, the Wyoming DEQ had been involved at Pavilion since 2005 but had been unable to tie the condition of local wellwater to Encana's drilling and fracking activities. This is hardly a case in which Encana's activities constituted "an imminent and substantial endangerment to the public health or welfare or the environment," certainly not on the scale of Love Canal or other notorious Superfund sites.[47] It looks more like a case in which the EPA went looking for a smoking gun to tie fracking to groundwater pollution.

This impression is reinforced when one considers that CERCLA explicitly exempts contamination from petroleum products; if the EPA had discovered that Encana contaminated the drinking water at Pavilion with methane, then there would have been nothing they could do about it under CERCLA. It is clear that the EPA intervened at Pavilion expressly to look for contamination of the groundwater by fracking fluids, and it appears that they were overly eager to interpret what evidence they uncovered as supporting the conclusion that such contamination had indeed occurred.

On December 7th, 2010 Al Armendariz, EPA's Region 6 administrator, sent out an email to anti-fracking groups in Texas urging them to watch "Tivo channel 8." "We're about to make a lot of news," he wrote excitedly.[48] The EPA had just accused Range Resources Corporation and Range Production Company of allowing gas to escape from one of its wells into an underground aquifer that supplied local drinking water. The Range well is in the Barnett Shale in Parker County, Texas. The EPA would require Range, among other things, to develop a plan to identify the path by which the gas, located some 5000 feet below the aquifer, had found its way into the aquifer, to stop the gas flow, and to clean up the aquifer.

The EPA acted under emergency powers granted under the Safe Drinking Water Act, powers which only come into play when state authorities fail to take the appropriate actions. In the Range case, the Texas Railroad Commission had been working on the case for months but had been unable to ascertain that the

Range well had caused the contamination of the aquifer. The EPA, however, claimed that the contamination "was likely due to impacts from gas development and production activities in the area," and they resisted calls for proof that Range had caused the contamination. It appears that, for all practical purposes, the EPA hoped to establish *by fiat* that contamination of groundwater had occurred as a result of gas drilling and fracking at Parker County. In the end, the EPA failed to produce proof of their claims, and in March of 2012 they dropped their case against Range.

This was a case in which Texas authorities were handling the situation, the EPA was not needed, and, after they had inserted themselves, events proved that the EPA had nothing to contribute beyond what the State authorities has already accomplished. It is difficult not to suspect that the EPA saw in Parker County a chance to tie fracking to groundwater pollution. Mr. Armendariz's apparent glee in announcing the EPA's intervention does little to dispel this impression, especially when one considers a statement he made in a 2010 video about how he intended to make an example of oil and gas companies:

> *The Romans used to conquer little villages in the Mediterranean. They'd go into a little Turkish town somewhere, they'd find the first five guys they saw and they would crucify them. And then you know that town was really easy to manage for the next few years.*[49]

In 2009 a homeowner's water well exploded in the small town of Dimock, Pennsyvania. Dimock is located on the Marcellus Shale, and Cabot Oil and Corporation had begun drilling and fracking for gas in Dimock the year before. After investigating the explosion, the Pennsylvania Department of Environmental Protection concluded that Cabot's drilling or fracking operations had contaminated the local groundwater with methane and other substances. Cabot denied responsibility but, in December, 2010, signed on to a $4.6 million consent agreement that required them to correct leaking or over-pressured gas wells and to supply fresh water to some homes. By this time, Dimock had become ground zero in the fight against fracking, thanks to the award-winning documentary *Gasland*.

By late 2011, the Pennsylvania DEP permitted Cabot to cease its deliveries of water to the affected residences, since Cabot had offered to provide methane removal systems to the homeowners and had offered each of the homeowners twice the value of their homes as required by the consent agreement. In the summer of 2012, the DEP would permit Cabot to resume fracking, though it would not allow them to resume drilling new wells until methane levels in the groundwater had returned to what the DEP considered normal.

Despite the fact that the DEP was very much in control of the situation, the EPA showed up in Cabot in November, 2011 to interview residents and to review the well-water test data of the DEP and of Cabot. On December 2nd they announced that Dimock's well water posed no threat to human health. But then additional well-testing data that had been under court seal became available, and the on December 7th the EPA requested the Agency for Toxic Substances and Disease Registry to review the data. On December 19, the EPA commenced a new round of interviews with residents of Dimock, and on Jan 19, 2012 they announced an Action Plan that called for the testing of 61 wells. Finally, on July 26th, 2012 the EPA announced once again that the well water in Dimock did not constitute a threat to human health.

As at Pavilion, Wyoming, the EPA had intervened at Dimock under CERCLA, the Superfund law. This precluded their doing anything about methane contamination, as it had at Pavilion. So, as at Pavilion, it was contamination by fracking fluid that the EPA was aiming to find. As if to emphasize this point, Richard Fetzer, the EPA Regional Administrator, referred to the Dimock Action Plan as "nationally significant or precedent setting." [50] (Bypassing the Pennsylvania DEP, Fetzer had extended an offer to the Susquehanna County Emergency Management Agency as early as February of 2010 for the EPA to get involved at Dimock, but the agency refused his offer, stating that the DEP was handling matters adequately.[51])

Three times in recent years, at Pavilion, Wyoming, Parker County, Texas, and Dimock, Pennsylvania, the EPA has intervened in a dispute involving possible groundwater contamination by fracking. In each case, state authorities were already involved and had the situation under control. In each case, the EPA exercised emergency powers to get itself involved, and in each case it turned out that there had not been any real need for the EPA to intervene. It is hard not to suspect that, in the absence of explicit statutory authorization to regulate fracking, the EPA has been attempting to insert itself into the fracking debate in any way it could in order to establish the environmentalists' long-sought connection between fracking and groundwater pollution.

The Marcellus Shale contains an estimated 84 trillion cubic feet of natural gas.[52] Nothing illustrates the environmentalist response to the Shale Revolution as vividly as recent events on the Marcellus. Situated in New York, Pennsylvania, Ohio, and West Virginia, it lies beneath some areas, especially in New York, where environmentalists are especially influential. At least 140 communities on the Marcellus have formally opposed drilling and fracking through zoning controls, road regulations, drilling moratoria, or outright bans. The town of Marcellus, New York, for example, from which the Shale derives its name, has banned gas drilling and fracking outright.[53] Anti-drilling and fracking lawsuits, like the one at Dimock, Pennsylvania, abound. In December of 2014, after a four-year moratorium on fracking, the New York became the second state to ban it outright.

In an editorial, the *Times Union* of Albany, New York asked, "What's the risk to human health and the environment? We're not convinced the state Department of Environmental Conservation knows—especially when some scientists and physicians are saying they aren't sure." [54] The implication of the Times Union, that a technology should be proven "safe" before it may be employed, harkens back, of course, to Rachel Carson. But how safe must a technology be? Fracking has been used on more than a million wells over six decades, and environmentalists are still searching for proof that it is unsafe. But, of course, no technology can be shown to be absolutely, positively safe, because there is always the possibility of some unforeseen occurrence—in the case of fracking, for example, some new and unusual combination of geology and groundwater, that could result in contamination. A case of groundwater contamination by fracking fluid is bound to occur at some time in the future.

A technology is not born either safe or unsafe. A technology is *made* safe, through trial and error. If some government agency had followed the Times Union's logic and had prohibited airline travel until it had been proven safe, we might still be waiting to get off the ground. Even today, airline travel carries *some* risk, though it is a great deal safer than it was in the 1930's—and it was *made* safer through the experiences of those who engaged in it.

The Shale Revolution offers America the chance to regain a measure of energy self-sufficiency, and, at the same time, an opportunity to reduce our dependence on a Middle East that largely wishes us ill. It also offers a chance to begin to rebuild our industrial strength, and to create jobs and begin to restore our economic health. It also will generate much-needed revenue for federal, state, and local governments. All of these things will happen if our governments at all levels simply allow them to happen. All our governments have to do is stay out of the way.

Indeed, our governments have a special incentive to allow all of these things to happen because, at the present time, unemployment is high, our national economic growth rate is anemic, and our federal government and many of our state, and local governments are deeply in debt and in dire need of revenue. The State of California, for example, could vastly improve its fiscal situation simply by allowing fracking on the Monterey-Santos Shale.[55] The American people, and their federal, state, and local governments, have every incentive to allow the Shale Revolution, which is still in its early stages, to run its course.

Why, then, would it not be allowed to run its course? Because environmentalists want to prevent the *mere possibility* of groundwater contamination, and because they want to preserve the quiet, rural ambience of certain places where the Shale Revolution is occurring, and because they want to prevent the growth of domestic energy production, re-industrialization, and the economic expansion that would follow.

The Green River Shale

The largest single reservoir of oil in the U.S. is the Green River Shale, which lies beneath portions of Utah, Colorado, and Wyoming. Rather than shale oil, which can be pumped out of the ground, the Green River contains oil shale, which must be mined as rock and then processed to produce oil. Some attempts have been made to produce oil from the Green River Shale, but none to date has proven sufficiently profitable. (The scarcity of water in the region adds to the cost of producing oil from shale there.) But unlike the privately owned Barnett Shale, where George Mitchell could spend many years discovering how to exploit it profitably, the Green River is mostly owned by the federal government, so access has been limited and environmentalists have had a great deal of influence over deciding what may or may not be done there.

In 2005 the Rand Corporation estimated there to be 800

billion barrels of technically recoverable oil in the Green River Shale.[56] At the United States' present rate of usage of 20 million barrels per day, the Green River could supply enough to last over 100 years. In 2008 the Bush Administration announced regulations to govern the opening up of almost 2 million acres of the federal land in the west, much of it on the Green River Shale, to oil development. In February, 2012 the Obama Administration reduced this to 676,000 acres.

Utah Tar Sands

Tar sands are a mixture of sand, clay, water, and bitumen, the latter of which is a thick form of oil. Tar sands are usually mined in open pits, and then the bitumen is separated from the rest and processed to produce a useful form of oil. Canada has immense tar sand deposits in Alberta, from which they produce significant quantities of oil. The U.S. has no major tar sands oil production at present, but there are substantial tar sand deposits in eastern Utah. The USGS estimated in 2006 that there are 12 billion to 19 billion barrels of in place oil in Utah's tar sands (and anywhere from 36 billion to 54 billion total in U.S. tar sands).[57] The same Bush Administration plan that would have opened up 2 million acres of the Green River Shale to oil development would also have opened 430,000 acres of Utah tar sands. An Obama Administration plan announced in late 2012 would reduce these 430,000 acres to some 129,000 acres.

Coalbed Methane

About 7.5% of the natural gas produced in 2012 in the United States came from coal beds. The USGS estimated in 1997 that there are 700 trillion cubic feet of coal-bed natural gas in the U.S., 100 trillion of which they estimate to be economically recoverable (as opposed to technically recoverable).[58] 100 trillion cubic feet of natural gas would supply U.S. needs for about four years.

Coal-bed natural gas is a relatively new source of natural gas in the U.S., and its development is still in the early stages, but it is beginning to attract attention from environmentalists. Coal-beds tend to be permeated with water, which is mixed in with the methane. Before the gas can be brought to the surface, the water must first be pumped out. As the USGS explains, "This water, which is commonly saline but in some areas can be potable, must be disposed of in an environmentally acceptable manner. Surface disposal of large volumes of potable water can affect streams and other habitats." Yes, the disposal of large volumes of water that is clean enough to drink "can affect streams and other habitats;" the danger is, quite literally, "pollution" by clean water. The disposal of this coal-bed water, some of it allegedly contaminated with methane, is beginning to generate opposition from environmentalists around the world.[59]

Methane Hydrates

Methane hydrate is a form of frozen methane. It exists in arctic regions and on the ocean floor. According to the USGS, "The worldwide amounts of carbon bound in gas hydrates is conservatively estimated to total twice the amount of carbon found in all known fossil fuels on earth."[60] That is twice the amount of oil, natural gas (other than in methane hydrates), and coal known to exist in the world. There are extensive methane hydrate deposits in the United States, including, for example, two areas off the coasts of North and South Carolina, "each about the size of the state of Rhode Island," which contain "intense concentrations of gas hydrates. USGS scientists estimate that these areas contain more than 1,300 trillion cubic feet of methane gas," which amounts to 54 years' worth of natural gas for the U.S. at 2011 consumption rates."[61]

If entrepreneurs in the United States ever begin to produce natural gas from methane hydrates, there will likely be a great hue and cry from environmentalists, who will argue, on the one hand, that burning methane will exacerbate global warming, and, on the other, that harvesting methane hydrates will do immeasurable damage to the arctic tundra or to sea life and the ocean floor.

Refineries and Pipelines

The last major oil refinery built in the United States opened in 1977.[62] Nothing illustrates environmentalists' hostility to the development of oil and gas resources so clearly as this simple fact. There have been numerous proposed refineries that were defeated by local opposition (as at Eastport, Machiasport, Trenton, and Searsport, Maine, Durham, New Hampshire, and Del, Maryland, to name just a few). Environmentalists are quick to point out that U.S. oil refining capacity is significantly greater today than it was in 1977. This is true, but only because oil companies have added capacity to their existing refineries.

In a free economy, companies tend to do things in the most cost-effective way. In many cases in the last few decades, oil companies' first choice would have been to build a new refinery rather than add to the capacity of their existing refineries. (At Eastport, Maine, for example, the Pittston Company wanted to take advantage of the deepwater port.) But because all these proposed new refineries were defeated, the companies had to fall back on less economically efficient alternatives. Americans are paying higher prices for their oil and gas because these companies were not free to build new refineries.

There have been two attempts recently to build major new oil refineries in the U.S.. One, in the desert outside Yuma, Arizona is "on hold," and the other, in Union County, South Dakota, has stalled as environmentalists challenge its air emissions permit in court. Even when environmentalists lose a battle, they often manage to make it a Pyrrhic victory for their opponents, by dragging out the construction process and increasing the cost of the project. They also make it clear to other, potential builders of new refineries that their costs will ultimately be exorbitant.

As for new oil pipelines, the proposed Keystone XL pipeline, which would run from Alberta, Canada to Steele City, Nebraska, would be part of a system intended to increase the flow of oil from western Canada and from the Bakken in North Dakota to the Gulf of Mexico. But the so-far successful environmentalist opposition to the Keystone would prevent that increased flow at just the time when the supply of oil is increasing dramatically in Canada and North Dakota.

In their last estimate in 2009, The United States Energy Information Agency put total U.S. crude oil resources (proved reserves plus unproved technically recoverable resources) at 220.2 billion barrels. This would last 32 years, at the 2011 rate of consumption. It did not include the 800 billion barrels in the Green River Shale, which would last another 116 years.

In the same report, the USEIA put total natural gas reserves (including shale gas and coal bed methane) at 2,203.3 trillion cubic feet. This would last 92 years at the 2011 rate of consumption. It did not include the estimated 1,300 trillion cubic feet of methane hydrates, which would last another 54 years.

The United States is not running out of oil and gas. The reason that it has been producing less and less oil since the beginning of the environmental age in 1970 is that access to oil has been restricted by government regulation. Natural gas production also peaked in 1970, but it recently exceeded that peak because of the shale gas revolution. Indeed, the Shale Revolution has shown what is possible when the market is left free enough for entrepreneurs and technological innovators to do their jobs.

Coal

The United States Energy Information Agency has estimated U.S. coal reserves to be 760 billion short tons.[63] From 2009 through 2011 the U.S. consumed about 1 billion short tons per year.[64] At this rate, the United States has enough coal to last over 700 years. But coal is an especially repugnant fuel to environmentalists. I described earlier in this chapter how then-presidential candidate Barack Obama predicted that new coal-fired electric power plants would be rendered economically impracticable under the cap-and-trade legislation he favored, and how, once that legislation failed to win passage, his EPA set about achieving the same end by regulatory means.

In July of 2012, the EIA reported that 27 gigawatts of coal-powered electric generating capacity would be retired over the next five years (8.5 % of the United States' total generating capacity of almost 318 gigawatts in 2011). This is about five times more gigawatts of coal-fired generating capacity than were retired in the previous five years, and it reflects the influence of the EPA's new rules, the Cross State Air Pollution rule and the Mercury Air and Toxics for Power Plants rule.[65] Most recently, in June on 2014, the Obama Administration announced a new requirement that fossil-fuel burning power plants reduce their carbon emissions by 30 percent by 2030, a measure that undoubtedly will require the closing of still more coal-fired plants.

As coal-fired power plants close down, so must the coal mines that supply them. Alpha Natural Resources, for example, decided to close eight mines in Virginia, West Virginia, and Pennsylvania and to lay off 1,200 workers. Alpha CEO, Kevin Crutchfield, laid part of the blame on a "regulatory environment that's aggressively aimed at constraining the use of coal." [66]

Nuclear energy, offshore oil and gas, oil and gas in ANWAR and the National Petroleum Reserve-Alaska, shale oil and gas, the

Green River oil shale, the Utah tar sands, coalbed methane, methane hydrates, and coal. The United States has many hundreds of years' worth of fossil energy available to it, but in every one of these cases (except methane hydrates, which have not yet begun to be developed), environmentalists have found reason to object to the development and use of these resources. It is hard not to conclude that environmentalists are opposed to any form of energy generation that promises to keep America growing in productivity and wealth.

Then there is the matter of global warming.

Notes

1. Energy Policy Staff of the White House Science Advisory Committee (with the Atomic Energy Commission, the Department of Health, Education, and Welfare, the Department of the Interior, the Federal Power Commission, the Rural Electrification Administration, the Tennessee Valley Authority, and the Council on Environmental Quality), "Electric Power and the Environment," August, 1970, xi. Emphasis added.
2. Paul Ehrlich, *"An Ecologist's Perspective on Nuclear Power"* in the *Federation of American Scientists Public Interest Report* 28, 5-6 (1975): 5.
3. "Nuclear Energy: Just the Facts," Nuclear Energy Institute, published online November, 2012, accessed October 29, 2014, http://www.nei.org/Master-Document-Folder/Publications-and-Brochures/Brochures/Just-the-Facts, 5.
4. Wolfram Krewitt, et al, "Health Risks of Energy Systems," *Risk Analysis* 18, 4 (August, 1998): 337. Krewitt and his colleagues did assign 2.4 cases of non-fatal cancer per terawatt to nuclear power generation and zero cases to any of the other methods.
5. Jon Palfreman, "Frontline: Why the French Like Nuclear Power," Public Broadcasting Service, accessed October 29, 2014,

http://www.pbs.org/wgbh/pages/frontline/shows/reaction/re adings/french.html.

6. "Annual Energy Review 2011," Energy Information Agency, United States Energy Information Administration, published online September, 2012, accessed October 29, 2014, http://www.eia.gov/totalenergy/data/annual/pdf/aer.pdf, 9.

7. Ibid., 7, 9.

8. Tim McMahon, "Oil Prices 1946-Present," published online March 6, 2014, accessed October 29, 2014, Inflationdata.com. Annually averaged and adjusted for inflation.

9. "Assessment of Undiscovered Technically Recoverable Oil and Gas Resources of the Nation's Outer Continental Shelf, 2011," U.S. Bureau of Ocean Energy Management, accessed Oct 10, 2013, http://www.boem.gov/National-Assessment-Map-2011/. The outer continental shelf extends from a line 3 nautical miles from land to a line 200 nautical miles from land.

10. Ibid.

11. Ibid.

12. Ibid.

13. Ibid. Proven reserves and undiscovered technically recoverable resources.

14. See "Report to the Secretary, US Department of the Interior: Survey of Available Data on OCS Resources and Identification of Data Gaps," 2009, accessed October 29, 2014, http://www.doi.gov/archive/ocs/report.pdf, figure II-7, p. II-16.

15. Dr. Bernard L. Weinstein, "The Outlook for Energy Production in the U.S. Gulf of Mexico: How the Regulatory Risk Premium is Restraining Production," Maguire Energy Institute, Cox School of Business, Southern Methodist University, May, 2012, accessed October 29, 2014, http://www.cox.smu.edu/c/document_library/get_file?p_l_id =68463&folderId=229433&name=DLFE-6063.pdf, 4-5.

16. "Assessment of Undiscovered Technically Recoverable Oil and Gas Resources," U.S. Bureau of Ocean Energy Management.

17. Ayesha Rascoe and Sofina Mirza-Reid, "U.S. Gives Shell OK to Begin Oil Drilling Prep in Beaufort Sea," *Chicago Tribune* (September 20, 2012), accessed October 29, 2014, http://articles.chicagotribune.com/2012-09-20/classified/sns-rt-us-shell-arctic-permitbre88j18e-20120920_1_top-hole-wells-beaufort-sea-arctic-oil.

18. "Shell's Beaufort Drilling Plans Dealt Costly Setback by Court," Anchorage Daily News, Aug. 16, 2007. Cited by Svend A. Brandt-Erichsen, "Despite the Challenges, Renewed Interest in Oil and Gas Development Focuses on Alaska," Marten Law, published online January 13, 2009, accessed October 29, 2014, http://www.martenlaw.com/newsletter/20090113-alaska-oil-development, n27.

19. The Resource Conservation and Recovery Act, the Comprehensive Environmental Response, Compensation and Liability Act ("CERCLA"), the Toxic Substances Control Act, the Surface Mining Control and Reclamation Act, and the Endangered Species Act all contain citizen suit clauses.

20. Jonathan A. Adler, "Stand or Deliver: Citizen Suites, Standing, and Environmental Protection," prepared for the symposium, "Citizens Suits and the Future of Standing in the 21stCentury:From Lujan to Laidlaw and Beyond," March 2-3, 2000, sponsored by the Duke Environmental Law and Policy Forum, accessed October 29, 2014, http://scholarship.law.duke.edu/cgi/viewcontent.cgi?article=1146&context=delpf, 50.

21. Undiscovered technically recoverable resources. "Analysis of Crude Oil Production in the Arctic National Wildlife Refuge," Energy Information Agency, U.S. Energy Information Administration, published online May, 2008, accessed October 29, 2014.

22. Ibid. This area is identified as Section 1002 in the 1980 Alaska National Interest Lands Conservation Act.

23. "Natural Gas Facts," US Dept of Energy, Office of Fossil Energy, National Energy Technology Laboratory, Strategic Center for Natural Gas and Oil, published online June, 2004, accessed October 29, 2014, http://www.netl.doe.gov/publications/factsheets/policy/policy 006.pdf.

24. "2010 Updated Assessment of Undiscovered Oil and Gas Resources of the National Petroleum Reserve in Alaska," Fact Sheet 2010-3102, USGS, posted online October 25, 2010, accessed October 29, 2014, http://pubs.usgs.gov/fs/2010/3102/.

25. Letter from Senators Mark Begich and Lisa Murkowski and Representative Don Young to the Secretary of the Interior, Ken Salazar, August 22, 2012. Cited in "Obama's Great Alaska Shutout," Wall Street Journal, updated online October 14, 2012, accessed October 29, 2014, http://online.wsj.com/articles/SB10000872396390443768804578040873921142716.

26. David W. Houseknecht, et al, "Assessment of Potential Oil and Gas Resources in Source Rocks (Shale) of the Alaska North Slope 2012 - Overview of Geology and Results," Fact Sheet 2012-3013, USGS, last modified online January 9, 2013, accessed October 29, 2014, http://pubs.usgs.gov/fs/2012/3013/.

27. This shale oil and gas is in addition to 10.03 billion barrels of oil equivalent contained in conventional deposits in the National Petroleum Reserve - Alaska.

28. "Review of Emerging Resources: U.S. Shale Gas and Shale Oil Plays," Energy Information Agency, United States Energy Information Administration, July, 2011, accessed October 29, 2014,

http://www.eia.gov/analysis/studies/usshalegas/pdf/usshaplela ys.pdf, 5.

29. Ibid. Undeveloped technically recoverable resources.

30. Ibid. Actually, the EIA estimate is not 424 trillion but 750 trillion cubic feet of *undeveloped* technically recoverable resources, a number which includes an estimated 410 trillion cubic feet for the Marcellus Shale, an estimate that the USGS reported as 84 trillion cubic feet of *undiscovered* technically recoverable resources and which I have used here. In addition to this 750 trillion, the *AEO2011* includes 35 trillion cubic feet of proved reserves reported to the Securities and Exchange Commission and the EIA, 20 trillion cubic feet of inferred reserves not included in the INTEK shale report, and 56 trillion cubic feet of undiscovered resources estimated by the USGS (total 18.50 BOE).

31. David Brooks, "The Shale Gas Revolution," Op-Ed, New York Times, Nov. 3, 2011.

32. "Natural Gas Consumption by End Use," Energy Information Agency, United States Energy Information Administration, published online September 30, 2014, accessed October 29, 2014,
http://www.eia.gov/dnav/ng/ng_cons_sum_dcu_nus_m.htm.
"Natural Gas Prices," Energy Information Agency, United States Energy Information Administration, published online September 30, 2014, accessed October 29, 2014,
http://www.eia.gov/dnav/ng/ng_pri_sum_dcu_nus_m.htm.

33. "Review of Emerging Resources: U.S. Shale Gas and Shale Oil Plays," Energy Information Agency, United States Energy Information Administration, published online July 8, 2011, accessed October 29, 2014,
http://www.eia.gov/analysis/studies/usshalegas/.

34. "Natural Gas Gross Withdrawals and Production," Energy Information Agency, United States Energy Information

Administration, published online July 8, 2011, accessed October 29, 2014, http://www.eia.gov/dnav/ng/ng_prod_sum_dcu_NUS_m.htm

35. "Rise in U.S. Oil Production Largest Since 1859, Output in 2013 Seen 7 Million BPD," press release, Energy Information Agency, United States Energy Information Administration, December 11, 2012, accessed October 29, 2014, http://www.eia.gov/radio/transcript/steo-oil-production-12112012.pdf.

36. "Widespread Economic Growth across States in 2011," press release, Bureau of Economic Analysis, U.S. Department of Commerce, June 5, 2012, accessed October 29, 2014, http://bea.gov/newsreleases/regional/gdp_state/2012/pdf/gsp 0612.pdf.

37. "State Employment and Unemployment, November, 2012," Bureau of Labor Statistics, U.S. Department of Labor, published online December 31, 2012, accessed October 29, 2014, http://www.bls.gov/opub/ted/2012/ted_20121231.htm.

38. "Average Application for Permit to Drill Approval Timeframes: FY 2005-FY 2011," Bureau of Land Management, updated online March 19, 2014, accessed October 29, 2014, http://www.blm.gov/wo/st/en/prog/energy/oil_and_gas/stat istics/apd_chart.html; "Memo to Colorado Oil and Gas Conservation Division," Colorado Department of Natural Resources, April 25, 2011, accessed October 29, 2014, http://cogcc.state.co.us/announcements/CommissionLtr4_25_ 11.pdf; "Ohio Oil and Gas Summary 2011," Division of Oil and Gas Resources Management, accessed October 29, 2014, https://oilandgas.ohiodnr.gov/portals/oilgas/pdf/oilgas11.pdf, 1.

39. Gary D. Libecap "Three Cheers for Fracking," *Defining Ideas*, Hoover Institution, March 5, 2014, accessed October 30, 2014,

http://www.hoover.org/publications/defining-ideas/article/170026.

40. "Hydraulic Fracturing Primer," American Petroleum Institute, July, 2014, accessed October 30, 2014, http://www.api.org/oil-and-natural-gas-overview/exploration-and-production/hydraulic-fracturing/~/media/Files/Oil-and-Natural-Gas/Hydraulic-Fracturing-primer/Hydraulic-Fracturing-Primer-2014-highres.pdf, 7.
41. In 2012, the Wall Street Journal reported that Dow Chemical was planning a new plant to take advantage of lower energy prices in the U.S., and Bloomberg reported that Austrian steel company Voestalpine AG was considering building a plant in the U.S., as were Nucor and four other American steelmakers. Bloomberg also reported that chemical producer LyondellBasell Industries NV was planning new plants along the Gulf Coast, as was fertilizer manufacturer CF Industries Holdings, Inc. Formosa Plastics Corp. USA will spend almost $2 billion to expand a plastics plant in Texas. Shell Oil will build a new plant in Pennsylvania near the Marcellus Shale gas deposits, and while Chevron Phillips Chemical Co. will build on the Gulf Coast. Westlake Chemical Corp. and Nova Chemicals Corp. also will expand at existing plants on the Gulf Coast and in western Canada. Daniel Gilbert, "Chemical Makers Ride Gas Boom," Wall Street Journal, April 18, 2012, accessed October 30, 2014, http://online.wsj.com/articles/SB10001424052702304331204577352161288275978. Sonja Elmquist, "Shale-Gas Revolution Spurs Wave of New U.S. Steel Plants: Energy," Bloomberg News, December 31, 2012, accessed October 30, 2014, http://www.bloomberg.com/news/2012-12-31/shale-gas-revolution-spurs-wave-of-new-u-s-steel-plants-energy.html. Frank Esposito, "Dow Chemical Picks Freeport, Texas for New Ethylene Cracker," Plastics News, April 19, 2012, accessed October 30, 2014,

http://www.plasticsnews.com/article/20120419/NEWS/3041
99960/dow-chemical-picks-freeport-texas-for-new-ethylene-
cracker.

42. The Sierra Club ardently supported the development of shale
gas resources until it came to light that the club had secretly
accepted $26 million in gifts from shale gas producer Chesapeake
Energy, whereupon the grass roots opposition to shale gas within
the Sierra Club prevailed and the club became an opponent of the
Shale Revolution.

43. "Vermont First State to Ban Fracking," CNN, May 17, 2012,
accessed October 30, 2014,

http://www.cnn.com/2012/05/17/us/vermont-fracking.

44. *"Draft Investigation of Ground Water Contamination near Pavillion,
Wyoming,"* U.S. EPA, December 8, 2011, accessed October 30,
2014,

http://www2.epa.gov/region8/draft-investigation-ground-
water-contamination-near-pavillion-wyoming.

45. Thomas E. Doll, State Oil and Gas Supervisor, Wyoming Oil
and Gas Conservation Commission, testimony on "Fractured
Science: Examining EPA's Approach to Ground Water Research
in Pavillion, WY," House Subcommittee on Energy and the
Environment of the Committee on Science, Space and Technology
during the 112th Congress, February 1, 2012, accessed October
30, 2014,

http://science.house.gov/sites/republicans.science.house.gov/fi
les/documents/hearings/HHRG-112-SY20-WState-TDoll-
20120201.pdf, 3.

46. 42 U.S.C. section 9606(a) (CERCLA).

47. The EPA wrote, "While CERCLA allows for groundwater
plumes, it is generally based on a point source or sources,
exposure pathways and targets. Therefore a plume with no
documented source or location presents a poor fit." "Site
Inspection—Analytical Results Report: Pavilion Area

Groundwater Investigation Site, Pavilion, Fremont County, Wyoming," U.S. EPA, August, 2009, accessed October 30, 2014, http://www2.epa.gov/sites/production/files/documents/Pavill ion_GWInvestigationARRTextAndMaps.pdf, p. 3.

48. Website of Congressman Pete Olson of Texas, April 27, 2012, accessed October 30, 2014,
http://olson.house.gov/2012-press-release/olson-colleagues-urge-dismissal-of-epa-region-vi-administrator-armendariz/.

49. Ibid.

50. "Because this response action could be considered nationally significant or precedent setting, it requires the prior concurrence of the Assistant Administrator, Office of Solid Waste and Emergency Response (AA-OSWER)." Richard Fetzer, "Action Memorandum - Request for Funding for a Removal Action at the Dimock Residential Groundwater Site, Intersection of PA Routes 29 & 2024, Dimock Township, Susquehanna County, Pennsylvania, January 19, 2012, accessed October 30, 2014, http://www.fossil.energy.gov/programs/gasregulation/authoriz ations/Orders_Issued_2012/58._EPA_III.pdf, 8.

51. Email from Richard Fetzer to colleagues, February 18, 2010, accessed October 30, 2014,
http://eidmarcellus.org/marcellus-shale/epa-dimock-distraction-ignoring-pollution-while-chasing-tips/13970/.

52. "USGS Releases New Assessment of Gas Resources in the Marcellus Shale, Appalachian Basin," USGS, August 23, 2011, accessed October 30, 2014,
http://www.usgs.gov/newsroom/article.asp?ID=2893&from=r ss_home.

53. Heesun Wee, "In Some Regions, Fracking Opponents Push Back Hard," CNBC, June 20, 2012, accessed October 30, 2014, http://www.cnbc.com/id/47279934/In_Some_Regions_Fracki ng_Opponents_Push_Back_Hard

54. Editorial, "Drill Deeper, New York," (Albany) Times Union,

January 14, 2013, accessed October 30, 2014,
http://www.timesunion.com/opinion/article/Editorial-Drill-
deeper-New-York-4193731.php%20%20%201/14/13.
55. Under present circumstances, presenting this fiscal windfall to
the State of California would be like giving an alcoholic the key to
the liquor cabinet. It would simply delay an inevitable reckoning.
56. "Gauging the Prospects of a U.S. Oil Shale Industry," Rand
Corporation, 2005, accessed October 30, 2014,
http://www.rand.org/pubs/research_briefs/RB9143/index1.ht
ml. The USGS estimates there to be 1.53 trillion barrels in the
Piceance Basin of the Green River, 1.32 billion in the Uinta Basin,
and 1.44 billion in the Green River Basin of the Green River
Formation, for a total of 4.29 trillion barrels.
http://pubs.usgs.gov/dds/dds-069/dds-069-y/,
http://pubs.usgs.gov/fs/2011/3063/) These are "in place," as
opposed to "technically recoverable" estimates, which means that,
given present technological capabilities, only a fraction of these
amounts could be recovered and used as oil. But if only one third
of these amounts could be used, that would supply the U.S., are
the present rate of usage, for almost 600 years.
57. "Natural Bitumen Resources of the United States," National
Assessment of Oil and Gas Fact Sheet, USGS, 2006, accessed
October 30, 2014,
http://pubs.usgs.gov/fs/2006/3133/pdf/FS2006-
3133_508.pdf. 12 billion barrels of oil would fuel the U.S. for
almost two years.
58. "Coal-bed Methane: Potential and Concerns," USGS,
October, 2000, accessed October 30, 2014,
http://pubs.usgs.gov/fs/fs123-00/fs123-00.pdf.
59. See, for example, Mary Griffiths, "Coalbed Methane Sparks
Debate: Who's to Blame for Gassy Water?" Calgary Herald,
March 12, 2006, Pembina Institute, accessed October 30, 2014,
http://www.pembina.org/op-ed/1223.

60. "Gas (Methane) Hydrates—A New Frontier," USGS, September, 1992, Cal State Los Angeles, accessed October 30, 2014, http://web.calstatela.edu/academic/natsci/zzstuff/Transfer/Ur ban_Geology/357_Lectures/357_Lecture14/meth_hydrates_sto ry.htm. By October of 2014, for some reason, the USGS had removed this fact sheet from their website.

61. Ibid.

62. A dozen small refineries have opened since 1977. Frequently Asked Questions, "When was the last oil Refinery built in the United States?" U.S. EIA, June 25, 2014, accessed October 30, 2014, http://www.eia.gov/tools/faqs/faq.cfm?id=29&t=6.

63. As of Jan., 1, 2012. "Coal Production, Selected Years, 1949-2011," U.S. EIA, accessed October 30, 2014, http://www.eia.gov/totalenergy/data/annual/pdf/sec7_7.pdf.

64. "Coal Consumption by Sector, Selected Years, 1949-2011, U.S. EIA, accessed October 30, 2014, http://www.eia.gov/totalenergy/data/annual/pdf/sec7_9.pdf. As of the end of the third quarter of 2002, the U.S. was on a pace to consume only .885 billion tons for the year.

65. That the U.S. Court of Appeals in Washington, D.C. declared the Cross State Air Pollution rule unconstitutional in August of 2012 does not invalidate it as an expression of the environmentalist ideology.

66. Joseph, Weber, "Obama Policies Hammered Following Coal-Mine Closings, Layoffs," Fox News, September 19, 2012, accessed October 30, 2014, http://www.foxnews.com/politics/2012/09/19/romney-pro-business-groups-blame-obama-polices-on-recent-mine-closings/.

9: Global Warming

THE BURNING OF fossil fuels, such as oil, natural gas, and coal, produces gasses that accumulate in the atmosphere. The quantity of fossil fuels burned every day in the world has increased dramatically since the start of the Industrial Revolution, and so have the concentrations of the resultant gasses in the air. Next to water vapor, the second most plentiful such gas in the atmosphere is carbon dioxide. In 1958, at a facility of the National Oceanic and Atmospheric Administration at Mauna Loa, Hawaii, the Scripps Oceanic Institution began measuring and recording CO_2 levels in the atmosphere. These measurements, which have continued to this day, show that atmospheric CO_2 levels have risen every year since 1958, that the rate of increase has risen steadily over that period, and that the concentration of CO_2 in the air is now about 25 percent greater than it was when the measurements began.[1]

Scientists believe that CO_2 and other gasses trap heat from the sun, much as a greenhouse does, and thereby increase the earth's ambient temperature. There is general agreement that the earth has undergone some warming since 1900. The UK Weather Service, which monitors world climate, estimates the increase to be about 1 degree Fahrenheit.[2] One possible consequence of rising worldwide temperatures, according to environmentalists, is that sea levels would also rise as polar ice melts. The UK Weather

Service reports that sea levels worldwide have risen about 6.75 inches since 1900.[3]

But the question of whether an increase in atmospheric CO_2 and other "greenhouse" gasses is indeed causing worldwide temperatures to rise is a matter of scientific dispute. First, it is not clear that temperatures are indeed rising. The data show that there has been no net gain in world-wide temperatures since 1997.[4] Second, it is not at all clear that even if some longer range rise in temperatures is occurring, from which the leveling since 1997 is a temporary departure, it is being caused by human action.[5] Dramatic, naturally caused changes in worldwide temperatures, some quite rapid, are a matter of historical record. The Medieval Warm Period, for example, which peaked about 1000 AD and was similar in intensity to the current warming, had to have been caused by some natural phenomenon, since it clearly was not caused by elevated CO_2 levels produced by industrial economies.[6]

If further warming of the earth were to occur, the question whether someone would see it as a positive or a negative development might well depend on his circumstances. A person living in northern Minnesota might welcome warmer temperatures, while one living just above sea level in the Maldives Islands of the Indian Ocean might not. Regardless of whether an increase in atmospheric CO_2 causes higher temperatures, it does cause plants to grow better. It might well improve world food supplies, a development that much of the world would no doubt welcome.

There is no answer to the question whether global warming would be good or bad for the people of the world considered collectively; it would be good for some individuals and perhaps bad for others, and whether it would be good or bad for any given individual would depend on his circumstances and on what he values. (Obviously, a warming severe enough to wipe out half the people on earth would be objectively bad. But there is no more evidence that we are headed for such a warming, much less that man is causing it, than that mankind will be wiped out by an asteroid next week.)

But Environmentalists see increased concentrations of CO_2 and other gasses in the atmosphere as a disruption of the balance of nature, and they see global warming as categorically bad. Considering their view of global warming, one would expect them to embrace nuclear power as the perfect solution to the global warming problem, *if they valued economic growth*. It would provide an economically feasible method of generating almost unlimited amounts of electricity without emitting any "greenhouse gasses." But even in the face of a global warming "crisis," environmentalists continue to oppose the development and use of nuclear power.

There are other ways that environmentalists could work to bring about reductions in carbon emissions; for one, they could try to persuade people voluntarily to drive smaller or hybrid cars, to use mass transit, to live in smaller houses, to recycle, to consume "organic" food, or to purchase fewer clothes and furniture and electronic goods. It would be especially appropriate for environmentalists to base their campaign to reduce carbon emissions on voluntary compliance, because many people simply do not agree that reducing carbon emissions is either desirable or necessary; they do not see it as desirable because, like the northern Minnesotan, they would enjoy warmer temperatures, or they do not see it as necessary because they find the environmentalists' science unpersuasive. Either way, there are perfectly rational and morally defensible reasons to choose not to reduce one's carbon emissions. In a free society, individuals should be left to make up their own minds about such matters.

Environmentalists have certainly not neglected to urge people voluntarily to do all these things. Indeed, they have raised

Americans' consciousness of carbon emissions more successfully than anyone would have imagined possible thirty years ago. But despite environmentalists' best efforts, most people continue to burn fossil fuels and pursue consumer life-styles with abandon. So, having failed to win over more people by persuasion, environmentalists have resorted to force.

One way to force everyone to reduce their carbon emissions would be for the government specify how much CO_2 and other gasses may be released within a given geographic area over a given period of time. Users of fossil fuels would then be able to purchase, and exchange, the right to emit a certain fraction of that total amount of CO_2 and other gasses. The State of California decided in 2012 to enact such a cap-and-trade program; as part of a plan to reduce total "greenhouse gas" emissions in the state to 1990 levels, it would reduce such emissions "from regulated entities" by 16 percent by 2020.[7] On the federal level, the House of Representatives approved a cap-and-trade bill in June of 2009, but the Senate failed to act on it. In response, the EPA decided to limit "greenhouse gas" emissions by fiat. In December of 2009, Administrator Lisa Jackson signed an endangerment finding officially declaring CO_2 and five other gasses to be pollutants, thereby enabling the EPA to regulate their emission.[8] In June of 2014 the Obama Administration mandated a 30 percent decrease in carbon emissions from fossil fuel-burning power plants by 2030.

One practical problem with placing limits on the release of so-called greenhouse gasses is that it forces users of fossil fuels either to capture the gasses and then transport and store them, or to reduce or eliminate their use of fossil fuels altogether and substitute other sources of energy. The process of capturing, transporting, and storing waste gasses is expensive; recall presidential candidate Barack Obama's 2008 comment to the San Francisco Chronicle that if newly-constructed electric power plants chose to burn coal, it would bankrupt them. But if the users of fossil fuels choose instead to reduce or eliminate their use of those fuels, it will prove enormously expensive for them to substitute other fuels; the reason they are using fossil fuels in the first place is that they are much less expensive than anything else. (Nuclear fuel would be economically competitive if it were freed of the burdens of over-regulation and litigation.)

The primary candidates to replace fossil fuels are wind and solar power. But wind is an unreliable source of power; ten wind farms in Kansas, for example, average about 37 percent capacity, whereas nuclear power plants produce at 90 percent of capacity.[9] Wind farms tend to take up a great deal more land than do conventional power plants; the Arkansas Nuclear One plant covers 1,100 acres and generates 1,800 megawatts. It would require 720 wind turbines covering 108,000 acres to generate 1,800 megawatts, and that is only if the wind turbines operated at 90 percent of capacity.[10] And because they must be built where the wind blows strongest, such as near mountain passes, wind farms are often situated far from the cities where the power is needed, thus necessitating the construction of long transmission lines.

These factors make wind power much more expensive than power generated by the burning of fossil fuels. After decades of government subsidies to wind power companies, they produced only 2.9% of the electricity generated in the United States in 2011.[11] Beyond these economic arguments, though, wind farms tend to offend environmentalists, precisely because they require that vast swaths of the landscape be covered with large machines and transmission lines. The Cape Wind Project, for example, which developers hope to build in the ocean between Cape Cod and Nantucket Island, has engendered fierce opposition from environmentalists in Massachusetts.

Solar farms also must cover huge areas in order to produce

significant amounts of electricity. In order to produce the 1,800 megawatts of the Arkansas Nuclear One plant, a solar farm would require 13,320 acres, if it produced at 90 percent capacity.[12] (Solar reflectors operate at only 25 to 33 percent of capacity on average.[13]) Thus, solar farms also tend to engender opposition from environmentalists, as has been the case, for example, with BrightSource's Ivanpah Solar Electric Generating System, First Solar's Desert Sunlight Solar Farm, and the Amargosa Farm Road solar project (pending as of 3/15/13), all in remote locales in the California desert. Solar power also cannot compete economically with fossil fuels. In 2012, again after decades of government subsidies, solar power generated only .1 percent of the electricity produced in the United States.[14]

Energy is the lifeblood of an industrial economy. To authorize governments to limit emissions of CO_2 and other gasses would be to give them the power of life and death over the economic activities of their citizens. Given the absence of economically viable alternatives to fossil fuels, to limit carbon emissions would be to limit productive human endeavor altogether. We simply could not manufacture as much as we do, produce as much food and clothing, construct as many buildings and roads and bridges, nor heat as many homes as we do now. Entrepreneurship and technological invention would diminish, and economic growth would come to a halt.

The meaning of California's plan to reduce carbon emissions to 1990 levels is, in effect, that the growth of productivity must stop, if not be reduced, a process that has already begun as California's producers move to other states.[15] To limit carbon emissions would be to impose a lower standard of living on the people of the United States—and keep it there. It would also diminish our competitiveness in international trade, and, because economic strength is the foundation of national defense, it would render us more vulnerable to foreign enemies in a very dangerous

Thomas McCaffrey

world. No nation subject to such limitations could remain a free people for long.

By way of analogy, imagine authorizing the federal government of the United States to limit the number of books and magazines and newspapers that could be printed, or the number of TV and radio stations that could broadcast, or the number of websites that could be put on the internet, or the number of churches there could be. Because of our commitment to the free exchange of ideas, we would not dream of giving such power to government (at least not yet), because to do so would be to invite tyranny.

For the same reason, it should be unthinkable that we would authorize our government to limit the productivity of our people by restricting carbon emissions. "Ah," the environmentalists would say, "but books and newspapers and churches, that is, *ideas*, cannot render portions of our planet uninhabitable." Tell that to the millions upon millions who perished as a result of bad ideas in Nazi-occupied Europe, in Soviet Russia, and in Communist China. Indeed, tell it to the millions who have died of malaria in third world countries since environmentalists conceived the idea of outlawing DDT. Or to the citizens of Detroit, whose post-apocalyptic landscape stands as a monument to modern liberal ideology.

Bad ideas are dangerous things, capable of immeasurable harm. It is not because the free exchange of ideas can have no devastating consequences that we protect it; it is because the only hope of reaping the benefits of good ideas is to make them compete with bad ideas; to do otherwise would be to surrender to tyranny at the outset. When we choose to be a free people, we commit ourselves to risking the consequences of bad ideas. Bad ideas and their sometimes terrible consequences are an occupational hazard of living in a free society.

If, in the case of carbon emissions, environmentalists would

422

choose freedom over tyranny, then CO_2 and other gasses would continue to accumulate in the atmosphere, at least for a while, and other consequences might ensue that some would prefer not to happen. But freedom would also mean prosperity, and material abundance, and leisure, and good health, and long life, and countless other benefits. And, because freedom promotes entrepreneurship and technological innovation, it would hold the promise, long-range, of new technological means to reduce carbon emissions without sacrificing individual freedom. We already have, in nuclear power, the ability to produce most of our energy without emitting greenhouse gasses. A free and prosperous and technologically innovative economy would offer the best hope of one day developing new modes of transport, and perhaps a whole economy, that emit would little or no greenhouse gas. The price for all this would be continued carbon emissions for the time being. The alternative would be to choose tyranny at the outset.

The Puritans believed God gave them the right to force people to behave righteously. Environmentalists believe science gives them that right. Indeed, they have succeeded remarkably in selling the idea that the question of whether government should limit carbon emissions is essentially a *scientific* question; if the earth is warming, and if that warming is being caused by the accumulation of CO_2 and other gasses in the atmosphere—both scientific questions, to be sure—then *obviously* government should impose limits on the emission of those gasses. But this is a textbook example of the preservationist premise at work; two matters of scientific fact, *is* human-caused global warming occurring, and what *will be* the consequences if it is occurring, are packaged together with two moral imperatives, that the alleged problem of human-caused global warming *ought* to be dealt with collectively, which means, politically, rather than by each of us acting individually and voluntarily, and that acting thus collectively we *ought* to halt the increase of greenhouse gasses and, as much as possible, return the atmosphere to its natural state.

The first two questions, the scientific ones, are certainly important; those who would impose tyranny on the rest of us in the name of science should at least be made to prove the validity of their science. But the second two questions, whether we should deal with the alleged human-caused global warming collectively or individually, and what, if anything, we should about it are not questions of science, but of morality. Environmentalists implicitly recognize this, of course, in their condemning of anyone who disagrees with them on the question of global warming as morally depraved.

But they treat it as a foregone conclusion, nonetheless, that if global warming is occurring and if human activity is a contributing cause, then governments ought forcibly to limit human carbon emissions. They argue that the problem of global warming is so serious, the consequences potentially so devastating, that the normal rules and principles that have governed our political life no longer apply. We are like the occupants of a lifeboat who must decide whom to throw overboard first. Constitutional principles and individual rights simply cannot be allowed to get in the way of what science tells us we must do in this emergency.

What environmentalists are really saying here is that freedom does not work. We tried it, and it failed. We allowed people to be free, and they brought the planet to the brink of destruction. Now it is time to bring freedom to an end. This is the meaning of allowing the global warming question to be decided by "science." There is good science and there is bad science. There is the science that put a man on the moon, and there is the "science" that stifled the development of nuclear power and that banned DDT. And the point at which the science of a question becomes "settled" is often very difficult to discern. A free people should no more be ruled by the government's idea of scientific truth than by the government's idea of God's Will.

All the more reason, then, that we should resist the argument of some environmentalists, the so-called Precautionary Principle, that the consequences of global warming are so potentially devastating that we should not wait until all the science is worked out before we take forceful steps to curtail carbon emissions. To follow the Precautionary Principle and authorize the government to place limits on human productive endeavor would be to choose tyranny *on the mere chance* that elevated CO_2 concentrations might cause global warming and that such a warming would have harmful consequences for man. We should no more do this than we should abolish free speech because someone *might* advocate an idea like socialism. In a free society, the risk of unwelcome consequences, both in the realm of ideas and in the realm of economic endeavor, comes with the territory.

But, environmentalists argue, we in the U.S. have a moral obligation to the Maldives Islanders and other people living near sea level, including Americans, to reduce our carbon emissions. Even if it were proven someday that elevated CO_2 levels do indeed cause some degree of global warming, and that such warming will cause the oceans to rise, it would no more be true that we have such an obligation to the Maldives Islanders than that I have a moral obligation not to put a second story on my house because it would block my neighbor's view and diminish his property value. In the latter case, as far as my neighbor's property is concerned my only moral obligation is not to violate his rights by invading or physically damaging his property.

In the case of the Maldives Islanders, if I emit CO_2 over my land, and if it migrates to the atmosphere, and if it contributes in some miniscule way to causing world-wide temperatures to rise (a causal relation so far unproved), and if that causes sea levels to rise (also unproved), and if that chain of events should then be taken to constitute a violation of the rights of the Maldives Islanders, then virtually anything I do can be construed as a violation of someone's rights, and my own freedom is at an end.

The idea that I have such a moral obligation to the Maldives Islanders is another example of the morality of selflessness, a distant echo of the Christianity that so many environmentalists eschew, but which lies at the root of much of their thinking. (Another such echo is the asceticism implicit in the environmentalist position on global warming, the idea that to get by with less material wealth, a smaller car, a smaller house, or fewer possessions, is morally admirable.)

If anyone wanted to stop economic growth, to halt the building of power plants and factories and skyscrapers and shopping malls and housing developments, he could not have devised a crisis better suited to the purpose than global warming. It provides environmentalists with seemingly plausible grounds for curtailing our use of fossil fuels, which are our primary source of energy. Without energy, no other economic activity is possible, at least not on an industrial scale.

Combined with all the other ways environmentalists have fought to limit our access to fossil fuels, and combined with their continued stifling of nuclear power, which is the one truly viable alternative to fossil fuels in power generation, their position on global warming makes it difficult to avoid the conclusion that they really do not want to there to be cheap, abundant energy made available to fuel the American economy. The forms of energy generation that environmentalists do promote, such as wind, solar, and geothermal, are so economically infeasible at present that they could not come close to fueling the American economy on its present scale, and even these forms of energy are opposed by some of the more ideologically consistent environmentalists.

If we knew nothing about environmentalists' positions on other issues, if all we knew about them were their positions on energy generation, we would be fully justified in concluding that they are actively opposed to economic growth, and that they would prefer to see the remainder of undeveloped America left as it is. We would be fully justified in concluding that environmentalists—all of them, not just the radicals—are essentially nature *preservationists*.

Notes

1. "Recent Monthly Mean CO_2 at Mauna Loa," National Oceanic and Atmospheric Administration, September, 2014, accessed October 31, 2014,

http://www.esrl.noaa.gov/gmd/ccgg/trends/.

2. "Global Average Temperature Records," Met Office (UK Weather Service), October 2, 2013, accessed October 31, 2014, http://www.metoffice.gov.uk/climate-

change/guide/science/explained/temp-records. This estimate is based on data from the UK Weather Service in collaboration with the Climatic Research Unit at the University of East Anglia, from Goddard Institute for Space Studies, which is part of NASA, and from the National Climatic Data Center, which is part of the National Oceanic and Atmospheric Administration.

3. "Climate Monitoring and Attribution," Met Office (UK Weather Service), March 14, 2014, accessed October 31, 2014, http://www.metoffice.gov.uk/climate-change/guide/how#Sea-level-rise. For a discussion of why a modest increase in global temperature might cause a *drop* in sea levels, see S. Fred Singer's excellent discussion on PBS in 2000 at:

http://www.pbs.org/wgbh/warming/debate/singer.html.

4. "RSS Global TLT Temperature Anomalies," Remote Sensing Systems, March, 2015, accessed April 28, 2015, http://data.remss.com/msu/monthly_time_series/RSS_Monthly_MSU_AMSU_Channel_TLT_Anomalies_Land_and_Ocean_v03_3.txt. When these data are plotted, the trend line established by least-squares regression analysis has a slope of zero. See, for example, the graph at:

http://o.b5z.net/i/u/10152887/f/Irish_document_what_went
_wrong_with_settled_science.pdf, which includes data through
December, 2014.

5. It is also possible, of course, that elevated CO_2 levels would be
causing global warming right now, except that some as yet
inadequately understood natural (or man-made) factor is
mitigating it.

6. See "The Medieval Warm Period," National Oceanic and
Atmospheric Administration, August 20, 2008, accessed October
31, 2014,
http://www.ncdc.noaa.gov/paleo/globalwarming/medieval.ht
ml.

7. "California Cap and Trade," Center for Climate and Energy
Solutions, accessed October 31, 2014,
http://www.c2es.org/us-states-regions/key-
legislation/california-cap-trade.

8. The other gasses are methane (CH_4), nitrous oxide (N_2O),
hydrofluorocarbons (HFCs), perfluorocarbons (PFCs), and sulfur
hexafluoride (SF_6).

9. "Kansas Wind Farms—Capacity Factors and Monthly Average
Power Generation," Kansas Energy Information Network, 2013,
accessed October 31, 2014,
http://www.kansasenergy.org/documents/Kansas_wind_capacit
y_Factors_122012.pdf. This website used data provided by the
U.S. Energy Information Administration, "Annual Electric Utility
Data."

10. This assumes 2.5 MW turbines covering 60 acres each. A
compilation of 107 wind farms by the American Wind Energy
Association shows an average acreage per megawatt of about 90
acres. "Acres of Industrial Wind Facilities," American Wind
Energy Association, accessed October 31, 2014,
http://www.aweo.org/windarea.html.

11. "Table 1.2. Summary Statistics for the United States, 2002-

2012," Tables 3.1.A. and 3.1.B, U.S. Energy Information Administration, accessed October 31, 2014,
http://www.eia.gov/electricity/annual/html/epa_01_02.html.
12. This assumes 1 Megawatt per 7.4 acres of photovoltaic solar panels. A recent estimate put the figure at about 1 megawatt per 8 acres. Dave Levitan, "Report Counts Up Solar Power Land Use Needs," Institute of Electrical and Electronics Engineering Spectrum, August 7, 2013, accessed October 31, 2014,
http://spectrum.ieee.org/energywise/green-tech/solar/report-counts-up-solar-power-land-use-needs.
13. "Table 6.7.B. Capacity Factors for Utility Scale Generators Not Primarily Using Fossil Fuels, January 2008, August 2014," U.S. Energy Information Administration, accessed October 31, 2014,
http://www.eia.gov/electricity/monthly/epm_table_grapher.cf m?t=epmt_6_07_b.
14. "Table 1.2. Summary Statistics for the United States, 2002-2012," U.S. Energy Information Administration, accessed October 31, 2014,
http://www.eia.gov/electricity/annual/html/epa_01_02.html, Tables 3.1.A. and 3.1.B.
15. See "The Price of Green Virtue," Wall Street Journal, updated July 9, 2012, accessed December 18, 2014,
http://www.wsj.com/articles/SB1000142405270230487030457 7491152903293004. "The first study … estimates the price tag for … cap-and-trade taxes on carbon emissions, a 'low carbon fuel standard,' and a stringent 33% renewable mandate for electricity production. Together these policies raise energy costs and are expected to reduce state GDP by between 3.5% and 8.9% by 2020."

10: Conservation of Natural Resources

ENVIRONMENTALISM IS COMPRISED of three broad categories of endeavor, the war on pollution, the effort to conserve natural resources, and the effort to preserve land and wildlife. We have seen that the war on pollution is essentially preservationist; in the name of eradicating pollution, environmentalists seek to limit the productive use of land and other resources in order to preserve them in their natural states. As for the second category, the effort to conserve natural resources such timber and mineral wealth, recall that Gifford Pinchot had believed that such resources exist for man's use and that the purpose of conservation was to ensure that America's natural resources would be used in a sustainable way, but *used* nonetheless, for the greatest good of the greatest number of human beings.

But since the rise of environmentalism, the old Pinchotian belief in making use of America's natural resources has increasingly been superseded, in the name of "conserving" them, by efforts to lock them up, public and private alike, and to transform the lands and waters where they are found into nature preserves. The effort to turn Pinchot's utilitarian "conservation of resources" in a preservationist direction dates from the very beginning of the environmental movement and is another manifestation of its essentially preservationist nature.

Resources and Man

In the late 1960s, the National Academy of Sciences and the National Research Council assembled a committee, funded in part by the federal government, to consider the adequacy of America's—and the world's—supplies of natural resources in relation to demand, both present and future.[1] The committee issued its report, *Resources and Man*, in 1969. Their view of natural resources was explicitly Malthusian; the amount of space on the surface of the earth that is occupiable by man, as well as the carrying capacity of that space, have "finite limits." [2] The period of rapid economic and population growth that has characterized the last few centuries, wrote contributor M. King Hubbert, will turn out to have been a remarkable aberration in human history.[3]

A Malthusian view of the world tends to lead to calls for conserving resources either by restricting access to them or by reducing the demand for them. One way to restrict access to natural resources is to preserve the lands where they are found. On page one of *Resources and Man*, the committee signaled their preservationist leanings with their version of the preservationist premise: "[C]an man approach a kind of dynamic equilibrium with his environment so as to avert destructive imbalances?" [4] As for reducing the demand for natural resources, what better way than to restrict economic growth, which the committee described as a threat to man's quality of life? [5]

In accord with their hostility toward economic growth, and in the spirit of Rachel Carson, the committee favored censorship of new technologies: "The gains from technological development must always be balanced ... against its costs" (implying that government must do the balancing, because only government can *enforce* whatever determinations are made as to the potential gains that new technologies promise as compared with their costs).[6] The report's anti-Baconism was explicit: "Our goal should not be to conquer nature, but to live in harmony with it." [7] Malthusianism easily lends itself to nature preservationism.

Like Gifford Pinchot, the committee saw natural resources as the collective property of "the people," but, whereas the old utilitarian conservationists meant the American people, these new preservationists meant "all mankind." The "overconsumption or waste [of natural resources] for the temporary benefit of the few who currently possess the capability to exploit them cannot be tolerated." [8] Note that "the few who currently possess the capability to exploit" natural resources are mostly the citizens of the industrialized countries. Note also that "the few" includes the owners of the private lands where many of those resources are found; this statement is an implicit attack on capitalism.

As advocates of collective ownership of natural resources, the committee had little concern for the rights of individuals, as their position on population control evinces:

> *Our Departments of State and Health, Education, and Welfare should adopt the goal of real population control both in North America and throughout the world. Ultimately this implies that the community and society as a whole, and not only parents, must have a say about the number of children a couple may have.* [9]

In contrast to the committee's apparent sympathy for nature preservation, its list of formal recommendations sounds surprisingly utilitarian, calling consistently for further exploration for and development of natural resources. The contradiction is illusory, though. Recommendation 19 advocates the formation of an additional, follow-up committee to study the "various social, psychological, legal, medical, religious, and political aspects of the problems of resources and man," such as "How can cultural preferences be altered so as to relieve demand on resources?" [10]

The committee seems to have accepted as their specific mission the job of identifying how the U.S. could best solve the problem of increasing the supply of natural resources, and this is the subject of most of most of their recommendations.[11] But, as Recommendation 19 suggests, they recognized that the long-range solution to the problem would also necessitate reductions in the demand for resources, reductions that would require fundamental changes in the values of the industrialized peoples, such as their preference for a consumer lifestyle and their belief in the possibility of perpetual economic growth. To a great extent, the committee believed, the long-term solution would need to be a preservationist one.

The President's Commission

The question of population growth is relevant to all three of the broad categories of environmentalist endeavor. More people mean more pollution, increased use of natural resources, and increased loss of wild lands and wildlife. Nevertheless, the topic of population growth tends to come up most often in discussions of conserving natural resources.

The President's Commission on Population Growth and the American Future, which was chaired by John D. Rockefeller III, sent its report to President Nixon on March 27, 1972. At the time, concerns about population growth were a more prominent part of American environmentalism than they are today. The commission identified three different approaches to solving the population "problem." The first advocated contraception and abortion as ways to reduce birth rates. Over the course of the next generation, contraception and abortion would contribute to a dramatic decrease in birth rates throughout the industrialized West. The second approach called for efforts to combat racism, which the committee said placed "undue pressure toward childbearing and child-rearing." [12] The third approach viewed population growth from an ecological perspective.

"Human life … is supported by intricate ecological systems that are limited in their ability to adapt to and tolerate changing conditions." [13] To put it differently, there exists a balance of nature, which man should refrain as much as possible from upsetting. A growing population threatens to upset that balance because it entails "more rapid depletion of domestic and international resources" and "greater pressures on the environment." [14] Indeed, "[h]uman culture, particularly science and technology, has given man an extraordinary power to alter and manipulate his environment," a power which he has used "to plunder and destroy rather than to conserve and create." [15]

So the commission advocated nothing less than "a basic recasting of American values." [16] (Recall Recommendation 19 of *Resources and Man* in the preceding section.) "Mass urban industrialism is based on science and technology, efficiency, acquisition, and domination [of nature] through rationality. The exercise of these same values now contains the potential for the destruction of humanity." [17] In other words, man can not command nature on an industrial scale without destroying nature and himself, so he should stop trying. "[N]othing less than a different set of values toward nature, *the transcendence of a laissez-faire market system*, a redefinition of human identity in terms other than consumerism, and a radical change if not *abandonment of the growth ethic*, will suffice." [18] This is environmentalism in a nutshell; it is hostile to science (in the service of extending man's technological command of nature), to industrialism, to economic freedom, and to economic growth, and it favors *preserving* nature over putting it to human use.

One Third of the Nation's Land

The federal government owns one third of all the land in the United States. In the late 1960s, Congress established a commission to review the laws governing how that land should be used. The 1970 final report of the Public Land Law Review Commission showed the influence of environmentalism. Among their recommendations, the commission proposed that "the enhancement and maintenance of the environment, with rehabilitation where necessary," be given equal priority with all other uses of the federal lands.[19] Note the proposed transition from using the nation's public lands to preserving them: the first step would be to give to the restoration and maintenance of the natural environment equal priority with logging, mining, oil and gas drilling, grazing, waterway management (including flood control and hydroelectric power generation), and recreation.

The commission acknowledged the central premise of ecology, "Everything is connected to everything," they wrote.[20] They recommended that, on federal lands, permission to engage in productive endeavors that entailed "severe, often irreversible impacts," such as the building of transmission lines, roads, dams, and open pit mines, as well as timber harvesting, "extensive chemical control programs," oil drilling on the outer continental shelf, and "high density recreational developments," should be conditioned upon "a detailed study of their potential impact on the environment."[21] This, of course, is the preservationist premise in action: as much as possible, man must refrain from upsetting the balance of nature. (This recommendation would be realized in the requirement that prospective users of federal lands first produce an Environmental Impact Statement, which was a first step in the direction of government censorship of land use.)

Perhaps recognizing the threat that all this ecologizing posed to Americans' ability to continue to make productive use of the federal lands, the commission included a number of recommendations intended to ensure that certain productive uses, such as timber cutting and mining, would continue on federal lands. For example, they called for a statutory requirement that "those public lands that are highly productive for timber be classified for commercial timber production as the dominant use." [22] In the event, the commissioners' fears that all this concern for the environment might eventually lead to reductions in the productive use of the public lands would prove well founded.

Forced Conservation of Resources

To Gifford Pinchot and the utilitarians, "conservation" did not mean curtailing the use of natural resources. It meant government management of those resources to ensure that they would be used in a sustainable manner. Today's environmentalists, though, explicitly advocate restricting the use of natural resources in the name of "conserving" them. The reductions in use, though, usually serve the purpose of nature preservation as much as (or more than) they serve to conserve resources.

Consider the campaign to replace fossil fuels with "renewable" energy sources such as solar and wind power. If environmentalists could eliminate the use of all fossil fuels tomorrow and replace them with wind and solar power, it would indeed serve to conserve fossil fuels such as oil, gas, and coal. But it would also make it impossible to maintain our industrial economy on anything like its present scale. Solar and wind power would simply be too expensive to allow us to maintain anything close to our current levels of productivity. The resultant collapse of our productivity would cause economic growth to come to a halt and with it the building of new homes and shopping centers and office buildings and highways. The effect of the current campaign to replace fossil fuels with wind and solar power, to whatever extent it succeeds, will be profoundly preservationist.

Consider, as another example of preservation in the name of

conservation, government-prescribed auto fuel efficiency standards. Although these have succeeded in forcing auto makers to produce cars that get dramatically better gas mileage than cars of the pre-environmental era, they do not "conserve" fuel in the sense of causing drivers to consume less of it. The reason is that, as the cost of driving drops because of improved fuel efficiency, car owners simply drive more miles. Between 1978 and 2008, the CAFE standard for passenger cars rose 51 percent, from 18.8 mpg to 27.5 mpg.[23] During the same time period, however, the average per capita miles driven per year in the U.S. rose 50 percent.[24] The CAFE standards have failed to conserve fuel, but, by physically shrinking the size of the gasoline-powered auto, they have moved us closer to the preservationist goal of abolishing the automobile altogether (a goal clearly contemplated, for example, by California's "Advanced Clean Car" regulatory package).

Another category of government-imposed conservation measures is intended to make buildings more energy efficient. Requirements for increased insulation, double-paned windows, and energy efficient heating and air conditions systems are all intended to conserve fossil fuels used to heat and cool buildings. But their effect is the same as it is for fuel-efficient automobiles; as the cost of heating and cooling goes down, the usage of heating and cooling goes up, thus resulting in little if any fuel conservation. As in the case of CAFE standards for cars, these measures do not really conserve resources, but they do serve another purpose, a preservationist one. They make homes more expensive, so that, in the long run, homes will necessarily become smaller. They also inure homeowners to thinking of their homes as a morally problematic imposition both on the environment and on their fellow citizens. They are a useful beginning toward the eventual abolition of the single-family home.

Then there is "conservation" of resources by government manipulation of prices. California, for example, employs "tiered" pricing of residential electricity to reduce power usage. Under a proper pricing system, it usually profits producers of a commodity to sell at lower prices to those who buy larger quantities. But the California Public Utilities Commission, which sets electricity rates, requires lower prices for residential users who buy smaller quantities of electricity and progressively higher rates for buyers of larger quantities. This system of tiered rates might "conserve" electricity, but it could only arise under a quasi-socialist system like that which the Utilities Commission oversees. It is based on the premise that the supply of electricity is limited for the foreseeable future and must, in effect, be rationed.

Now, in a free economy prices automatically "ration" goods in short supply and cause producers to increase the supply. As the supply of any commodity decreases relative to demand, the price rises, and, in response, consumers use less of it. Eventually, though, the high price will encourage producers to increase their production in order to fulfill the demand of those consumers who curtailed their usage.

But under a system of tiered pricing, which encourages consumers to purchase no more than a minimal amount of electricity, producers have little incentive to increase the supply, thus ensuring that electricity will be kept in short supply indefinitely, and thus guaranteeing that its *average* cost (as opposed to the price charged to customers who purchase just the minimum amount) will remain high. Though advertised as a measure to conserve energy, tiered pricing of commodities like electricity and water, has the distinctly preservationist effect of discouraging economic growth in California.

The National Forests

The flagship of the original conservation movement was Gifford Pinchot's system of national forests. These embodied his vision of a federal domain in which natural resources would be administered by a technocratic elite for the greatest good of the greatest number of Americans. The original purposes of the national forests, as identified in the Organic Act of 1897, were timber production and forest and watershed protection. More than any other endeavor, logging in the national forests exemplified the belief among many early conservationists that natural resources were meant to be *used* for productive purposes.

As it turned out, timber production in the national forests was fairly light until the demand for lumber soared after World War II. The timber harvest went from about 3.1 billion board feet in 1945 to about 11 billion in 1974.[25] For a full generation after the war, timber production would be the primary occupation of the Forest Service. But by 1960, even before *Silent Spring*, a reaction against this industrial-like use of the forests had already taken form. That reaction issued in the Multiple Use–Sustained Yield Act of 1960. The Act identified five uses of the national forests, timber production, livestock grazing, watershed protection, recreation, and wildlife preservation, and, in an important change, it accorded equal priority to all five. But it would require a court decision and two more major statutes before the Forest Service would seriously begin to give nature preservation and other uses equal priority with timber production.[26]

Although the Forest Service did eventually begin to devote more attention to other uses, logging continued apace. From 1963 through 1990, lumbermen in the National Forests harvested, on average, 10.8 billion board feet per year. It took the Endangered Species Act and the northern spotted owl to finally eliminate most of the logging in the national forests. After a court shut down all logging on the owl's range in the forests in 1991, the volume of logging dropped precipitously. From a peak of 12.7 billion board feet in 1987, the harvest fell to 2.5 billion in 2012.[27]

The transformation of productive logging lands to spotted

owl habitat represented the triumph of nature preservation over productive use on a vast expanse of national forest lands. Then, in 2001, President Clinton's Roadless Area Conservation Rule went into effect. It prohibited most logging and road building on over 58 million acres of national forest and national grasslands. (The Multiple Use–Sustained Yield Act of 1960 had sanctioned the "establishment and maintenance of areas of wilderness" in the National Forests.) President Bush subsequently scaled back Clinton's roadless rule, but in 2011 a federal appeals court in Denver in effect reinstated the original rule.[28] This was another huge victory for preservation over use, since it closed almost a third of the national forests and grasslands to most productive uses.

Nowadays, logging in the national Forests is primarily a function of ecological restoration and maintenance; it is largely limited to "thinning," which is done either to restore sections of the forests to a more healthy and "sustainable" state or to enable foresters to maintain them in such a state. In other words, logging in the National Forests today is primarily a function of *preserving* the forests, rather than making productive use of them. The typical dispute between loggers and environmentalists—and they are numerous, since environmentalists challenge a great many of the timber harvests that manage to win approval—involves the charge that a timber company intends to engage in out-and-out logging, rather than in thinning. Both sides in the dispute have accepted the premise that only thinning is and ought to be permitted, and that in most cases outright logging for profit is no longer a morally defensible activity in the national forests.

Environmentalists consider livestock grazing to be an inappropriate use of the national forests (and of most other federal lands). But there is no such thing as a use that is categorically appropriate or inappropriate to a given parcel of land without regard to the purposes of its owner. If my purpose were to spend my life earning a living from my land, then livestock grazing might or might not be an appropriate means by which to achieve that purpose. If the parcel simply was not large enough to carry enough animals to support me and my family, then grazing would be an inappropriate use. On the other hand it might be that, through judicious determinations as to the intensity of grazing to be carried on, combined with diligent efforts to maintain the land in good working condition, the parcel might be made to accommodate enough grazing to support me and my family indefinitely.

Many families throughout the American west have supported themselves for generations on grazing lands, including lands within the national forests. There is nothing intrinsic to a parcel of land, nothing "ecologically predestined," that determines how it ought to be used. If my purpose were to earn the maximum return possible on the money I used to purchase a parcel of land, then it might be that a steel mill would be the most appropriate use. Note that this would be an economic judgment, and it would have little or nothing to do with the ecology of the parcel itself or of the surrounding terrain.

While environmentalists seem to treat grazing as categorically inappropriate to the national forests, in fact they assume a certain specific purpose for the land (on behalf of its owners, the American people)—that being to enjoy the land in a near-natural state for an indefinite period of time—and then they judge prospective uses of the land, such as grazing, by how well they would serve that purpose. Of course, grazing consistently comes up wanting because, like most human uses of land, it does tend to alter land from its natural state.

John Muir called sheep "hoofed locusts." In 1922 the Izaak Walton League sued the United States Forest Service to halt all sheep and cattle grazing in the national forests of California. Ranchers, who have been besieged by opponents of grazing for decades, tend to view them as environmental extremists. But the term "environmental extremist" is redundant; to be an environmentalist is to be an extremist. For someone who believes that man has a moral obligation to refrain from upsetting the balance of nature, it is perfectly logical to oppose grazing in the national forests, which is always a potential, and often an actual, source of natural upset.

The question whether we should judge grazing by the standard of how much it upsets the natural *status quo* is, of course, more complicated when the grazing takes place on public land. The national forests were, after all, set aside as a kind of nature reservation, though expressly intended for human use, most notably logging.[29] If the land were privately owned, there would be no basis for opposing its use for grazing. Since it is public land, there is no objective way to decide how it "should" be used, other than to say that it would be better to put it to productive use than to no use at all. As for the ranchers, it will always be a challenge for them to defend the use of public land for private profit. On the other hand, for all the criticism of "welfare ranchers" on public land, it was not the ranchers who decided that 43 percent of the range land in the United States should be owned by the federal government.[30]

In a fully free society, all the grazing land would be privately owned, and grazing decisions would be entirely up to the landowners. As long as a rancher's use of his own land did not cause harm to his neighbor's land, no one would have any grounds for complaint. The question of how these privately owned lands should be used—whether for grazing or nature preservation—or for steel mills—would be left to the free market to decide.

But as ranching developed in the West, a great many ranches came to consist of a base of private land supplemented by much larger grazing allotments on federal lands, without which the ranches would not have been economically viable. Many of these ranches came into being when the federal lands were simply open range. But when the federal government began to regulate these public grazing lands, some of them in the national forests, the ranchers themselves came to exercise a great deal of influence over the regulatory process. So they have been, and continue to be, remarkably successful at resisting environmentalist efforts to abolish grazing in the national forests. (Today, with hundreds of millions worth of mortgages secured by grazing allotments on federal lands, banks also have an incentive to resist environmentalist attacks on public lands grazing.)[31] As a result, although grazing on public lands has been under attack by environmentalists for decades, there is no statute or regulatory measure that threatens to wipe out national forest grazing in the immediate future.

This is not because environmentalists have not tried to enact such a statute. In 2005 members of Congress sympathetic to environmentalist interests sponsored the Multiple-Use Conflict Resolution Act. This act would have authorized the use of taxpayer funds to purchase grazing permits from willing ranchers and then retire the permits. Advocates of the free market would normally favor allowing the holders of federal grazing permits to sell them freely to anyone of their choosing. (At present, they may sell only to ranchers who will make use of them.)[32] But the ranchers fear that lands on which the permits are allowed to be retired will be lost to grazing forever and that allowing permits to be bought out and retired will spell the end of public lands ranching altogether. The Multiple-Use Conflict Resolution Act never became law, but enacting some sort of buyout scheme has become the primary legislative goal of those who would abolish livestock grazing on federal lands.

"For $1.6 billion the scourge of livestock grazing—not only within the National Park and Wilderness Preservation Systems, but on all public lands, can end," anti-grazing activist Andy Kerr has written.[33] Congress has authorized permit buyouts on limited tracts of land, such as the California Desert Conservation Area in 2012, but it has yet to authorize buyouts on all federal lands. In late 2012, Democrat Adam Smith of Washington sponsored a new buyout scheme, the Rural Economic Vitalization Act, which would authorize privately-funded buyouts from willing ranchers anywhere on federal lands.

Where legislation has failed to dislodge the ranchers from their grazing allotments in the national forests, however, lawsuits are succeeding. The National Environmental Policy Act of 1969 requires that the Forest Service conduct an environmental impact study before renewing a grazing permit.[34] (Permits must be renewed every ten years.) These reviews have complicated the renewal process and have made ranching more expensive to carry on. NEPA has also expanded the opportunities for the opponents of grazing to mount regulatory and legal challenges to it. And here the Endangered Species Act enters the picture. Environmentalists have yet to find a species that will end most grazing in the national forests as the spotted owl ended most logging there, but it is not for lack of trying. The experiences of the Chilton family in New Mexico illustrate the difficulties of public lands ranching in the era of the Endangered Species Act.

In 1997, the Center for Biological Diversity and the Forest Guardians of New Mexico, taking advantage of the citizen suit clause in the Endangered Species Act, filed suit to enjoin grazing on the Chilton's 21,500-acre federal allotment on national forest land. This was followed in 1999 by a similar suit to enjoin grazing on their 15,000-acre allotment. In both cases, the plaintiffs alleged that the U.S. Forest Service had failed to consult the U.S. Fish and Wildlife Service about federally listed endangered species (none of which were known to inhabit either of the Chilton's allotments).

"The Center [for Biological Diversity] followed its established pattern of suing the federal agencies, not the rancher, but actually targeting the ranchers and their essential grazing leases." [35] Said Jim Chilton:

> The Forest Service settled the 1997 lawsuit behind closed doors without our knowledge or agreement or the agreement of the Arizona or New Mexico Cattle Growers associations who were interveners in the suit. The settlement paid the Center [for Biological Diversity] substantial sums of U.S. taxpayer money. Outrageously ... the Forest Service in the settlement agreed to withdraw important portions of our grazing allotment and take water rights we had under Arizona law. [36]

In 1999 the Chilton's successfully defended themselves against the second lawsuit, but at a cost of $400,000. Environmentalists are waging a war of attrition, and the ranchers cannot afford to defend themselves indefinitely.

As a result of the limited permit buyouts, the litigation, and increased regulation of federal grazing allotments, the number of grazing permit holders in the national forests declined by about 61 percent between 1980 and 2012. Over that same span, the number of animal unit months for cattle declined 17 percent, for horses and burros 42 percent, and for sheep and goats 37 percent. [37] Grazing is gradually disappearing from the national forests.

Mining is another productive land use that is under siege in the national forests. The General Mining Act of 1872 still governs mining on federal lands in the United States. The act entitles anyone to stake a mining claim on any federal lands except those that have specifically been withdrawn from mining and other productive uses. Much of the land in the national forests is thus open to mining.

Environmentalists have tried to change important parts of the 1872 act, with different degrees of success. For example, the act entitles anyone who successfully stakes a claim on federal land to patent (in effect, purchase) that land for a nominal price. In 1994 Congress suspended the issuing of mining patents, and it has yet to lift the suspension. Without a patent, miners are still able to carry on mining operations, but, as occupants of federal land, they must ask permission for much of what they do, something that would not be necessary if they owned the land under a patent.

Another element of the 1872 Mining Act entitles miners to the full proceeds of what they extract from federal lands, without the need to pay royalties to their landlord, the federal government. Two recent bills in Congress, one in 2007 and another in 2009, would have required miners to pay royalties on what they extract, though neither bill became law.

In 2012 Interior Secretary Ken Salazar imposed a 20-year ban on mining on a million acres of federal lands, including portions of the Kaibab National Forest, because these lands abut Grand Canyon National Park. This is a common and long-standing tactic of environmentalists, to treat it as an obvious moral imperative that the prohibitions against productive land use within the national parks should be extended to adjoining lands outside the Parks simply because those lands are near the parks.

The government does not keep records of mining production on federal lands, so it is not possible to produce statistical evidence that the amount of mining in the national forests has declined since the start of the environmental era. But mining in the United States as a whole has declined substantially during that time. Copper production peaked in 1998 at 2.14 million metric tons. In 2012 just under a million tons were produced, a decline of 54 percent.[38] Iron ore production declined 42 percent between 1966 and 2012.[39] Lead production dropped 81 percent from its peak year of 1973. Silver dropped 52 percent between 1997 and

2012.[40] Zinc declined 88 percent between its peak year of 1969 and 2012. And potash dropped 68 percent from 1967 to 2012. There are still substantial quantities of all these minerals in the ground in the U.S., so the declines in production are not due to exhaustion of supplies, but to government restraints on mining, as exemplified by two recent events.

In July of 2014, the EPA announced that it would place severe restrictions on the proposed Pebble Mine in a remote area in the southwestern part of the state; the Pebble would have been the largest copper and gold mine in North America. And in July of 2013 President Obama announced the prohibiting of mining on over 300,000 acres of BLM land in six states for twenty years. The amount of mining throughout the United States has declined dramatically since the start of the environmental era, so in all probability it has declined—at least as dramatically—in the national forests.

The U.S. Department of Agriculture's Strategic Plan for 2010 to 2015 lays out four primary objectives for the national forests: to restore and conserve the nation's forests, farms, ranches, and grasslands; to lead efforts to mitigate and adapt to climate change; to protect and enhance America's water resources; and to reduce the risk from catastrophic wildfire and restore fire to its appropriate place on the landscape. Note that all four objectives are preservationist to one degree or another, and none of them is directly concerned with the productive use of national forest lands.

Then consider the "Ecological Restoration Plan" of the Pacific Southwest Region of the National Forests, which includes California, small portions of Nevada and Oregon, Hawaii, and the Pacific Islands.

From this point forward, Ecological Restoration will be the central driver of wildland and forest stewardship in

*the Pacific Southwest Region, across all program areas
and activities.* [41]

In 1906, Gifford Pinchot attempted to have the national
parks placed under his authority so that he could put them to
productive use, as he was doing with the national forests. He
failed, and the parks continued to function essentially as nature
preserves and recreation areas. Today, in a development that
would have horrified Pinchot but delighted John Muir, the Forest
Service is itself abandoning its original, utilitarian mission and the
national forests are instead being transformed into the functional
equivalent of nature preserves and recreation areas, much like the
national parks.

The Morality of Conservation

As stated earlier, in a capitalist economy, natural resources such as
timber, minerals, and fossil fuels are "conserved" by prices. If the
supply of copper runs short, then the price of copper rises, and
consumers use less of it. The price will stay high as long as the
supply remains low, thus ensuring that copper will be used
sparingly.

Prices also bring about increases in the supplies of natural
resources; as the price of copper rises, it becomes profitable to
search out new deposits or to develop lower grade ones. High
copper prices also spur the search for copper substitutes. But
despite the tendency of the price system to conserve resources in
short supply, over the long run capitalism encourages the use of
resources. Copper producers, for example, have an economic
interest in producing and selling as much copper as they can. This
rankles environmentalists, who argue that we should *dis*courage
the use of precious resources such as copper, rather than hastening
the day when we will run out of it.

Environmentalists treat natural resources the way survivors

in a lifeboat would treat their remaining food and water, as a fixed supply that must be strictly rationed until it runs out, as it inevitably must. But the supply of natural resources on earth differs from the supply of food and water in a lifeboat in important ways. First, there is no immediate danger of our running out of most "non-renewable" resources.

Consider metals, many of which exist in stupendous quantities in the earth. Aluminum constitutes over 8 percent of the earth's crust, iron 5 percent; for every 10-mile by 10-mile square that is a mile deep on the surface of the earth, there are eight cubic miles of aluminum and five cubic miles of iron.[42] Man likely will never be able to put so much aluminum and iron to use. Even copper, which is scarce enough to be used as money, exists in such quantity (.0068 percent of the earth's crust) that there is enough in the top mile of just the landed areas of the globe to last over fifteen hundred years at the current world rate of usage.[43]

The more scarce a metal becomes, the greater the economic incentive to recycle it. Gold is recycled at virtually 100 percent. Copper, at present, is being recycled, or, rather, ripped out of street lights and buildings, before its owners are even finished using it. (The copper "shortage" is being caused by government restrictions on producing new copper, such as at the Pebble Mine in Alaska.) The reason we do not recycle many metals is that they are simply not scarce enough, and therefore valuable enough, to justify recycling them. When governments force recycling or subsidize it, it means that wealth that people would have preferred to spend in other ways is directed toward recycling instead. A great deal of metals recycling used to be carried on by junkyards without the need for government compulsion, but junkyards have become an endangered species in the age of environmentalism.

What most distinguishes the scarcity of natural resources on earth from that of food and water in a lifeboat is technology.

Technology enables man to continue to locate new deposits of metal ores and to devise ways of extracting them from formerly inaccessible places. It also enables man to develop substitutes for resources that are in short supply. And metals do not cease to exist once they have been used and discarded. Iron might rust away, but rust is simply iron in a different form, from which it can be transformed back into useable iron.

Someday technology might enable man to locate and process all the metal that has ever been used, discarded, and then scattered to the four corners of the earth. It might even transform metal from a non-renewable resource into a renewable one. Technology is the engine of plenty. It transforms what appears to be a fixed supply of a natural resource, in the way food and water on a lifeboat are fixed, into a supply that can be made to last indefinitely.

What about genuinely non-renewable resources, such as fossil fuels? As we have seen earlier in this chapter, we have supplies of oil, gas, and coal sufficient to last hundreds of years at today's rates of consumption. There is no immediate need to replace rationing-by-price with rationing-by-government-fiat (not that there ever would be such a need). But what about the long-range possibility that we might someday run out of these fuels?

The answer to this question is, again, technology. We already possess an energy source, in nuclear power, that would enable us to conserve substantial amounts of fossil fuels by ceasing to use them to generate electricity.[44] For anyone who seriously wanted to conserve fossil fuels *without diminishing either our capacity to generate energy or our economic growth*, nuclear power would be indispensable. Absent nuclear power, though, our best hope of finding substitutes for fossil fuels, again without diminishing either our capacity to generate energy or our economic growth, lies in our ability to develop new technologies. This ability depends, in turn, on maintaining a vigorous capitalist industrial economy.

But environmentalists consider capitalism to be the problem rather than the solution. It was capitalism that gave us "gas guzzling" automobiles that got eight miles to the gallon, which environmentalists consider wasteful. To conserve gasoline they would force auto manufacturers to produce more fuel-efficient cars, and they would restrict access to oil still in the ground. But by replacing rationing-by-price with a system that rations natural resources by government fiat, environmentalists are destroying the capitalist system. And as the capitalist system is destroyed, so also will be the capacity to generate new technologies, and along with it man's ability to replenish indefinitely his supplies of natural resources. By destroying man's capacity to generate new technologies, environmentalists are placing him in a position very much like that of the man in the lifeboat, in which a fixed supply of natural resources must be carefully rationed to last as long as possible.

This will also mean a lowering of the standard of living, and this will not be an accidental side-effect of conserving resources. Environmentalists believe that the period of prosperity we are living through is an aberration, a mistake. No people should ever be allowed to make such profligate use of the earth's resources as we do in the U.S. today. It is not that we need a few small adjustments, such as switching from incandescent to high tech light bulbs. What is needed is deep, radical change.

"No one seems to have examined carefully enough," wrote Samuel H. Ordway for the Princeton Symposium in 1956, "the causes responsible for our current fantastic consumption of raw materials ... No one has assumed to analyze the basic philosophy, indeed religion, of modern man, that makes us what we are: a race working, struggling, inventing, fighting, living *to create an ever higher standard of living for all mankind* ... and it may turn out to be our great illusion."[45] For the sake of conserving natural resources, say the environmentalists, we simply must give up the idea that we are entitled to an ever rising standard of living, or even to the standard we now enjoy. Automobiles will have to give way to mass transit, single-family homes to mass housing, and wide-screen TV's and power boats and recreational vehicles and the whole panoply of luxury "things" that we aspire to must cease to be. This is the real meaning of "conservation of resources."

Environmentalists treat it as self-evident that morality requires us to conserve natural resources. The man who drives a Prius is considered a person of heightened moral sensibilities, while one who drives an SUV is a philistine. Environmentalists argue that we in the West use more than our fair share of the world's natural resources, and that we owe it to the peoples of the undeveloped countries, and to "future generations," to reduce dramatically our use of resources.

As I have said, in treating all this as morally self-evident, environmentalists are relying on a pair of ideas borrowed from the Christian tradition, a tradition that most serious environmentalists assiduously eschew. The first is the idea that it we are morally required to place the welfare of others, in this case the peoples of the undeveloped countries or "future generations," ahead of our own welfare. The second is the idea that it is morally commendable to make do with less material wealth rather than with more.

Capitalist industrialism has brought longer, more healthful lives to many peoples of the earth, rich and poor alike. It has brought to the common man material wealth, comfort, and leisure time that in earlier ages were known only to a wealthy elite. In advocating forced conservation of natural resources, environmentalists believe they are on the side of the angels. But what they are really advocating is that Americans surrender their freedoms, on the one hand, and submit to working longer and harder throughout their lives, on the other hand, only to end up poorer, less healthy, and dead at a younger age. No proper morality would sanction such a life-destroying bargain.

Notes

1. The report was funded, in part, by the Department of Health, Education, and Welfare and the U.S. Department of the Interior.

2. Committee on Resources and Man of the Division of Earth Sciences, National Academy of Sciences—National Research Council, *Resources and Man: A Study and Recommendations* (San Francisco: W.H. Freeman and Company, 1969), 2.

3. M. King Hubbert, "Energy Resources," in *Resources and Man*, 238. Hubbert was probably correct that, in the absence of some catastrophic reduction in world population yet to occur, the world can only undergo one population explosion like the one it has experienced in the last few hundred years.

4. *Resources and Man*, 1.

5. Ibid., 3. "The quality of life, which we equate with flexibility of choices and freedom of action, is threatened by the demands of an expanding economy and population."

6. Ibid.

7. Here the report quotes Roger Revelle, "United States Participation in the International Biological Program," U.S. National Committee for the International Biological Program, Report No. 2, 1.

8. *Resources and Man,* 5.

9. Ibid., 11. Emphasis in the second sentence added.

10. Ibid., 16.

11. Huppert saw nuclear power as the answer to man's future energy needs. Ibid., 238.

12. "Population and the American Future: The Report of the Commission on Population and the American Future," 1972, Center for Research on Population and Security, accessed November 1, 2014,
http://www.population-

security.org/rockefeller/001_population_growth_and_the_amer
ican_future.htm.
13. Ibid., chap. 1.
14. Ibid., chap. 5. The question of population growth bears on all
three of categories of environmentalist endeavor, the combating
of pollution, the conservation of natural resources, and the
preservation of land and wildlife.
15. Ibid., chap. 1.
16. Ibid.
17. Ibid.
18. Ibid. Emphasis added. In his book, *Environmentalism: Ideology
and Power*, Donald Gibson has assembled a mass of evidence that
elements of America's wealthy WASP establishment, Rockefellers
prominent among them, were instrumental throughout the
conservation era and into the environmental era in ginning up
opposition to industrialism and economic growth.
19. United States Public Land Law Review Commission, "One
Third of the Nation's Land: A Report to the President and to the
Congress ," Oregon State University Library, June, 1970,
accessed November 1, 2014,
http://ir.library.oregonstate.edu/xmlui/handle/1957/11183, 8.
20. Ibid., 88.
21. Ibid., 80.
22. Ibid., 91.
23. "Summary of Fuel Economy Performance," National Highway
Traffic Safety Administration, accessed November 2, 2014,
http://www.nhtsa.gov/fuel-economy.
24. Doug Short "American Vehicle Miles Driven Haven't Been
This Low since the 1990's" Business Insider, December 22, 2012,
accessed November 2, 2014,
http://www.businessinsider.com/vehicle-miles-driven-2012-12.
Since 2005 the average per capita miles driven has declined about
14% of where it stood in 1971. This is not because CAFÉ

standards have suddenly begun to cause Americans to drive less. It probably has more to do with rising gas prices and the aging of the baby boomers than anything else.

25. "Number of Sales, Volume, Value & Price Per MBF of Convertible Timber Cut & Sold," U.S. Forest Service, accessed November 1, 2014, http://www.fs.fed.us/forestmanagement/documents/sold-harvest/documents/1905-2012_Natl_Summary_Graph.pdf.

26. In the 1973 case of *Izaak Walton League v. Butz*, Judge Robert E. Maxwell found that the Forest Service's permitting of extensive clear-cutting in the Monongahela National Forest violated the requirement of the Organic Act of 1897 that only "dead, physiologically mature, and large growth" trees that had been individually marked for harvesting could be sold. There followed the Forest and Rangeland Renewable Resources Planning Act of 1974 and its amendments, known as the National Forest Management Act of 1976. These gave effect to the equalization of priorities called for in the Multiple Use–Sustained Yield Act of 1960.

27. Ibid.

28. *Wyoming v. USDA*, Nos. 08-8061, 09-8075 (10th Cir. 10/21/11).

29. Grazing was not explicitly included among the purposes of the National Forests. The Organic Act of 1897, which established the Forests, did stipulate that "[t]he Act is not intended to authorize the inclusion within national forests of lands that are more valuable for mineral or agricultural purposes." (16 U.S.C. 475)

30. James Lester (Political Science Department, Colorado State University), "Livestock Grazing on Public Lands," Colorado State University, accessed November 1, 2014, http://www.colorado.edu/economics/morey/8545/student/livestock/grazing.html.

31. [1] "$450 Million in Bank Loans to Public Lands Ranchers

Guaranteed with National Forest Grazing Permits," Forest Guardians, June 24, 2002, accessed November 1, 2014, http://www.publiclandsranching.org/htmlres/plr_escwaiver.ht m.

32. Current law prohibits the retiring of a grazing permit that is otherwise valid; all federal land that may legally be grazed must be grazed.

33. Andy Kerr and Mark Salvo, "Livestock Grazing in the National Park and Wilderness Preservation Systems," *Wild Earth* 10 2 (Summer, 2000), accessed November 1, 2014, http://www.andykerr.net/grazing-in-wilderness-parks/, 53-56.

34. In 2005 Congress authorized the Forest Service to grant "categorical exclusions" which exempt grazing permit applicants under certain circumstances from having to do an environmental impact study.

35. "Statement by the National Cattlemen's Beef Association & Public Lands Council on Livestock Grazing on Public Lands," submitted by Mr. Jim Chilton to the Subcommittee on Forests and Forest Health of the House Committee on Resources, April 13, 2005, Property Rights Foundation of America, accessed November 1, 2014, http://prfamerica.org/2006/LivestockGrazingOnPublicLands.ht ml.

36. Ibid.

37. The 1980 statistics are from Rachel Thomas, "Forest Service Grazing Statistical Report, 1980-2002," January 1, 2004, accessed November 1, 2014, http://www.propertyrightsresearch.org/2004/articles/forest_se rvice_grazing_statistic.htm. The 2012 statistics are from "Grazing Statistical Summary – FY 2012," US Forest Service, accessed November 1, 2014, http://www.fs.fed.us/rangelands/ftp/docs/GrazingStatisticalSu mmary2012.pdf.

38. "Mineral Commodity Summaries 2013," USGS, accessed November 2, 2014,
http://minerals.usgs.gov/minerals/pubs/mcs/2013/mcs2013.pdf.
39. Iron ore production peaked in 1953, well before the start of the environmental era.
40. Silver production peaked in 1916 at 2450.
41. "Region 5 Ecological Restoration Implementation Plan," US Forest Service, accessed November 2, 2014,
http://www.fs.usda.gov/detail/r5/landmanagement/?cid=stelprdb5409054.
42. "List of Periodic Table Elements Sorted by Abundance in the Earth's Crust," Israel Science and Technology Homepage, accessed November 2, 2014,
http://www.science.co.il/PTelements.asp?s=Earth.
43. "Copper," *New World Encyclopedia*, accessed November 2, 2014, http://www.newworldencyclopedia.org/entry/Copper. World copper usage in 2011 was 1908 million tons. "[Copper] Production and Consumption," London Metal Exchange, accessed November 2, 2014, http://www.lme.com/metals/non-ferrous/copper/production-and-consumption/. It illustrates the difficulty, and the technological challenge, of finding and extracting the copper that, although it is so plentiful in the earth's crust, it is also so scarce in human commerce as to be useful as money.
44. 39% of U.S. fossil fuel usage in 2010 was for electricity generation. "Americans Using More Fossil Fuels," Lawrence Livermore National Laboratory, November 9, 2011, accessed November 2, 2014,
https://www.llnl.gov/news/newsreleases/2011/Nov/NR-11-11-02.html.
45. Samuel H. Ordway, Jr., "Possible Limits of Raw-Material Consumption," in Thomas, ed., *Man's Role in Changing the Face of the Earth*, 990.

11: Land and Wildlife Preservation

THE FIRST TWO broad categories of environmentalist
endeavor, the war on pollution and the conservation of natural
resources, have been essentially preservationist from the start of
the environmental era. There remains the third category, land and
wildlife preservation. Since these are preservationist by definition,
what needs to be shown is that they have expanded dramatically in
range and importance since the mid-1960s.

The Preservation of Public Lands

When the federal government took control of the unsettled
western lands of the United States following the ratification of the
Constitution, its primary responsibility toward those lands was to
sell them off as quickly as practicable. For almost a century after
the founding, the lands under federal jurisdiction were used freely
by Americans, without government oversight, for such uses as
grazing, mining, and logging. The establishment of Yellowstone
National Park in 1872 marked the first time that anyone had
attempted to *preserve* federal lands from productive use.[1] After the
conservation movement began to influence policy, the federal
government stopped selling off the remaining unsettled lands and
instead went into the business of being a permanent landlord.
They also got serious about preserving more of the remaining
federal lands from human use. Since Yellowstone, fifty-eight more

national parks have been created, and the pace of park creation quickened after the environmental era began in the mid-1960s.

In 1906, President Roosevelt established the first national monument, at Devil's Tower in Wyoming. A national monument is similar to a national park except that, whereas a park requires an act of Congress to create it, a monument can by created by presidential proclamation. Roosevelt also established the first national wildlife refuge, at Pelican Island in Florida in 1903. Today there are 108 national monuments (although some, like the Statue of Liberty, are not nature preserves), over 560 national wildlife refuges, and thirty-eight wetland management districts.[2] A great many of these were established before the environmental era, but the pace at which all these types of federal nature preserves were being created increased significantly after start of the environmental era. In addition, a number of new varieties of federal nature preserve have sprung up since the mid-1960s.

The national seashores and lakeshores are almost entirely limited to recreational uses.[3] The first national seashore was established at Cape Hatteras in 1953. Almost all of the remaining nine national seashores and four national lakeshores came into being in the environmental era. There are eighteen national recreation areas, all of which are, in effect, nature preserves. The first of these was established at Lake Mead following the completion of the Hoover Dam in the 1930s, but all the rest were established in the environmental era. There are eighteen national preserves and three national reserves. These are similar to national parks, except that some oil and gas extraction, some mining, and hunting are allowed. The preserves and reserves all date from the environmental era.

As of April 2012, the National Wild and Scenic Rivers System protected 12,598 miles of 203 rivers in 39 states and in Puerto Rico.[4] This system dates from the passage of the Wild and Scenic Rivers Act signed by President Johnson in 1968. There are also five national rivers administered by the National Park Service. According to the park service, national rivers preserve lands alongside free-flowing streams. They also provide opportunities for outdoor hiking, canoeing, and hunting. All of the national rivers date from the environmental era. And finally, there are nine national scenic areas, all administered by the U.S. Forest Service. These are specially protected lands, mostly within the national forests, although the Columbia River Gorge National Scenic Area encompasses, and places restrictions on, a great many private holdings. The first of the national scenic areas, the Mono Basin National Forest Scenic Area in California, was established in 1984.

As this proliferation of federal nature preserves shows, the pace at which the federal government has been preserving its land holdings (and, in some cases, such as the Columbia River Gorge National Scenic Area, preserving *private* lands) from human uses has increased markedly since environmentalism made its appearance in the mid-1960s.

What of the rest of the federal lands? We have already seen that in the national forests, once the utilitarian workhorse of federal lands, logging, grazing, and mining have all been dramatically curtailed during the environmental era. The national forests are now well on their way to becoming just another category of federal nature preserve. The last category of federal lands to consider, and the largest, are those managed by the Bureau of Land Management, which lands comprise about 40 percent of all federal landholdings.

These are the leftovers among federal lands, the ones that have not been scenic enough or important enough to qualify as a national park or a national forest or for some other federal designation. (They are, of course, the lands from which most newly designated entities, such as national monuments, are drawn.) These also have, in the past hundred years, been highly utilitarian lands; logging, grazing, mining, oil and gas drilling, and recreation have all been extensively carried on on BLM lands. But since the start of the environmental era, the use of these lands for most productive purposes has diminished markedly.

The statistics for BLM lands tell the story. Consider logging. In 1996, BLM lands produced $75,354,821 worth of forest products; in 2012 they produced $20,317,108 worth.[5] What about coal mining? In 1990, there were 489 leases in effect involving 730,427 acres; in 2011 there were 304 leases in effect involving 472,871 acres.[6] As far as other types of mining are concerned, in 1989, 503 plans of operation were reviewed; in 2012, 109 were reviewed.[7] Grazing has declined much less than these other categories of productive use. In 2011, there were 94 percent of the animal unit months logged on Section 3 lands, and 88 percent of the animal unit months logged on Section 15 lands, as were logged in 1996.[8]

The statistics for oil and gas drilling on BLM lands require explanation. In 1988, 9,234 leases were issued involving 12,215,573 acres for oil and gas drilling on all Federal lands.[9] In 2012 there were 1,729 leases issued involving 1,752,060 acres, or 19 percent of the leases issued in 1988 involving 14 percent of the acreage. On the other hand, the amount of oil and gas being produced on Federal lands increased dramatically between 1988 and 2012, thanks to the fracking revolution. Whereas in 1988 there were 1,772 drilling permits approved on Federal lands, in 2012 this number increased to 4,256. In other words, in 2012 there was far more drilling going on, on far less land, than had been the case in 1988. Nevertheless, it is also true that the *amount* of BLM land being used for oil and gas drilling in 2012 was only 14% of what it had been in 1988.

Finally, the number of recorded visitor days on BLM lands, a measure of recreational uses of those lands, rose from 18 million in 1996 to almost 67 million in 2012. In the environmentalist view, recreation is an acceptable human use of most lands.

It is also worth noting that 8.7 million acres of BLM land, or 3 percent of the total of BLM lands in the lower 48 states, are now preserved as wilderness. In addition, The Bureau of Land Management administers sixteen national conservation areas and six "similarly designated lands." These are specially protected lands within the BLM domain, and they range in size from the 18-acre Piedras Blancas Light Station in California to the 1.2 million-acre Black Rock Desert-High Rock Canyon Emigrant Trails National Conservation Area in Nevada. The BLM lands are headed in the same direction as the national forests; both are being used less and less for productive purposes and more and more for nature preservation and recreation.

	1996	2012	2012/1996
Wood Products	$75,354,821	$20,317,108	27%
	1990	2011	2011/1990
Coal Mining	489 leases	304 leases	62%
	730,427 acres	472,871 acres	65%
	1989	2012	2012/1989
Hard Rock Mining	503 plans reviewed	109 plans reviewed	22%
	1996	2011	2011/1996
Grazing	8.42 m AUM, sec 3	7.83 m AUM	93%
	1.315 m AUM, sec 15	1.151 m AUM	88%
	1988	2012	2012/1988
Oil and Gas	9,234 leases	1,729 leases	19%
	12,215,573 acres	1,752,060 acres	14%
	1,772 permits	4,256 permits	240%
	1996	2011	2011/1996
Visitor Days	18,412,754	66,950,000	364%

The Forced Preservation of Private Lands

More remarkable than environmentalists' success at preserving government lands has been their success at having preservationist restrictions placed on private lands. One way that governments preserve private lands from development is through simple zoning. In California it is common for planning departments to zone a rural section of a county so that only one house can be built on every twenty acres of land. This ensures that that part of the county will be kept mostly undeveloped.

Zoning, of course, contributes to "sprawl." As a city grows, the demand for land near the center increases, and the value of that land rises accordingly. As the value rises, so does the intensity of use that such land will support; land that in the early days of the city was economically appropriate for single-family homes, later on might be more appropriate for apartment buildings, and still later for commercial sky scrapers. But when zoning locks in single-family home usage forever, then the city is unable to grow upward and is forced to grow outward. Sometimes, however, a city will change a zoning designation in response to economic or political pressures, so environmentalists have looked for other ways to freeze development on private lands.[10]

A device that has become popular in California is growth mitigation fees. Throughout most of the United States, when growth occurs in an area and it becomes necessary to build new roads to accommodate it, the city or county pays for the new roads out of the general tax revenues. But in some California counties, before breaking ground on a new home a builder must pay a specified amount to the government in growth mitigation fees. This money is then put into a fund that will eventually be used to pay for the construction of new roads and other infrastructure as they become necessary. The cost of the fees is passed along to the homebuyer, of course.

In parts of El Dorado County, the total cost of the various mitigation fees and building permits can approach $75,000 for a single family home. Since the fee system was put into place, the construction of new homes by individual landowners in the county has come to a virtual halt. Most people looking to build a home now go to other counties that do not impose such fees.

This, of course, is precisely what the proponents of the fees intended, to halt economic growth in the county. (As it is, almost 75 percent of the El Dorado County consists of national forest land.)

There is a certain political logic to shifting the responsibility to pay for new roads away from the taxpayers of a locality and toward new home buyers who make the construction of new roads necessary. California has taken to charging fees for more and more things that formerly were "free," such as admission to state parks. The premise behind the fees is that the users of amenities such as parks ought to pay for their maintenance. But to impose such fees without a corresponding decrease in the taxes paid by the general public—a decrease that the public never has and never will see—amounts to an increase in taxes, and, in the case of growth mitigation fees, the full burden of the increase unjustly falls on the buyers of new homes.

There was a time in America when economic growth was seen as a good thing, something to be encouraged. Growth paid for itself through the increased economic productivity it generated. Today in California, growth is seen as something to be discouraged, a problem whose effects must be mitigated by imposing a tax on those who commit growth.

In California, each county is required by the state to establish a general plan, which identifies which types of land use will be permitted in every part of the county over a twenty-five year period. Based on this general plan, the county then identifies in more specific terms what types of land use will be permitted on each parcel of land. It is instructive to examine some of the preservationist measures that are applied to private land under the general plan of El Dorado County.

- In order to plow up one acre or more of "undisturbed vegetation," a farmer must obtain a permit from the

County (unless he uses the "best management practices" approved by the County and does not change the contour of the land).[11]

- "Where practical and when warranted … parking lot storm drainage shall include facilities to separate oils and salts from storm water." [12]
- "[D]evelopment projects … shall be designed to avoid disturbance or fragmentation of important [wildlife] habitats to the extent reasonably feasible. Where avoidance is not possible, the development shall be required to fully mitigate the effects of important habitat loss and fragmentation." [13]
- "Where critical wildlife areas and migration corridors are identified" on a development site, the County must preserve them as natural areas and, if necessary, require that buildings be grouped together.[14]
- On lands with "high wildlife habitat value," the following will apply:
  1. the landowner will be required to retain a higher than normal percentage of the tree canopy;
  2. there will be greater protection for rare plants;
  3. there will be additional limitations on the height of buildings and on how much of the land may be covered by them;
  4. more wetlands and riparian values will be retained than on non-habitat lands;
  5. "no hindrances to wildlife movement (e.g., no fences that would restrict wildlife movement)" will be permitted;
  6. there will be requirements that large expanses of non-oak plants, should they exist, be retained.[15]
- For new development projects on over an acre that have

at least one percent canopy coverage by "woodlands habitat," the landowner is required to retain a certain percentage of the tree canopy. For example, if the canopy covers 80 to 100 percent of the site, then the landowner must retain 60 percent of canopy; if the canopy covers 10 to 19 percent of the site, he must retain 90 percent of canopy. In lieu of retaining the canopy, the landowner may choose to contribute to the County's Integrated Natural Resources Management Plan conservation fund.[16]

- "Except under special exemptions, a tree removal permit shall be required by the County for removal of any native oak tree with a single main trunk of at least 6 inches diameter at breast height, or a multiple trunk with an aggregate of at least 10 inches dbh.[17]

- "Any person desiring to remove a native oak shall provide the County with the following as part of the project application:[18]

> 1. a written statement by the applicant or an arborist stating the justification for the development activity, identifying how trees in the vicinity of the project or construction site will be protected and stating that all construction activity will follow approved preservation methods;
> 2. a site map plan that identifies all native oaks on the project site; and
> 3. a report by a certified arborist that provides specific information for all native oak trees on the project site."

One other County requirement deserves mention: a developer of a large, residential "planned development" in El Dorado County must leave 30% of the entire site in its natural state. This last requirement has two indirect consequences: it raises the cost of building a house, and it exacerbates "sprawl."

Just as the National Environmental Policy Act of 1969 requires the preparation of an Environmental Impact Statement before work can begin on a new Federal project, so the California Environmental Quality Act requires that an Environmental Impact Report be completed before work can begin on public *and private* projects. Below is a list of the categories of criteria that must be satisfied before a project can be approved.[19] With so many possible grounds for refusing approval to a project, it is a wonder that anything ever gets built in California.

Aesthetics
Agriculture Resources
Air Quality
Biological Resources
Cultural Resources
Geology and Soils
Hazards and Hazardous Materials
Hydrology and Water Quality
Land Use and Planning
Mineral Resources
Noise
Population and Housing
Public Services
Recreation
Transportation/Traffic
Utilities, Energy and Service Systems
Growth-Inducing Impacts
Cumulative Impacts
Significant Irreversible Changes
Alternatives

No free people should ever submit to land-use censorship like this any more than they should submit to any other kind of censorship.

In 1973 the Oregon Legislature passed a statute requiring that every urban area in the state establish an "urban growth boundary" around itself in order to preserve agricultural and forest land from development. The plan required that within each boundary there should be twenty years' worth of vacant land for new development. As the vacant land within the boundaries was developed, the boundaries would expand to encompass additional vacant land. Thus the plan did not prohibit new growth; it just directed growth away from suburban and into urban areas. (A second purpose of the boundaries was to reduce auto traffic by concentrating homes and jobs within the urban areas. The boundaries, therefore, were another manifestation of the war on the automobile.)

Like zoning, urban growth boundaries are unjust to individual property owners. At the moment when the urban growth boundary is put into place, landowners outside the boundary are deprived of their rights to develop their land, suffering what amounts to an uncompensated taking. Also, by artificially restricting the supply of vacant land urban growth boundaries cause the price of land—and, therefore, of all real estate—to rise. Wendell Cox reports that in a normally functioning American city, the ratio of median home value to median income (called the median multiple) is about 3.0; in June of 2013, the median multiple for Oregon's largest city, Portland, was 4.8, meaning that home prices in Portland were inordinately high.[20] Among the many deleterious consequences of high home prices, companies tend not to locate where their employees will face high housing costs.[21] By causing home prices to rise, Oregon's urban growth boundaries thus tend to inhibit economic growth, in spite of the planners' intention that growth inside the boundaries should continue unabated.

By prohibiting upward growth, zoning forces it outward, thereby exacerbating "suburban sprawl." Urban growth boundaries counteract the effects of zoning by restricting outward growth. San Jose, California instituted an urban growth boundary in 1973, but without the built-in growth flexibility that characterized the Oregon boundaries. As a result of this boundary, both upward and outward growth were thwarted. One result has been exorbitantly high house prices; in 2013, San Jose had a median multiple of 7.9, which tied with Santa Barbara for third highest in the U.S., behind Honolulu at 9.3 and San Jose's neighbor to the south, Santa Cruz, at 8.2.[22]

When home prices reach this level, poor people, blue collar workers, and even lower middle class families are priced out of the market, as are young families, retired persons, and some ethnic minorities, such as Hispanics and Blacks. Thus the one-two combination of zoning and urban growth boundaries can severely derange not just the economy of a city but also its demography and its socio-economic structure.

Between 2000 and 2010, the population of Santa Clara County, where San Jose is located, grew by just 5.9%.[23] Over that period of time more new residents moved into growth-friendly but mostly rural Placer County, located in the Sierras east of Sacramento, than moved into Santa Clara County.[24] Is Santa Clara County "full"? With a density of 1,366 persons per square mile, it is far less densely populated than, say, Bergen County, New Jersey, at 3,664. At least seventeen California localities have followed San Jose's lead and instituted urban growth boundaries. But land use control is about to go to a whole new level in the Bay Area.

In 2006, the California Legislature passed Assembly Bill 32, which requires that Californians reduce their emissions of "greenhouse gasses." There followed, in 2008, Senate Bill 375, which mandates that each of the state's eighteen metropolitan areas develop specific plans to reduce greenhouse gas emissions from automobiles. In the San Francisco Bay Area, per capita emissions must be reduced by 7% by 2020 and by 15% by 2035. Plan Bay Area is the name of the grand strategy to achieve these emissions reductions. The essence of the plan is to channel future growth in the Bay Area into "Priority Development Areas" near mass transit in already existing urban areas. This mass transit-mass housing strategy is intended to get people out of their cars, on the one hand, and to preserve the region's remaining undeveloped areas, of which there are vast expanses in the 7,000 square miles that comprise the Bay Area, on the other.

The old zoning policy of prohibiting upward growth in favor of outward growth will now be reversed throughout the region in favor of a policy that promotes (a certain degree of) upward growth and discourages outward growth. The plan is a major step toward what seems to be the environmentalist goal of putting people onto reservations in order to preserve the remaining natural landscape. It is noteworthy for the new distance it places between the individual landowner and those who decide what he may or may not do with his land. Zoning is administered at the city level; this plan is being put together at the "Bay Area" level, which consists of nine counties, 101 cities, and 7 million people.

In 1972 Congress passed the Coastal Zone Management Act. The Act encourages states to put mechanisms into place for regulating development along the coastlines of the United States. The rationale for the act is stated as a series of Congressional findings. Finding "a" states that "There is a national interest in the effective management, beneficial use, protection, and development of the coastal zone." [25] This is the only mention among the findings of "development" as being something desirable. Eight of the remaining eleven findings spell out the need to protect the coastal zone *from* use and development. The purpose of the Act is not to encourage commercial and industrial development within the coastal zones, but the control and restrict them; it is to *preserve* the coastal zones as much as possible.[26]

Pursuant to the act, California established a Coastal Zone Commission in 1976 to regulate the use of land in the "coastal zone."[27] This is an additional layer of regulation on top of all the usual general plans, zoning, and building codes that landowners are subjected to. Why this additional layer? The State of California has determined that a) the Pacific Ocean and the six feet of beach above mean high tide that is owned by "the public" must be made physically and visually accessible to the public; b) the waters and lands of the coastal zone are especially deserving of protection from man; and c) it is important that as much agricultural land as possible in the coastal zone be preserved from more intensive human uses such as residential, commercial, or industrial. The Commission is a State super-authority with the power to judge the adequacy of land-use regulation by local authorities and, when it deems necessary, to step in and regulate land use itself.

The California Coastal Act of 1976 stipulates in general terms what shall and shall not constitute appropriate development within the coastal zone. It specifies that local governments may develop their own "local coastal plans" to administer the Act within their respective jurisdictions, but these local plans may not take effect until they have been certified by the Coastal Commission. Even then, any member of the public, whether an interested party or not, and any member of the commission, may appeal a decision of a local planning authority to the commission for final administrative adjudication. This makes the commission the final administrative authority on questions of coastal land use.

One of the commission's most useful tools is the building permit. Often the commission will require, as a condition of allowing a building permit to be issued, that a landowner cede a portion of his land to the public. Usually these cessions are demanded as a way to facilitate public access to the beaches. Long ago, the original rationale for requiring owners of private property to seek government permission to build new improvements or to alter existing improvements on their own property was safety; the idea was that unless government made sure that buildings and other improvements on private land were properly constructed, innocent people would be injured or killed by shoddy construction. But now that the idea has become widely accepted that it is proper for property owners to have to seek government permission to improve their own property, the permit process is being used literally as a means of extortion.

Consider the 1985 case of *Grupe v. California Coastal Commission.*[28] Grupe sought a permit to build a single-family home on a beachfront parcel he owned. The parcel was bordered on both sides by state beaches, so there was no lack of public access to the shoreline. But the Commission conditioned issuance of the permit on Grupe's granting an easement across *two thirds of his land* for public access and recreation. The California Court of Appeals upheld the easement on the grounds that, although Grupe's house would not restrict access to the beach, shorefront development *can* restrict access to the shoreline, so any new development may be treated as though it *would* limit access.

In the 1985 case of *Remmenga v. California Coastal Commission*, Remmenga wanted to build a house on land he owned within the Hollister Ranch development.[29] There was no public access through Hollister Ranch to the beach, so Remmenga's proposed house would not have blocked access in any way (and Remmenga's house was one mile from the beach). So instead of extracting the typical public access easement across Remmenga's land, the Commission required that he pay a fee "in-lieu" of an access easement. The California Court of Appeals reasoned, as it had in Grupe, that because development along the coast can block access to the shoreline, all development may be treated as though it does block access.

The regulatory landscape changed with the case of *Nollan v. Coastal Commission of California* in 1987.[30] Nollan had challenged the Commission's right to condition the issuance of a building permit on Nollan's agreeing to cede a public access easement across his land. The U.S. Supreme Court ruled that for the commission to require an access easement when Nollan's proposed improvements would not restrict access amounted to an unconstitutional taking of private property. This case established that government exactions from property owners would only be constitutional if the action of the property owner directly caused the need for the exaction, as in the case where a new housing development would block off previously existing access to the shoreline.

But, following the Nollan decision, the commission continued to show what can only be called contempt for the rights of coastal property owners. They refused to vacate easements or return in-lieu payments that had been unjustly required of property owners in the years before Nollan.[31] In one case involving a permit to build a home, they simply refused to act on the request until ordered to do so by a court.[32] In another case, they arbitrarily extended the "coastal zone" to encompass a 20-acre parcel in order to block construction of a private elementary school that had been approved for the site.[33] In still another, they denied a landowner permission to erect gates and to post "no trespassing" signs on his private land that the public had begun to make use of.[34] And in another, the commission required, in return for approval to build a home, that a landowner agree to record an agricultural easement on all of his 143 acres except the 10,000 square feet where the house was to be built. The easement would have required the landowner to go into the farming business. When a court disallowed this requirement, the commission required that the owner agree to record an open space deed restriction on the 143 acres, which would have precluded virtually all use of the parcel except the building envelope. This again was disallowed by a court.[35]

The requiring of building permits is an exercise of the government's police power (the power to regulate). Recall that the police power arose from the common law of nuisance. If an activity by a landowner could be construed as causing a nuisance to the public in general—think of placing a rancid smelling rendering plant in a residential neighborhood—then the government could legally prohibit such activities before the fact, i.e., before any specific individual had proven in court that his rights had been violated by the landowner's action. Police power ordinances can be enforced by enjoining a landowner from initiating or continuing his offensive behavior or by fining him or punishing him in some other way.

When the police power is used to force a landowner to part with a portion of his wealth in consequence of some action by him, he is, in effect, being punished for that action. (It was the levying of such regulatory "punishments" without the benefit of a court action that lay at the root of the doctrine of "substantive due process," which I discussed in Chapter 3.) To require all landowners who apply for building permits to part with a portion of their property as an "access easement" regardless of whether the proposed development would block access is, in effect, to treat all such development as a public nuisance. This was precisely the policy of the Coastal Commission before the Nollan decision, and their continued assaults on property rights since Nollan suggest that they would have continued that policy if they could have. The ideas of the man who headed the commission for most of its existence give us no reason to think otherwise.

Peter Douglas was responsible for the state-wide initiative that created the Coastal Commission in 1972. He served as Executive Director of the Commission from 1985 until 2011. At an environmental law conference at Yosemite Valley in 2002, he said:

> *Active advocacy for the conservation of Nature … is [in part] about struggle against dehumanizing, amoral corporate capitalism and imperialism at all levels" around the planet, and environmental destruction resulting from greed and materialism. It is about the loss of human rights, human health, community values, and livelihood. It is about the degradation of home for all life.*[36]

These are the words of the head, not of the Soviet Politburo, but of the California Coastal Commission. Douglas has elaborated on the idea, held by conservationists from the very beginning, that natural resources are the collective property of "the people."

> *We argue that certain environmental resources and values associated with human and natural communities are held in common and cannot be owned or harmed by an individual. We speak of environmental commons, such as ecosystems, that know no property boundaries but whose biological health and vitality affect many people and other life.*[37]

What constitutes an environmental common? An environmental common can be an "eco-system, watershed, waterbody, coastal bight, air basin, public park, neighborhood, village, town,[or]city." It can include "wilderness areas, nature reserves, environmentally sensitive habitat and neighborhoods where people live, work and play. The environmental commons of a residential neighborhood, for example, are those places in it whether on public or private land, the uses of which directly or indirectly impact the quality, functionality and safety of the human community (i.e., streets, sidewalks, playgrounds, schools, park and open spaces)." [38]

As defined by Peter Douglas, the "environmental commons" could include every square foot of the earth's surface. "Over time we have privatized the commons and enclosed them to the exclusion of the general public. This is the tragedy of the commons in our time." [39] To a great extent, if not entirely, private property is immoral in Peter Douglas's world. This goes a long way to explain the unending conflicts that the California Coastal Commission has had with property owners. [40]

Critics dismiss Peter Douglas and the Coastal Commission as wild-eyed environmentalist radicals of the peculiarly California variety. But Douglas's view of who should own the earth is identical to Gifford Pinchot's. Douglas just advanced it a bit further toward its logical conclusion. Pinchot said that natural resources belong to "the people" collectively, but he never attempted to apply the idea outside of the national forests. Douglas applied it to the seacoast; his tenure with the commission amounted to a moral crusade to establish "the people" as the rightful owners of as much of the California coastline as he could.

Douglas is well aware that the kind of land regulation he envisions must come *"at some not insignificant cost to individual rights and aspirations."* [41] He is also aware that the kind of regulation he advocates is fundamentally opposed to man's "commanding nature," as Francis Bacon called it.

> *American character has, for several hundred years, been identified with taming of wilderness and exploitation of land for human profit. Why cannot commitment to preservation and restoration of community environmental values to benefit all life exemplify our character as a people in decades ahead.* [42]

Again, this idea is not confined to the radical fringe of the environmental movement. It came up repeatedly at the Princeton Symposium in 1956. It is one of the defining ideas of environmentalism.

Peter Douglas was able to carry on his crusade against private property for thirty-four years (starting as Chief Deputy of the Coastal Commission) because he was operating in California, where environmental ideas enjoy wider acceptance than in most other states, and where the liberal elite enjoy an unusual degree of political control. But it would be a mistake to think that what environmentalists are achieving in California cannot be duplicated elsewhere. It would be more accurate to speak of *when* environmentalist policies will come to predominate in the rest of the U.S., rather than *if* they will.

California is ahead of the rest of the country on the matter of global warming, for example, because they already have their own cap and trade regimen. But Mr. Obama's EPA, by declaring carbon dioxide to be a pollutant that the EPA may regulate, has already enlisted the rest of the country in the war on global warming.

Opposition to economic growth and development has been institutionalized in California, as the Coastal Commission and Plan Bay Area attest, but "smart growth" is already being promoted by the EPA (in their stormwater runoff regulations, for example); it is the Trojan Horse that will bring anti-growth policies to the rest of the U.S. California already has a plan in place to phase out the gasoline-driven engine by the middle of this century, but they designed their plan to be consistent with the long-range policies of the EPA. What Peter Douglas and the Bay Area planners have been practicing is not some exotic form of radical environmentalism; it is standard issue environmentalism, carried on at a more advanced stage of development. And it is on its way to every town in the United States.

It is clear that the preservation of land, both public and private, by federal, state, and local governments has increased dramatically during the environmental era. What about wildlife preservation?

The Endangered Species Act

Among the most overtly preservationist of all environmental statutes is the Endangered Species Act. The first attempt in the environmental era to protect species from extinction was the Endangered Species Preservation Act of 1966. This act merely authorized the federal government to purchase land for the preservation of wildlife habitat. In 1969, the Endangered Species Conservation Act enlarged the government's authority to acquire land, expanded the range of protected species to include invertebrates, required the federal government to identify endangered species beyond the borders of the U.S., and prohibited the importation of those species. With the passage of the Endangered Species Act of 1973, species protection assumed it current, radically preservationist form.

Under the Endangered Species Act, if the continued existence of a species of animal or plant or insect is officially determined to be "threatened" or "endangered," then it becomes illegal to "take" a member of that species, except with special permission.[43] "Take" in this context includes altering its officially designated habitat. For a private landowner, or for a user of public land for private purposes, such as a logging company in a national forest, an official finding that a threatened or endangered species inhabits the land one owns or uses can result in the placing of severe, and costly, restrictions on one's ability to make use of the land.

It could mean that a farmer will be prohibited from farming his land, as happened to the Domenigoni family of Riverside County, California when the Stephen's kangaroo rat, which inhabited 800 acres their farmland, was added to the endangered species list in 1988. It could mean that irrigation water will be cut off to farmers, as happened to 1,400 of them in Oregon's Klamath Basin in 2001 when the U.S. Fish and Wildlife Service determined that the water was temporarily needed in the rivers for the Coho Salmon and a couple of species of sucker fish. Or it could mean that a livestock farmer will be prohibited from killing an endangered carnivorous predator that is attacking his livestock.[44]

The potential of the Endangered Species Act to wreak economic havoc first came to national attention with the snail darter controversy of the 1970's. The snail darter is a three-inch fish that inhabited a stretch of the Little Tennessee River where the Tellico Dam was nearing completion. A lawsuit contending that the completion of the dam and the filling of the lake behind it would wipe out the snail darter succeeded in halting work on the dam, despite the $100 million that had already been spent on it. In agreeing that the dam should forever remain uncompleted, the U.S. Supreme Court determined that Congress had indeed intended the act to halt extinctions whatever the cost.[45] In the event, Congress passed a special exemption that allowed the dam to be completed.

Farmers in California's San Joaquin Valley have not been so favored by Congress. Those on the west side of the valley rely for their irrigation water on Sierra snow melt that finds its way to the Sacramento River Delta, from whence it is moved by giant pumps into canals that deliver it to farmers to the south. In 2006, the Natural Resources Defense Council and other groups filed a lawsuit to reduce the amount of pumping being done, charging (1) that the removal of large amounts of fresh water from the Delta was increasing the salinity of the remaining water and, thereby, harming the threatened, three-inch delta smelt, and (2) that the pumps themselves were killing large numbers of smelt.[46] Subsequently, the U.S. Fish and Wildlife Service ordered severe restrictions on the volume of water being pumped out of the Delta to the farmers, restrictions that caused crop losses and high unemployment in the valley.

As of early 2014, farmers were still fighting in court about how much water may be denied them, but it was almost certain that some reduction from the *status quo ante* would be permanent. Meanwhile, Governor Jerry Brown has been promoting a major expansion of the system that moves Sacramento River water south to the San Joaquin Valley and beyond. Whatever the merits of this project, the delta smelt could well prevent it from happening.

Then there is the northern spotted owl. In 1992 the Federal Government designated 6.9 million acres of Federal land in Washington, Oregon, and northern California as critical habitat for the threatened owl. The habitat designation meant the closing of those lands to logging and, as a result, the virtual destruction of the logging industry, and a way of life, in the Pacific Northwest. An industry that had produced 12.7 billion board feet in the peak year of 1987 produced only 2.5 billion feet in 2012.[47]

The Endangered Species Act aims to avoid extinctions literally at all costs. Section 4 states that the listing of a species as threatened or endangered should be based "*solely* on the basis of the best scientific and commercial data available."[48] In *Tennessee Valley Authority v. Hill*, the U.S. Supreme Court interpreted this to mean that "Congress intended to halt and reverse the trend toward species extinction *whatever the cost*." [49] The ESA treats the preservation of a species, at least as far as designating it as threatened or endangered—and prohibiting takings thereof—is concerned, as invaluable. It does so regardless of whether very many individual human beings actually do value it.

In a free society the value of any good, even a species of wildlife, should be determined by the individual valuations of every citizen, just as the value of a parcel of land or a Rembrandt painting is determined. And, in the real world of individual valuations by free men, nothing is invaluable. But environmentalists consider the preservation of species to be just too important to be left to the valuations of free persons, so they have taken it upon themselves to force their fellow citizens to pay an exorbitant price to preserve them.

All things being equal, most of us would prefer not to cause the extinction of a species that would otherwise continue to live and to thrive. But all things are not equal. There are enormous costs associated with preserving many endangered species, at least as the Endangered Species Act goes about it. How even to measure the cost of preserving the northern spotted owl is a daunting task, considering the economic and social devastation it has caused. But such costs are dwarfed by the harm done to our system of individual rights, capitalism, and industrialism. The ESA runs rough-shod over private property rights by allowing the federal government severely to restrict the use of private land that has officially been declared "critical habitat." And such restrictions, both on private land and on huge portions of the nation's vast public land holdings, harm America's economic productivity.

In one sense, the authors of the Endangered Species Act (and environmentalists generally) might be right to blame capitalism and industrialism for an increase in the rate of the extinction of species. In 1800, at the dawn of the industrial era, there were about a billion people in the world. Today there are seven billion. More people mean less wildlife. And less wildlife probably means fewer species in the wild. On the other hand, the industrialized countries of the West have been experiencing low, and even negative, population growth for several decades. After a point, it turns out, capitalism and industrialism, and the ideas and attitudes they give rise to, lead to a leveling off of population growth rates. In the United States, the main reason that the population continues to grow is that we are importing people from the rest of the world.

In its retreat from capitalism and industrialism, though, the Endangered Species Act also represents a retreat from an active, utilitarian orientation toward nature. It implies that, to one extent or another, we cannot continue to grow economically and to alter our natural environment accordingly without causing so many extinctions that we will run the risk of making our planet uninhabitable by humans. Here again is the familiar repudiation of Francis Bacon's premises; the ESA implies that it is not possible for man to command nature on an industrial scale such as the United States enjoyed up to 1970, and that, therefore, he ought not try. As I have shown, this repudiation of Baconism has characterized environmentalism since that symposium at Princeton in 1956.

The rationale most often given for preserving every species possible is that each one contributes to the planet's "biodiversity." The biodiversity argument holds that the more species in a given ecosystem (or the more genetic diversity within a species), the better. The purported benefits of biodiversity are many.

- Farmers have found that some species of insect can be used to control other, agriculturally harmful species.
- Almost 80,000 species of edible plants exist, "of which fewer than 20 produce 90 percent of the world's food." [50]
- Plants and animals also provide man with a wide range of other products, including medicines, fertilizers, pesticides, adhesives, oils, cotton, and silk.

In each of these cases, the thinking goes, since we cannot know which species will prove useful in the future, we should endeavor to keep as many alive as we can. Note that all of these purposes could be served by preserving endangered species in captivity or in DNA banks. But there are other reasons commonly given for preserving biodiversity.

- Many species have proven useful as monitors of environmental conditions, like the canary in the coal mine. Certain lichens, for example, "are good indicators of excess ozone, sulfur dioxide, and other air pollutants." [51]

- Every species contributes, in one way or another, to the health of its ecosystem. We humans benefit from such "ecosystem services" as "air and water purification, detoxification and decomposition of wastes, climate regulation [and the] regeneration of soil fertility." [52]

- Biodiversity contributes to the long-term stability of ecosystems. (Biodiversity is a balance-of-nature idea.)

- Having a wide variety of life forms makes the world a more interesting and esthetically pleasing place.

Note that these four purposes are best served by preserving endangered species in the wild, rather than in captivity. [53]

If we preserved endangered species in captivity or in DNA banks, we would cause far less economic dislocation than the current practice of preserving them in the wild. But the Endangered Species Act pointedly aims to preserve biodiversity in the wild. It does so because biodiversity serves *all* of its useful purposes only when it is a characteristic of a functioning ecosystem. [54] This quality of the act makes it profoundly preservationist, leading, as it does, to the closing of millions upon millions of acres of American land to productive use.

But it also has an important implication for the future of species protection in the U.S. If the value of biodiversity does indeed lie in its role within functioning ecosystems, then it makes no sense to allow a species to disappear from any portion of its natural range, and to allow the biodiversity of that portion to diminish accordingly, before the Endangered Species Act is triggered. According to the biodiversity argument, the Endangered Species Act ought to prevent local "extinctions" as well as global ones.

In fact, a species does not need to be heading toward total extinction—disappearing from the face of the earth—before it qualifies for protection under the Endangered Species Act. From the beginning, the act was intended to prevent partial extinctions under some circumstances. The ESA of 1973 authorized the protection not just of full species, but of subspecies, as well. The meadow jumping mouse, for example, has never been at all threatened or endangered as a species. But a population of meadow jumping mice that inhabits parts of Colorado and Wyoming was found to be low in numbers. By virtue of the Fish and Wildlife Service's 1998 declaration, based on questionable science, that this population constituted a bona fide subspecies— the Preble's jumping mouse—the Service was able to extend protection to it as a threatened subspecies.[55] Much can ride on whether an animal or plant qualifies as a subspecies. When the Preble's was added to the threatened list, 31,000 acres of public and private land in Colorado and Wyoming were closed to most productive uses.[56] In 2005, about 25% of the entries on the threatened and endangered list were subspecies.[57]

But the ESA also authorizes the protection of populations (of vertebrates) of even narrower taxonomic classification than subspecies. In 1978, the act was amended to include "any *distinct population segment* of any species of vertebrate fish or wildlife which interbreeds when mature."[58] The Fish and Wildlife Service admits that "[a]vailable scientific information provides little scientific enlightenment in interpreting the phrase "distinct population segment."[59] But, in conjunction with the National

Marine Fisheries Service, Fish and Wildlife has worked out a set of criteria to identify what qualifies as a distinct population segment.[60]

The endangered Sacramento River winter run Chinook Salmon is a distinct population segment of the Chinook salmon species. The threatened Central Valley spring run Chinook Salmon is also a distinct population segment, and a separate entry, on the list, even though the range of this population segment consists of the Sacramento River and all its tributaries. So, a distinct population segment can be as narrowly defined as a portion of a certain species, which portion occupies a certain geographical location at a different time of year from other members of its species. (Fish are especially prone to form subspecies, as the Endangered Species Act defines them, since, like Darwin's birds in the Galapagos, fish populations that are geographically isolated from each other tend to evolve along slightly different genetic lines.)

There is yet another classification below the level of species that merits protection under the ESA. The original Act of 1973 defined an endangered species as one "in danger of extinction throughout all *or a significant portion of* its range." [61] The Fish and Wildlife Service defines a portion of a species's range as "significant" if "its contribution to the viability of the species is so important that, without that portion, the species would be in danger of extinction." The coastal California gnatcatcher, listed as threatened in 1993, was found to be threatened over a significant portion of its range though not over its entire range.

Until now, the Fish and Wildlife Service has applied the threatened or endangered status just to that "significant portion" of a species's range where it is in fact threatened or endangered. But in early 2014 the service proposed to change its policy and to apply the threatened or endangered status to the species's entire range, as though the entire species were on the verge of extinction. Such a change would greatly expand the amount of land subject to use restrictions under the Endangered Species Act.

One other strategy that involves protecting something other than a single, full species from extinction deserves mention. In July of 1994, the Fish and Wildlife Service announced a radical new plan for protecting threatened and endangered species. "Species will be conserved best not by a species-by-species approach but by an ecosystem conservation strategy that transcends individual species." [62] In 2008, the Service unveiled a plan to institute protections for no less than 48 separate species all at once by declaring 43 square miles of public and private land on the Hawaiian island of Kauai as critical habitat. This will amount, in practice, to establishing a permanent federal wildlife refuge, since it is highly improbable that even half of those 48 species, let alone every one of them, will merit removal from the threatened or endangered list in the foreseeable future.

In spite of these efforts to prevent not just the complete extinction of bona fide species, but also the extinction of subspecies and alleged subspecies, of "distinct population segments," and of species from a "significant portion" of their ranges, the Endangered Species Act does not prevent *all* disappearances of species from any part of their ranges. But if the act were consistent with its purpose of preserving biodiversity, it would indeed aim to prevent every last one of these "local extinctions." The more we succumb to the notion of biodiversity as an ace in the hole that trumps such core American values as property rights and economic productivity, the more difficult we will find it to resist the growing push to expand the ESA to cover every local extinction, an expansion that would severely harm what remains of Americans' property rights and economic vitality. [63]

As it happens, there is just such a push under way. On their website the Center for Biological Diversity, which initiates many of the lawsuits that are filed under the Endangered Species Act, argues that "most of biodiversity's benefits take place on the local level," and, therefore, "conserving *local populations* is the only way to ensure genetic diversity critical for a species' long-term survival." [64] Meanwhile, the center is seeking to add hundreds more species to the list of threatened and endangered species. In July of 2012 they reached an agreement with the Fish and Wildlife Service under which the service would expedite the review of 757 species of plants and animals which the center believes should be added to the threatened or endangered lists.

On another front, the *National Geographic* reports on their website, "[W]e are in the midst of the Sixth Great Extinction, an event characterized by the loss of between 17,000 and 100,000 species *each year*." [65] Unlike the last extinction, which wiped out the dinosaurs 65 million years ago, the one allegedly under way at present began during human times; some call it the "anthropocene" extinction. According to the New York Times, we are "on track to lose three-quarters or more of all species within a few centuries." [66] The extinction of species (and subspecies and pseudo-species) is shaping up to be the "Next Big Crisis," one that promises to rival the ongoing global warming campaign.

But given the present state of man's knowledge of extinction, claims that he is causing the "Sixth Great Extinction" are premature, at best. We do not know how many species inhabit the earth. Estimates range from several million to tens of millions.[67] The wide range of this estimate speaks volumes about the degree of our ignorance. We do know from the fossil record that a great many species that existed in the past do not exist today, so we know that extinction is a natural phenomenon. We also know that there have been several "mass" extinctions in the distant past, some or all caused by cataclysmic events such as an object from space striking the earth. And we know that there is an ongoing process of natural extinction at any given time. It is possible that the rate of extinctions has increased at least somewhat during the period of man's dominance on earth, and especially since the Industrial Revolution, if only because man's rapidly increasing numbers have been displacing wildlife populations.

But whether man has caused an ecologically harmful number of extinctions, much less a "Sixth Mass Extinction," simply has not been proven. We do not know what the rate of extinctions has been in the past. And since we do not know either the rate of natural extinction at present, nor the rate of extinctions caused by man, we have no way of knowing to what extent man has increased the overall extinction rate.

"[I]t would be scientifically unjustified to blame man for all species that have disappeared (or that seem to have become extinct) in the past few thousand years," writes A.J. van Loon.[68] Commenting on a resolution offered at the 2002 World Summit on the calling for "a significant reduction in the rate of biodiversity loss," van Loon says, "Since extinction is—as it has been ever since life originated on Earth—a natural phenomenon, this … statement is, obviously … ridiculous: one might question … how the advocated policy should be implemented and the results evaluated—if neither the past nor the present-day extinction rate is known or can be estimated reliably with modern techniques."[69] Says van Loon, "There are more questions than answers, which implies that we are still far away from understanding how biodiversity works, or is affected by man."[70]

Given the likelihood that man has increased the extinction rate to some extent, but given also the limited state of his knowledge at present, van Loon advises that we "monitor biodiversity worldwide."[71] This is a far cry from America's policy of suspending private property rights wholesale and shutting down productive uses on huge areas of public and private land.

It would be an easy matter to prevent extinctions such as that of the passenger pigeon, which was wiped out by commercial hunters in the U.S. in the late 19ᵗʰ and earth 20ᵗʰ centuries. Indeed, such egregious, human-caused extinctions have already been halted in the United States. But the Endangered Species Act goes far beyond preventing this type of extinction.

One thing that makes the act so radically preservationist is its failure to distinguish between species that are naturally rare and those that have been rendered rare and endangered by human actions. In western El Dorado County, California, there is a tract of land called the Pine Hill Preserve, where four endangered and one threatened plant species are preserved virtually in amber.[72] Since this region on the west slope of the Sierras is lightly populated, it is hard to argue that these plants have been made rare by human action. In all likelihood, they are simply rare plants, species uniquely tailored to a highly specific combination of soils and weather.

But in the case of many species like these, it is automatically assumed either that their rarity has been caused by human action and must therefore be countered, or that, simply because they are rare, they must be preserved from extinction, even though the ecological consequences of their passing would be nil. To prove man's responsibility for their rarity, by the way, would be difficult, if not impossible, and, at any rate, such proofs are rarely if ever considered necessary. (Van Loon has suggested that the ongoing warming from the last ice age might be the cause of continuing extinctions today.[73] We know this is the case for several populations of endangered desert pupfish that live in water-filled caves near Death Valley. These water holes are remnants of a great lake that covered the area after the last ice age. As the lake dried up, an entirely natural process, the fish were left isolated in these tiny shrinking environments.)

Since the Endangered Species Act became law in 1973, only

five species have officially been declared extinct within the fifty states.[74] This hardly qualifies as a wildlife Armageddon. One of the five, the Santa Barbara song sparrow, inhabited a tiny, 639-acre island off California. Island species are notoriously prone to extinction. Environmentalists blame the sparrow's disappearance partly on feral cats, which were brought to the island by man. But if we are going to blame man for extinctions caused by his introducing of harmful alien species into new environments, then we must also credit him for the dramatic *increase* in the number of species of fauna and especially flora that inhabit North America. Mark A. Davis has written that over 4000 alien species of vascular plants alone introduced to North America in the last four hundred years have become naturalized, "[y]et there is no evidence that even a single long-term resident species has been driven to extinction, or even extirpated within a single US state, because of competition from an introduced plant species." [75]

If we were to abolish the Endangered Species Act and reinstitute secure property rights and a genuinely capitalist industrial economy, is it likely that we would cause so many extinctions to occur that the United States would eventually become incapable of sustaining so many people as now live here? In addition to the doubts cited above as to whether man has been causing a catastrophic number of extinctions, there are other, important reasons to believe that we would not.

Freedom, capitalism, and industrialism give rise to forces that can mitigate the pressure that a large human population tends to place on wildlife. Environmentalism is itself a product of capitalism and industrialism, and not just because without the latter there would be no need for an environmental movement. There have been industrial countries, such as the Soviet Union and Red China, that were not capitalist. But it was in the free, capitalist countries of the West that environmentalism arose. It was in the wealthy West, and not in the poor Communist countries, that some citizens came to believe, however mistakenly, that the problem of producing the material wealth man needs to survive had been so thoroughly solved that we could begin to constrain men's productive efforts in the interest of cleaning up and restoring the health of the environment. And it was in the capitalist West that environmentalists have had the freedom (to freedom's detriment) to challenge the existing economic and political order as they have in the United States. While capitalist industrialism (or, more often, its distortions) have given rise to some environmental problems, it also has created the conditions under which men can afford to give thought to improving the quality of their environment, and to developing and implementing the means of doing so.

The science of ecology also is a product of the free, capitalist West. It is wealth that makes possible the pursuit of scientific knowledge. A people who had to occupy themselves entirely with keeping alive could not afford to support individuals who did nothing but study nature. Wealth also makes possible the rapid and easy dissemination of new ecological knowledge the world over, including to the general public. This would be important in a genuinely free society that would rely on the *voluntary* cooperation of its citizens to preserve wildlife. Wealth also enables people to carry on their own preservation efforts. Private individuals are today helping to preserve wildlife through such programs as the Audubon Society's certifying of backyard bird habitats and the National Wildlife Federation's certifying of private wildlife habitat.

But imagine a world in which wildlife, including endangered species, could be *owned* by private landowners. Imagine individuals collecting endangered species of plants, like rare coins, and taking on the challenge of restoring them to species health. Imagine greenhouses full of endangered plants, and commercial landscaping companies using them to landscape homes and businesses.

The best thing that can happen to an endangered species in a free society is for a valuable human use to be found for it. There is no reason why any fish that is of commercial value should ever become endangered if men could own them and farm them. (That four species of Pacific salmon are listed as threatened or endangered in various locales on the west coast, even though Salmon is one of the most farmed fishes in the world, attests to environmentalists' preference for preserving species in the wild.) Whales would never have become endangered if men could have owned them—perhaps by bidding for exclusive rights to their ranges, the way broadcasters purchase the rights to a radio frequency. The near-demise of the whales was a classic case of the tragedy of the commons; no one owned the whales, so everyone had an incentive to harvest as many as he could before someone else beat him to it.

Elephants provide an instructive example. In 1973, Kenya in effect made their elephants public property and prohibited the hunting of them. At the hands of poachers, and of villagers defending their crops and their lives, the elephant population declined from 167,000 to 16,000 by 1989.[76] No one had a direct stake in keeping the elephants alive. Zimbabwe, on the other hand, in 1989 placed its elephants under the control of local leaders, who treated them as communal property and instituted hunting quotas. Elephants were thus made a source of value to the local villagers. While this was not a case of instituting full private property rights in elephants, it was a step in that direction, and the population of elephants increased from 37,000 to 85,000.[77] "In Namibia, which allows hunting, more than 80% of all large wild mammals live on private and community lands, and those populations have increased by 70% in recent years."[78]

There is no reason why private citizens should not be free to own large carnivores like bears, wolves, and mountain lions.

(Legal liability would give them an incentive to keep the animals apart from humans. The current effort to restore large carnivores to former ranges that are now populated by humans is insane.) Freedom and prosperity would also enable individuals and groups, such as non-profits, and even corporations, to purchase habitat and preserve it for wildlife. Under the Endangered Species Act, in contrast, landowners have every incentive to avoid having endangered species found on their land.

Among the most powerful forces for good that freedom and prosperity give rise to are scientific and technological advancement. They have made possible dramatic increases in agricultural productivity per acre. According to Indur M. Goklany, "Between 1961 and 1993, global population increased 80% (from 3 to 5.5 billion), but cropland increased only 8%," yet over the same period, "food supplies increased in all regions except sub-Saharan Africa." [79] Such productivity increases make it possible to leave wildlife habitat intact on land that would otherwise have been needed for crops.

Science and technology have given us the ability to store the DNA of endangered species in DNA banks. All things being equal, storing it in banks would be less desirable than preserving species in the wild, but it is a nice insurance policy. We are close to having the ability to create viable clones from the DNA of a living member of an endangered species. We have even created a living clone from the DNA of a dead goat, although the clone did not live long. Such clones would lack the genetic diversity needed to reconstitute an entire species in the wild. But we also have the ability to engineer changes in DNA, so it is conceivable that we could someday restore an entire species to viability in the wild (though whether we should want to is a question worth asking). [80]

This would be expensive, to be sure, but what is the Endangered Species Act if not exorbitantly expensive? The day might not be far off when our wealth and technology will enable us to create whole new species from scratch, or even to create whole new natural communities made to order. We are also growing in leaps and bounds in our ability to *repair* natural communities that have suffered damage either from natural or human causes. And this last gives rise to an important point.

We know that natural communities can afford to lose some species without serious ecological consequences. There is hardly a place on earth that has not suffered the disappearance, at least locally, of one or more species because of human action, and yet we are not being overwhelmed by extinction-caused crises. Indeed, throughout history, man's well-being has required the extirpation of some species, such as large carnivores, from the vicinity of his settlements.

We also know that some species, such as the desert pupfish, will not cause any great ecological repercussions were they to go extinct. (Yet we spend public wealth to keep the pupfish alive. These would be a great candidate for private adoption.) We also know, by the time a species becomes officially threatened or endangered, what would be the ecological consequences of its extinction, because *it has already disappeared from part of its range*, and we can see the ecological consequences, if there are any, in those locales. If there have been no serious consequences, then there is no pressing human need to preserve such a species in the wild. So if voluntary efforts to preserve this species in the wild were to fail, such species would be good candidates for preservation in captivity or in a DNA bank.

On the other hand, in a genuinely free society there would likely be some extinctions that would entail undesirable consequences for man, and which voluntary efforts would fail to avert. This is where man's scientific and technological ability to repair ecological damage would come in. When industrial man undertakes a new technological venture, he does not know in advance all the problems that will arise as a result of it. But he embarks upon it in the confidence that whatever those problems will be, he will find a way to solve them. Already, the accumulation of carbon dioxide in the atmosphere, which a short time ago had seemed to environmentalists to be an insoluble side-effect of the industrial age, is giving rise to a number of possible technological solutions.[81] There is every reason to believe that if a human-caused extinction threatened to cause serious ecological problems for mankind, he would be able to engineer a solution.

If man were to abandon freedom, capitalism, and industrialism, to reduce the number of human inhabitants of the earth to a small fraction of its current seven billion, and to return to living in small pastoral groups, then he might succeed in eliminating most human-caused extinctions. But since the costs of this strategy would be so unacceptably high, only a handful of environmentalists openly advocate it today. Most do not advocate it openly because, for any rational person, it is not a realistic option. But it is the course represented, in principle, by the Endangered Species Act.

The act implies that capitalism and industrialism must—to one extent or another—be suspended until the current extinction "crisis" has passed. They must be suspended in every locale where their pursuit might cause the extinction of a species or a portion of a species. But based on the evidence of the last forty years, the chances are good that the implicit state of emergency that the Endangered Species Act is supposed to bring to an end will be permanent. In those forty years, only 30 threatened or endangered species have officially "recovered." And if the current effort to turn the extinction of species into the Next Big Crisis succeeds, then property rights will be abrogated beyond recognition and economic productivity will be severely diminished.

Yet environmentalists say we have no choice but to play it safe and to do all we can now to prevent further extinctions, even if it means partially shutting down capitalism and industrialism. But risk is part of living. Suspending our liberties and our prosperity to prevent extinctions is like deciding not to leave home for fear of being hit by a car. It is, for all practical purposes, a decision to stop living our lives. The choice that Americans face is whether the miniscule risk (if any) that we will suffer a major extinction-caused ecological catastrophe is outweighed by the benefits we all derive from individual liberty, capitalism, and industrialism, and all they entail. Even to the severely risk-averse, this has to be an easy choice.

Property Rights and Environmentalism

Imagine an America in which questions of free speech were determined on a case-by-case basis depending on the contents of the speech in question. Imagine, for example, that if the consequences of socialist ideas could be shown to be detrimental to the "public good," then advocating them in speech or print could be prohibited. Or imagine the same with regard to religious teachings.

In the United States, we do not decide on a case-by-case basis which ideas or religious teachings may be freely advocated. Whether some ideas or religious teachings can be shown in fact to have harmful consequences is legally irrelevant in a country that protects individuals' *rights* if free speech and freedom of religion. Yet by inducing us to decide questions of land use on the basis of "scientific" determinations of what would be the environmental consequences of one use or another, environmentalists have succeeded in doing to property rights precisely what deciding questions of free speech or religious freedom on a case-by-case basis would do to the rights of freedom of speech or freedom of religion. They have transformed into a question of "science" what had been, and should continue to be, one of rights.

Recall that ownership entails the rights to possess, to use, and to dispose of a thing. The environmentalist strategy for preserving private lands has been to leave the rights to possess and to dispose of these lands intact, but for government to usurp piecemeal the right to determine how these lands may be used. Environmentalists would preserve private ownership in form while eviscerating it of its substance. The resulting hybrid is a species of socialism—one that characterized the fascist regimes of Germany and Italy in the middle of the last century. Once the environmentalists and their liberal allies control land use, the rest of our rights will fall like dominos—a process that has already begun, and the Lockean republic founded in 1783 will be at an end.

Notes

1. The only use permitted in most national parks is recreation, although some of the parks permit limited oil and gas production.
2. As of summer, 2013.
3. Very limited oil and gas production has been conducted at Padre Island National Seashore.
4. "A National System," National Wild and Scenic Rivers System, accessed November 5, 2014,
http://www.rivers.gov/national-system.php.
5. "Public Land Statistics, 2012," U.S. Department of the Interior, Bureau of Land Management, accessed November 5, 2014,
http://www.blm.gov/public_land_statistics/index.htm, Table 3-9, p. 94. Adjusted for inflation.
6. "Total Federal Coal Leases in Effect, Total Acres Under Lease, and Lease Sales by Fiscal Year Since 1990," U.S. Department of the Interior, Bureau of Land Management, accessed November 5, 2014,
http://www.blm.gov/wo/st/en/prog/energy/coal_and_non-energy/coal_lease_table.html.

7. "Public Land Statistics (2012)," BLM, accessed November 5, 2014, http://www.blm.gov/public_land_statistics/index.htm, Table 3-20, p. 133 and Table 3-23, p. 143.

8. Ibid., Table 3.5, p. 76, Table 3-7a, p. 83. "Section 3" and "Section 15" refer to the Taylor Grazing Act of 1934.

9. Includes all Federal lands, not just BLM lands. "Number of New Leases Issued During Fiscal Year," BLM, updated December 23, 2013, accessed November 5, 2014, http://www.blm.gov/pgdata/etc/medialib/blm/wo/MINERA LS__REALTY__AND_RESOURCE_PROTECTION_/energy/o il___gas_statistics/data_sets.Par.62098.File.dat/table04.pdf; "Number of Acres Leased During Fiscal Year," BLM, updated December 23, 2013, accessed November 5, 2014, http://www.blm.gov/pgdata/etc/medialib/blm/wo/MINERA LS__REALTY__AND_RESOURCE_PROTECTION_/energy/o il___gas_statistics/data_sets.Par.80157.File.dat/table05.pdf; "Number of Drilling Permits Approved on Federal Lands, updated November 18, 2013, accessed November 5, 2014, http://www.blm.gov/pgdata/etc/medialib/blm/wo/MINERA LS__REALTY__AND_RESOURCE_PROTECTION_/energy/o il___gas_statistics/data_sets.Par.65795.File.dat/table08.pdf.

10. Zoning changes can carry their own injustices; in the case of a couple who invested their life savings in a dream home only to find later on that the undeveloped tract of rural land next door that was zoned for 20-acre residential use has been re-zoned to accommodate high density residential development, it is at least understandable that they would complain about their treatment by their government.

11. El Dorado County General Plan Conservation and Open Space Element, Policy 7.1.2.7.

12. Ibid., Policy 7.3.2.3.

13. Ibid., Policy 7.4.1.6.

14. Ibid., Policy 7.4.2.2.
15. Ibid., Policy 7.4.2.9.
16. Ibid., Policy 7.4.4.4.
17. Ibid., Polity 7.4.5.2.
18. Ibid., Policy 7.4.5.2 B.
19. "UC CEQA Handbook: 3.3 - Environmental Impact Report," University of California, accessed November 6, 2014,

http://www.ucop.edu/ceqa-handbook/chapter_03/3.3.html.
20. Wendell Cox ,"The Evolving Urban Form: Portland," NewGeography.com, August 3, 2013, accessed November 6, 2014,

http://www.newgeography.com/content/003856-the-evolving-urban-form-portland.
21. Compared to Oregon, the state of Texas has few restrictions on home building. In 1970, the median price of a home in Oregon was 28.3% higher than in Texas. Over the period of 2007-2011 (as measured by the U.S. Census), after a generation of land use restrictions in Oregon, the median home price in Oregon was 99.8% higher than in Texas. Between 1970 and 2010, Texas's population grew by 125%, while Oregon's grew by only 81%. "Historical Census of Housing Tables: Home Values," U.S. Census Bureau, June 6, 2012, accessed November 6, 2014, http://www.census.gov/hhes/www/housing/census/historic/values.html and "State and County Quick Facts: Oregon," U.S. Census Bureau, July 8, 2014, accessed November 6, 2014, http://quickfacts.census.gov/qfd/states/41000.html.
22. "9th Annual Demographia International Housing Affordability Survey: 2013", Demographia (Wendell Cox Consultancy)," accessed February 15, 2014,

http://www.demographia.com/dhi.pdf, 17.
23. "Bay Area Census," Metropolitan Transit Commission-Association of Bay Area Governments Library, accessed November 6, 2014,

http://www.bayareacensus.ca.gov/counties/SantaClaraCounty7 0.htm.

24. See "State and County QuickFacts: Placer County, CA," U.S. Census Bureau, accessed November 6, 2014, http://quickfacts.census.gov/qfd/states/06/06061.html.

25. Public Law 92–583, Approved Oct. 27, 1972, 86 Stat. 1280. Emphasis added.

26. Ibid. Finding "h" states, "In light of competing demands and *the urgent need to protect and to give high priority to natural systems in the coastal zone....*" Emphasis added.

27. Generally, the "coastal zone" extends 1000 yards inland, or to the nearest public road.

28. *Grupe v. California Coastal Commission* (1985) 166 Cal. App.3d 148 [212 Cal. Rptr. 578].

29. *Remmenga v. California Coastal Commission* (1985) 163 Cal. App 3d 623 [209 Cal Rptr. 628].

30. *Nollan v. California Coastal Commission* 483 U.S. 825 (1987).

31. See, for example, *Daniel and Hill v. County of Santa Barbara*, 288 F.3d 375 (9th Cir. 2002).

32. See *Healing v. California Coastal Commission*, 22 Cal. App. 4th 1158.

33. See *Encinitas Country Day School v. California Coastal Commission*, (2003) 108 Cal. App. 4th 575.

34. See *LT-WR, LLC v. California Coastal Commission*, (2007) 152 Cal.App.4th 770, 797.

35. See *Sterling v. California Coastal Commission*, No. CIV. 482448 (Cal. Super. June 18, 2010).

36. Peter Douglas, "Shades of Green: Buying and Selling Environmental Protection," Address to Yosemite Environmental Law Conference (Oct. 26, 2002). Quoted in J. David Breemer, "What Property Rights: The California Coastal Commission's History of Abusing Land Rights and Some Thoughts on the

Underlying Causes," *UCLA Journal of Environmental Law & Policy* 22, 2 (Winter, 2004): 289, n242.

37. Peter Douglas, "The Loss of Community Environmental Values: An American Tragedy," Remarks to the California Chapter of the American Planning Association, Oct. 13, 1997. Quoted in Breemer, "What Property Rights," 290.

38. Ibid.

39. Peter Douglas, "Shades of Green," Quoted in Breemer, "What Property Rights," 291, n247.

40. "The legal landscape relative to land use management for the purpose of protecting community environmental values changed dramatically in 1987. Two Supreme Court rulings that year (Nollan and First English) fundamentally changed the way in which land use decisions would henceforth be treated by inventing a new and expansive "regulatory takings" doctrine. At the urging of private land rights advocates, these and subsequent court rulings have effectively neutered government in its ability to protect important community environmental values. These decisions significantly redefined and expanded individual, private land rights and benefits at considerable public expense." Peter Douglas, "Minding the Coast: It's Everyone's Business," Paper presented to the Coastal Society 16th International Conference, July 12, 1998, California Coastal Commission, accessed November 6, 2014, http://www.coastal.ca.gov/coastsc.html.

41. Peter Douglas, "Minding the Coast," http://www.coastal.ca.gov/coastsc.html.

42. Ibid.

43. To "take" means to "harass, harm, pursue, hunt, shoot, wound, kill, trap, capture, or collect, or to attempt to engage in any such conduct." An exception is made for endangered insects officially considered to be a nuisance. See "Endangered Species Act," National Wildlife Federation, accessed November 6, 2014,

http://www.nwf.org/Wildlife/Wildlife-
Conservation/Endangered-Species-Act.aspx.
44. Some states compensate owners for livestock killed by a protected predator.
45. *TVA v. Hill*, 437 U.S. 153 (1978)
46. The winter (endangered) and spring (threatened) runs of the Chinook Salmon are also the cause of restrictions in the Sacramento Delta.
47. In 2008 the Bush Administration reduced the acreage of the critical habitat to 5.4 million acres, but in 2012 the Obama Administration increased it to 9.6 million acres.
48. ESA & 4(b)(1)(A); 16 U.S.C. &1533 (b)(1)(A) Emphasis added.
49. 437 U.S. 153 (1978). Emphasis added. The Fish and Wildlife Service is required to designate critical habitat for threatened or endangered species, and, when doing so, they *are required* to take into account the costs that their designations will entail.
50. "Why Save Endangered Species," U.S. Fish and Wildlife Service, July, 2005, accessed November 6, 2014,
http://www.fws.gov/nativeamerican/pdf/why-save-
endangered-species.pdf.
51. Ibid.
52. Ibid.
53. The Endangered Species Act states that "[t]hese species are of esthetic, ecological, educational, historical, recreational, and scientific value to the nation and its people." 16 U.S.C. 1531, Sec. 2, (a)(3).
54. The Endangered Species Act aims "to provide a means whereby the ecosystems upon which endangered species and threatened species depend may be conserved....") 16 U.S.C. 1531 (b).
55. Peter Aldhous, "Subspecies, Distinct Populations and the Fight for Land," New Scientist, July 15, 2006, accessed November 6,

2014,
http://www.newscientist.com/article/mg19125604.000-
subspecies-distinct-populations-and-the-fight-for-land.html.
56. M. David Stirling, *Green Gone Wild: Elevating Nature above
Human Rights*, (Bellevue, WA: Merrill Press, 2008), 119-134.
57. "Taxanomic Considerations in Listing Subspecies under the
U.S. Endangered Species Act"
By Haig SM, *et al*, *Conservation Biology* 20 6 (December, 2009):
1585.
58. "Interagency Policy Regarding the Recognition of Distinct
Vertebrate Population Segments under ESA," U.S. Fish and
Wildlife Service, accessed November 6, 2014,
http://www.fws.gov/endangered/laws-policies/policy-distinct-
vertebrate.html. Congress has stipulated that the authority to
protect distinct population segments be used "sparingly and only
when the biological evidence indicates that such action is
warranted." (Senate Report 151, 96th Congress, 1st Session,
1979, 6-7.).
59. "Interagency Policy," U.S. Fish and Wildlife Service,
http://www.fws.gov/endangered/laws-policies/policy-distinct-
vertebrate.html.
60. The population must be separated from the rest of its species
physically, physiologically, ecologically, or behaviorally, and it
must be important to the survival of the species in some way, such
as by being markedly different genetically from the rest of the
species, thus increasing the biodiversity of the species.
"Endangered and Threatened Species; Determination of Nine
District Population Segments of Loggerhead Sea Turtles as
Endangered or Threatened," U.S. Fish and Wildlife Service,
accessed November 6, 2014,
https://www.federalregister.gov/articles/2011/09/22/2011-
23960/endangered-and-threatened-species-determination-of-
nine-distinct-population-segments-of-loggerhead#h-9.

61. "Improving ESA Implementation: Significant Portion of Its Range," U.S. Fish and Wildlife Service, accessed November 6, 2014,
http://www.fws.gov/endangered/improving_esa/spr.html. Emphasis added.
62. "Endangered and Threatened Wildlife and Plants: Notice of Interagency Cooperative Policy for the Ecosystem Approach to the Endangered Species Act," Federal Register, July 1, 1994, accessed November 6, 2014,
http://www.nmfs.noaa.gov/pr/pdfs/fr/fr59-34274.pdf.
63. As of February, 2014, there were 1,515 species on the endangered or threatened list for the United States.
64. "The Extinction Crisis," Center for Biological Diversity, accessed November 6, 2014,
http://www.biologicaldiversity.org/programs/biodiversity/ele ments_of_biodiversity/extinction_crisis/. This is actually two different reasons for preventing local extinctions, joined to form a non-sequitur.
65. "The Sixth Great Extinction: A Silent Extermination," *National Geographic* "Newswatch," March 28, 2012, accessed November 6, 2014,
http://newswatch.nationalgeographic.com/2012/03/28/the-sixth-great-extinction-a-silent-extermination/.
66. Richard Pearson, "Protecting Many Species to Help Our Own," *New York Times* (June 1, 2012). The author cites a study published in the journal *Nature*.
67. A.J. van Loon, "The Dubious Role of Man in a Questionable Mass Extinction,"
Earth-Science Reviews 62 (2003): 180. Accessed 12/18/14, http://geoinfo.amu.edu.pl/wngig/ig/UAM_Ing/VanLoon/Eart hSciRev_62%282003%29.pdf, On the lower estimate, van Loon cites Goldschmidt, T., 2002. Wankele soorten. NRC Handelsblad. 32 (306), 33.

68. van Loon, "Dubious," 181.

69. Ibid., 180. Van Loon cites Pal, C., Papp, B., Hurst, L.D., 2003. "Rate of Evolution and Gene Dispensability," *Nature* 421, 496–497.

70. van Loon, "Dubious,"182.

71. Ibid., 181.

72. Pine Hill ceanothus, Stebbins' morning-glory, El Dorado bedstraw, Pine Hill flannelbush, and Layne's butterweed. "Rare Plants of PHP," Pine Hill Preserve, accessed November 6, 2014, http://www.pinehillpreserve.org/rare_plants/index.htm.

73. Van Loon, "Dubious," 181-182.

74. Santa Barbara song sparrow, longjaw cisco, Sampson's pearlymussel, blue pike, dusky seaside sparrow. "Delisting Report," U.S. Fish and Wildlife Service, accessed November 6, 2014,

http://ecos.fws.gov/tess_public/DelistingReport.do.

75. Mark A. Davis, "Biotic Globalization: Does Competition from Introduced Species Threaten Biodiversity?" *Bioscience* 53 5 (May 2003): 481. This is not to say that the number of species world-wide has not declined over the last 400 years.

76. Terry Anderson and Shawn Regan, "Shoot an Elephant, Save a Community," *Defining Ideas* (Hoover Institution), June 6, 2011, accessed November 6, 2014,

http://www.hoover.org/research/shoot-elephant-save-community.

77. Ibid.

78. Ibid.

79. "Saving Habitat and Conserving Biodiversity on a Crowded Planet," by Indur M. Goklany, *Bioscience* 48 11 (Nov., 1998): 941.

80. Van Loon suggests that restoring extinct species might be an inadvisable step backwards in the evolutionary process, van Loon, "Dubious Role," 183-184.

81. See, for example, Hugh Powell, "Fertilizing the Ocean with

Thomas McCaffrey

Iron," *Oceanus Magazine* 46 1 (January, 2008), published online November 13, 2007, accessed November 6, 2014, http://www.whoi.edu/oceanus/viewArticle.do?id=34167). See also Jens Hartmann, et al, "Enhanced Chemical Weathering as a Strategy to Reduce Atmospheric Carbon Dioxide, Supply Nutrients, and Mitigate Ocean Acidification," *Reviews of Geophysics* 51 2 (June, 2013): 113-149, published online May 23, 2013, accessed November 6, 2014, http://onlinelibrary.wiley.com/doi/10.1002/rog.20004/abstract.

12: Conclusion

ACCORDING TO THE preservationist premise, man has a moral obligation to refrain as much as possible from upsetting the balance of nature. The surest way to avoid upsetting the natural balance is to preserve the landscape in its natural state. Economic growth is the enemy of nature preservation; more wealth means more factories and office buildings and shopping malls and housing developments, and those mean more and more of the landscape plowed up, paved over, and built upon. The great engine of economic growth, in turn, is manufacturing. Anyone who wanted to slow down economic growth would want to slow down the growth of manufacturing. If environmentalism is indeed fundamentally preservationist, then we would expect to see environmentalists working to constrain industrial growth. This is precisely what the Clean Air Act, the Clean Water Act, the war on energy, the war on carbon emissions, and the rest of what constitutes environmentalism have been doing to U.S. manufacturing.

The Decline of U.S. Manufacturing

One could fill a small library with all the academic papers that purport to show that environmental regulation does not drive manufacturers out of business or cause them to relocate to other countries. But anything that drives up production costs in a given locale necessarily becomes an incentive for manufacturers to move

Thomas McCaffrey

someplace where costs are lower. A Sunday drive through the old mill towns of New England, whose looms once led the world in textile production but now stand silent, will attest that businesses most certainly do move elsewhere to take advantage of lower production costs.

California has become the exemplar of what happens when businesses are subjected to punishing levels of regulation. A study commissioned by the California Legislature in 2009 found that state regulation of small businesses alone costs Californians almost $5 billion per year, which is almost a third of the state's gross domestic product.[1] This equated to over $134,000 per small business in 2007.[2] In 2011, California's economy, the eighth largest in the world, ranked thirty-fourth among American states in economic growth.[3] Businesses are leaving California. In 2011, according to one consultant, 254 companies moved at least a part of their operations from California to other states.[4] Environmental regulations are not the only cause of these relocations, of course, but in environmentally-friendly California, they are an important one. A common complaint among California employers, for example, is the high cost of housing that their employees must pay, a cost that is a direct consequence of environmentalist land use restrictions and growth controls that dramatically restrict the supply of new housing in the Golden State.

On the federal level, a 2012 study found that the agency that imposed the highest number of regulations on the manufacturing sector is the EPA, and that it also imposed the highest regulatory costs—a total of $158 billion for the period from 1993 to 2011. (Second place went to the Department of Transportation, at a mere $30 billion.)[5]

So how has manufacturing fared in the United States since 1970? Manufacturing employment peaked in 1979, at 19,553,000 persons.[6] From there it trended downward, hitting 17,322,000 in July of 2000.[7] Soon after, it began to drop precipitously, until it

508

hit 11,981,000 in March of 2010, almost 39% below its peak of 34 years earlier.[8] These numbers suggest that American manufacturing has been in decline since the late 1970s. But many economists and academics say that these employment statistics are not indicative of industrial decline. They argue that, just as happened in American agriculture, increased productivity has enabled fewer workers produce more output than it took far more workers to produce in the past. This is the argument both of those on the political left, who are anxious the show that increased government regulation does not cause a diminution of industrial output, and of some on the political right, who are equally anxious to show that free trade among nations does not entail such a diminution for the U.S.

But a paper published in March of 2012 argues persuasively that this conventionally agreed-upon argument is incorrect, and that the growth rate of American manufacturing has indeed diminished markedly, just as the employment statistics suggest.[9] The authors argue that the government's estimates of manufacturing productivity, which show an increase in industrial productivity of 15.5 percent from 2000 to 2010, are incorrect, and that in fact output declined about 11 percent during that time.[10] Their reasoning on this point is technical, and I direct interested readers to their article.

But they go on to make an argument that is hard to refute: the way to improve productivity on a significant scale is to invest in new, technologically advanced machinery. They show that, in fact, manufacturers have not been buying new machinery since 2000 in anywhere near the quantities that would be necessary for America's drastically reduced workforce of 2010 to be able to produce as much or more than the larger workforce of 2000. To support this argument, the authors show that manufacturers' investments in plants and machinery as a share of gross domestic product peaked in the 1970s and have been declining ever since.

In the 1970s, these averaged 2.4 percent of gross domestic product per year.[11] That figure dropped to 2.2 percent in the 1980s, 2.1 percent in the 1990s, and 1.5 percent from 2000 to 2010.

The authors also point to how long it has been since the quantity of capital stocks, such as buildings and machinery, reached its peak within various industries. For example, the quantity of capital stock in the primary metals industry (steel and aluminum manufacturing, for example) peaked in 1981 and by 2010 had declined by 24.2 percent.[12] This statistic indicates, first, that primary metals manufacturers in the U.S. have not been improving their productivity by investing in new plants and machinery, and second, that the primary metals industry has been in decline since 1981 or so. (The quantity of raw steel produced in the U.S. peaked in 1973, aluminum peaked in 1980.)[13]

Wood products manufacturers' capital stocks also peaked in 1981 and by 2010 had declined by 13.1%.[14] (U.S. hardwood production peaked in 1991.[15] Think spotted owl.) Textile mills' capital stocks peaked in 1997 and had declined by 33.9 percent by 2010.[16] Paper products manufacturers' capital stocks peaked in 1996 and by 2010 had declined by 22.3 percent.[17] (Think of the Edwards Dam in Augusta, Maine, which was expropriated from its owners and demolished by the federal government. Not only was the dam used to generate hydroelectric power, it was also a collection point of pulp logs floated down the Kennebec for the paper mills in the region. The log runs were outlawed in the 1970s, the dam removed in 1999, and now the paper mills are disappearing.) Only three of 19 manufacturing sectors did not show a decline in their total capital stocks by 2010.[18] It is clear that these industries have not improving their productivity by investing in plants and machinery, so they could not today be producing greater quantities of goods with fewer employees than they were thirty years ago.

Then there is the percentage of capital that U.S. companies have been investing in their affiliates abroad rather than at their plants here in the U.S. In 1982 (the first year available), U.S. companies with operations abroad directed 11 percent of their capital expenditures to their foreign affiliates. But from 1982 onward they tended to direct more and more of their capital investment to their affiliates in other countries, so that by 2010 they devoted about 40 percent of their capital investment to countries outside the U.S.[19]

Finally, the authors look at manufacturing profits as a percentage of total domestic profits. In the 31 years from 1950 to 1981, these declined 9 percent, from 58 percent to 49 percent. But in the 29 years from 1981 to 2010, all within the environmental era, they declined a whopping 31 percentage points, from 49 percent to 18 percent.[20]

To show declines in manufacturers' investment in plants and machinery as a share of gross domestic product, and to show declines in manufacturers' profits as a share of total domestic profits, is to show that manufacturing is coming to comprise a smaller proportion of the economy of the U.S. Some experts argue that this is a perfectly natural and healthy development for an advanced economy in the modern world. According to this view, smokestack industries, like the old steel mills of Pittsburgh, are characteristic of a certain stage of development on the way to becoming an "advanced" economy. When an economy outgrows this stage, the mills move to a less advanced country where labor costs are lower, and the country that formerly hosted them transforms itself from a manufacturing economy into one with a smaller, high-tech manufacturing sector and an expanded service sector.

As Atkinson, *et al*, point out, though, this process of modernization need not entail a shrinking of a nation's industrial sector. Germany has seen a diminution of its smokestack industries, but at the same time it has seen its high-tech sector grow correspondingly, so that the overall size of its manufacturing sector has hardly diminished at all.[21]

It is natural and economically healthy for manufacturers to move from a country where production costs, including the cost of labor, are high to a country where they are lower, provided, of course, that those high production costs have not been artificially increased by government regulation. In the case of the United States, though, production costs have most certainly been increased artificially by such regulation. Virtually every new regulation increases the cost of doing business, and the myriad environmental regulations that have been imposed on U.S. manufacturers have increased their costs of doing business substantially.[22]

Of course, environmental regulation is only one of the factors that raised their costs, others being forced unionization, heavy taxation, and a whole range of other regulations. But environmental regulation fell most heavily on the smokestack industries, and it is no coincidence that they were the first to begin to disappear from the American landscape.[23]

And there was nothing economically healthy about their leaving the United States. Manufacturing jobs, as Atkinson, *et al*, point out, tend to pay considerably better than service sector jobs. A country that loses manufacturing jobs and replaces them with service jobs will be a poorer country for it. But a poorer U.S. would be a more environmentally friendly U.S. The de-industrialization of America is a direct—and intended— consequence of the environmentalist campaign to eradicate pollution, to conserve resources, and to preserve land and wildlife. Now the Shale Revolution is beginning to counteract this process of de-industrialization. If it continues to do so, it will be in spite of the efforts of environmentalists.

Environmentalism: Mainstream and Radical

To discover the nature of environmentalism, I have examined federal statutes such as the Clean Water Act, the Clean Air Act, and the Endangered Species Act, federal land and energy policies, and certain state and local policies, mostly in California. The ideas embodied in these statutes and policies represent mainstream environmentalism, as opposed to its radical fringe. The Clean Water Act, the Clean Air Act, and the Endangered Species Act have all been the law of the land for forty years or more. Although certain elements of those statutes are still contested in court from time to time, the premises underlying them are no longer seriously opposed. Those premises, and the ones underlying the other federal, state, and local policies I have referred to, constitute mainstream ideas by virtue of the widespread acceptance they enjoy among the American people.

Consider the idea that pollution is categorically evil. Rather than seeing some degree of pollution as an unavoidable concomitant of industrial productivity, environmentalists treat pollution as an absolute, unmitigated evil that must be extirpated completely—perhaps not right away, if only because to do so would be politically impossible—but eventually. As I have shown, this view of pollution manifests itself in the Clean Water and Clean Air Acts, specifically in the relentless tightening of pollution standards that they mandate. The idea that pollution is categorically evil is a radical idea, since it necessarily entails a severe diminution of our heavy industry. Yet it enjoys almost universal support in the United States today.

Another radical idea that has won general acceptance in the United States is the Malthusian one that natural resources are running scarce and must be "conserved." This idea is embodied in such policies as the federal fuel mileage standards that are imposed on auto manufacturers, the energy efficiency standards imposed on the manufacturers of home appliances, and the requirements of some states that new homes be built with dual-paned windows, substantial amounts of insulation, and water-conserving plumbing. The imposition of such requirements has already helped reduce the physical size of the American automobile, and it could one day bring about its demise, as it could the demise of the detached single-family home. Yet the idea that we must conserve energy resources is widely accepted in the United States today, as is the idea that government should enforce the conserving.

The early conservationists, epitomized by Gifford Pinchot, combined this idea of Malthusian shortages with another one now also widely accepted, the idea that private landowners, and especially commercial enterprises, tend to use natural resources wastefully and in other ways that do not conduce to the "public good." This idea is especially radical in that it strikes at the root of the American politico-economic system, private land ownership. (The first of Karl Marx's "Ten Points" called for the abolition of private land ownership.) The combination of these two ideas, the alleged imminence of resource shortages and the alleged profligacy of private landowners, enabled the early conservationists to make permanent the federal government's ownership of one third of the land in the United States. While the advocates of freedom and prosperity frequently contest the federal government's decisions as to how best to use—or not use—these federal lands, almost no one contests the *principle* that the government should own a third of the country.

The early conservationists' idea that private landowners tend to use land in ways contrary to the public good gave rise to another phenomenon that now poses a serious threat to America's politico-economic system, government regulation of *private* land use. It made its appearance in the Progressive era in the form of zoning, which, along with its more recent progeny, building codes and county-wide general land use plans, so inured Americans to having government dictate what they could or could not do with their land that it was a simple matter, in the 1970s, for our governments to begin to impose a plethora of new land-use restrictions for environmental reasons.

The threat that such regulation poses to the existence of the institution of private property is becoming clear in California. It is evident in the policy that developers of large new housing tracts may be compelled to devote a large percentage of their tracts to undeveloped "greenbelts." It is evident in the panoply of restrictions that counties such as El Dorado impose, such as the prohibitions on cutting down trees. It is evident in San Jose's urban growth boundary, and in the highly restrictive "Smart Growth" principles being forced upon San Francisco Bay Area municipalities by the One Bay Area plan. And it is evident in the draconian restrictions and exactions imposed on landowners along the coast by the California Coastal Commission. Note that, although there has been opposition to many of the Coastal Commission's rulings, the *principle* that government should regulate private land use is universally accepted today.

Two other ideas, now widely held, also work in favor of preserving land as opposed to developing it for productive uses. The first is the idea that a natural landscape is esthetically superior to a man-made one, especially one made over for commercial or industrial use. This idea is almost unanimously agreed upon in the United States today. The second idea derives from the preservationist premise. In any conflict over whether to develop a tract of land, it manifests itself as a presumption that to leave the land in its natural state would be the morally superior course.

California, again, is leading the way. In many parts of the Golden State, it is becoming rare for a proposed real estate development of any size to go forward unopposed, and the opposition often comes not just from environmental groups but from private citizens, Democrat and Republican alike (including Tea Party activists), who are seeking to preserve the ambience of their neighborhoods. One is struck by the air of moral righteousness with which they trample their neighbors' rights to use their land as they choose.

When one lives in a free society, one gives up the prerogative to force one's neighbors to "keep the county rural," just as one gives up the prerogative to force one's neighbors to "keep the county gun-free," or to "keep the county Protestant," or to "keep the county White." Yet these two ideas, the alleged esthetic superiority of the natural landscape and the alleged moral superiority of keeping it natural, are making it increasingly difficult to make productive use of our lands, public and private alike, in the United States.

Then there are the waters of the United States. While there has been much wrangling over exactly what constitutes a "navigable waterway" for purposes of federal jurisdiction, there is virtually no opposition to the *principle* that the federal government should exercise control over the vast majority of the nation's surface waters, including most swamps, bogs, and wetlands.[24] The danger of ceding monopoly control of the nation's waterways to the government seems not to have occurred to anyone, but the farmers of California's giant San Joaquin Valley are learning that a government that controls the water controls everything. They are having to leave their fields unplanted and watch their orchards wither and die as they watch what should be their irrigation water be diverted into rivers for the sake of the fish there.

And while there has been some opposition to specific waterway restoration plans here and there, such as the Chesapeake Bay TMDL plan, the idea goes virtually unopposed that, by and large, the proper goal of federal control ought to be, as much as possible, to preserve those waters that are already in a close-to-natural state, and to restore the rest to as natural a state as possible. Again, the idea of government control of the nation's waters, and the idea that the purpose of that control ought to be the preservation of those waterways, are both radical ideas that now enjoy mainstream support in the United States.

Finally, there is the Endangered Species Act. Although there are frequent outbreaks of opposition to a specific application of the act, and although there is opposition to the way the act itself is written, with its protections of subspecies and "distinct populations," especially when private property rights or economic interests are harmed, there is widespread agreement with the *principle* that species in danger of extinction should be preserved and that government should do the preserving.

A common complaint about environmentalists is that they fail to limit their demands to reasonable ones. If ever there was a time for reasonableness on the part of environmentalists, it was in the spring of 2014 in California's Central Valley. After three years of drought, with reservoir levels in the Sierra foothills and the early-spring snowpack at near-historic lows, environmentalists cheered a decision by a federal appeals court that would keep water draining out of the reservoirs and into the Sacramento Delta for the sake of the threatened Delta smelt.[25] The decision meant that tens of thousands of acres of farmland would go dry, thousands of jobs would be lost, food prices would rise, and some municipalities would face severe shortages of water for household use.

These kinds of crises will become common as environmentalists transform America. Indeed, from the environmentalist perspective this is not a crisis but a step on the road to reclaiming for nature some of the farmland that never should have been plowed up to begin with. Nothing could express more eloquently the meaning, and the intent, of the ideology of environmentalism than this sacrificing of the interests of the human beings who rely on this water to the needs of the tiny Delta smelt.

The absence of "moderate" environmentalist voices criticizing the court decision further attests that such measures are not manifestations of environmental "extremism," but of environmentalism, *per se*. As this hijacking of water resources for the sake of a tiny fish attests, it is folly to expect reasonableness from someone whose purpose, the halting of economic growth, is inherently unreasonable. (The future of water in California, in the United States, and in the world lies in the desalinization of seawater. We could make deserts bloom everywhere if capitalism were allowed free rein.)

Mainstream environmentalism is, by its nature, a radical ideology, since it aims at a fundamental reordering of our economic and political institutions. To say this is not to deny the existence of a genuine radical fringe of the environmental movement. The environmentalist ideas that I have identified as mainstream are all justified by reference to *human* welfare, as opposed to the welfare of nature. (Recall that Rachel Carson once wrote that in *Silent Spring* she strategically chose to ground her criticism of synthetic pesticides on the requirements of human welfare.) But there is another class of environmentalist ideas that are validated by reference to the requirements of nature itself. "Animal rights" is an example of this class of ideas. The idea of animal rights is based on a fundamental misunderstanding of the concept of individual rights, as they are enunciated, for example, in the Declaration of Independence.

As I wrote in chapter 2, rights derive from man's nature as a rational being. They provide a moral and legal sanction for individual freedom from coercion. To speak of animals, who are capable neither of reason nor morality, as having rights is as nonsensical as arguing that animals have a moral obligation to refrain from harming each other.

The idea that there could exist a moral imperative, such as animal rights, that limit's man's freedom of action without benefiting him in any way, and which in fact works to man's clear disadvantage, harkens back to an older conception that viewed moral law as an arbitrary divine imperative. As I argued in Chapter 5, the concept of individual human rights is a highly *practical* one in that it provides a moral-legal basis for a society based on non-coercive relations among men. Men have every reason to subscribe to the idea of individual rights, since everyone is better off when men do not go about assaulting and robbing their neighbors. It is no accident that those who advocate animal rights tend to be mortal enemies of the genuine individual human rights of life, liberty, and property.

For anyone to posit that man should order his actions around the needs of nature in general rather than around his own interests would be to commit the same mistake as according rights to animals. It would severely diminish man's legitimate freedom of action without benefiting him in any way, and it would therefore be irrational and immoral. Those environmentalists who place the welfare of nature ahead of the welfare of man constitute the genuine radical fringe of the environmental movement.

It is common among the opponents of environmentalism to attribute good intentions to environmentalists themselves. But, as I have shown, environmentalism is essentially preservationist; it aims to limit economic growth and to halt it altogether wherever it can. One way to grasp what this would mean for America would be to consider what the United States would be like today if our forebears had halted economic growth in 1900.

At that time, most Americans lived outside the cities.[26] That means they had no electric lighting, no telephones, no central heating, no indoor plumbing, and no paved roads. No Americans had refrigeration, air conditioning, washing machines or dryers, radio, TV, computers, or internet. They had no automobiles and no airplanes. They traveled by horse, by boat, or by steam train. They bathed once a week. Public water supplies were untreated, and outbreaks of water-borne diseases, such as typhus, were common. Health care was primitive by today's standards. There was no penicillin or other antibiotics, and only the most rudimentary types of surgery. The influenza epidemic of 1918-1919 killed 675,000 in the U.S.[27] When people's teeth rotted, they pulled them out. The average life expectancy in 1900 was 47 years.[28] Today it is 77. In 1900, the average American worked 55 hours per week, and only 35 percent of men were able to retire by age 65.[29] By 2000, 82.5 percent were able to retire by then.[30]

If our forebears had somehow managed to halt economic growth in 1900 and had kept it halted ever since, then we would still be living much as they did then. Knowing as we do now all the good that has come from the economic growth that has occurred since 1900, would we be inclined to credit with "good intentions" a group who had tried to prevent that growth from occurring? And can we say that if they had succeeded, they would deserve our gratitude for having done well by "future generations"?

Knowing as they do all the benefits that have flowed from the economic growth of the last 115 years, can today's environmentalists really be credited with good intentions for wanting to deny the benefits of the next 115 years of growth to our descendants. At the very least, today's environmentalists want our descendants to enjoy no better a standard of living than we do today, to enjoy no better health care, to work no fewer hours, and to live no longer. By what stretch of the imagination can these intentions be called "good"?

Then there is the destruction of individual rights that is implied by a forced halting of economic growth. Can a group whose plans would necessarily entail the extinguishing of our fundamental rights of life, liberty, and property really be credited with "good intentions"? And, should they succeed, would our descendants look upon them with gratitude for having done right by future generations?

Ecology

The balance of nature idea originated in ancient Greece. According to Frank N. Egerton, it remained largely a background assumption in Western thought until the Swedish botanist, Carl von Linne, made it explicit in 1749, referring to it as the "economy of nature." [31] That the word "economy" derives from the same root as the word "ecology" is no accident. Linnaeus, the inventor of the system of taxanomic classification by pairs of Latin names, was an important progenitor of the science of ecology. For Linnaeus, a nature "in balance" meant a nature perpetually unchanging, an arrangement sanctioned, he believed, by divine providence. [32]

Linnaeus's conception of an unchanging nature encountered a challenge around 1800 as it became generally accepted, consequent to the discovery of fossilized remains, that a great many plants and animals that had once populated the earth no longer existed. [33] It was further challenged when Louis Agassiz became persuaded of something that Swiss farmers in the lower Alps already knew, that gravel moraines there had been deposited by glaciers long extinct. [34] And it received a fatal blow with the publication of Charles Darwin's "*On the Origin of Species*" in 1859.

In the early 1900s, a plant ecologist at the University of Nebraska developed a new, more dynamic conception of the balance of nature. Frederic Clements theorized that plant communities develop through a series of stages that culminate in a "climax" state that remains balanced and stable, like the grasslands of Clements's native Nebraska. Following any disturbance that wiped out this climax formation, nature would follow an orderly and predictable process of regeneration, one that would ultimately bring about a return to the balanced and stable climax formation. The balance thus achieved would be subject to changes in climate and to the forces of evolution, but otherwise, it would be largely stable and predictable.

In the 1930s, the British botanist, Arthur Tansley, conceived the idea of the "ecosystem." It incorporated Clements's plant community into a larger whole that included all the living and non-living things on a given expanse of earth, all functioning as an integrated system. Building on this idea, G. Evelyn Hutchinson and his student, Raymond Lindemann, pioneered the study of the "energy economy" within an ecosystem, beginning with the energy input from the sun. This focus on energy flows reached an important stage of development in the work of two brothers from North Carolina, Eugene and Howard Odum, important because theirs was the form in which the idea of the ecosystem, and the science of ecology more generally, entered the popular imagination in the formative years of the environmental movement.[35] The book they co-authored in 1953, *Fundamentals of Ecology*, entered its fifth edition in 2004.[36]

Before the rise of ecology as a formal science, men had divided the study of animate nature into such disciplines as biology, botany, and zoology. But some naturalists argued that man cannot truly understand an organism until he understands how it functions in its natural setting. Nor can he understand nature itself until he understands how it functions not as an assortment of discrete parts but as a unified whole. Thus was born the holistic science of ecology.[37]

Eugene Odum went so far as to suggest that in order to think ecologically, we must cease to think about the individual organism as nature's primary. "[I]n dealing with man and the higher animals," he wrote, "we are accustomed to think of the individual as the ultimate unit."[38] But this will not do for the ecologist. "[F]rom the standpoint of interdependence, interrelations, and survival," said Odum, "there can be no sharp break [between the individual organism and the community]." The individual organism, in Odum's conception, depends on its community just as an organ depends on its organism.[39]

The Odums' *Fundamentals of Ecology* was a kind of Declaration of Interdependence for nature.[40] It is no coincidence that the development of ecology as a science of interdependence in nature paralleled the rise of Progressivism in the United States as an ideology of interdependence among human beings. More than one ecologist hoped to find in nature support for a less individualistic, more collectivist morality for humans.

Back in the early years of Progressivism, Frederic Clements had first proposed the idea that a plant assemblage functions collectively, like a single organism. (Clements co-authored a study with a young fellow ecologist, Roscoe Pound, who would later abandon ecology for law and become, as I discussed in chapter 3, a pivotal figure in the development of Progressive jurisprudence in the United States.) In the same era, the Harvard entomologist, William Morton Wheeler, discovered altruistic behavior in organism-like ant colonies. According to Donald Worster, "While [Wheeler] might live, as he believed, in a time of social "disintegration," he could see in the example of organismic science the promise of "a biologically renovated" morality." [41]

A related idea, that plants and insects and even higher animals agglomerate into superorganisms, became an important theme of five Chicago scientists, whose *"Principles of Animal Ecology"* went through nine printings between 1949 and 1969.[42] At a 1941 symposium on "Levels of Integration in Biological and Social Systems," organized by the "Chicago Five," behavioral scientist Ralph W. Gerard wrote that, reflecting similar tendencies throughout the animal world, "societies are evolving into more integrated [organisms].... [T]his means a greater influence of the whole on its parts, a greater subordination of the individual man to the larger group." [43] Nothing could express more clearly than this the expectation, common among ecologists, of finding in nature scientific grounds for advocating greater degrees of political collectivism among human beings.

Continued Gerard, "Perhaps the most significant of these gradual changes in individual man, considered collectively, is the progressive growth of altruism." If egoism is the morality of American-style individualism, as epitomized most clearly by American capitalism, then altruism is the morality of political collectivism, as epitomized by socialism.[44] (Arthur Tansley, the inventor of the idea of the ecosystem, was himself a Fabian socialist.)

The ecosystem would play an important role in the rise of the ideology of environmentalism. The man in the street would come to believe that the word "ecosystem" designates a patch of the earth's surface, such as a pine forest or a meadow, which is readily distinguishable from neighboring ecosystems by the composition of plant species that inhabit it. This conception would prove invaluable to environmentalists as a rationale for preserving lands and waters from development. Note that ecosystem boundaries do not conform to private property boundaries, so in order to preserve ecosystems intact environmental legislation and regulation would often need to override private property rights. The decision of the Fish and Wildlife Service to enforce the Endangered Species Act by setting aside whole "ecosystems," for example, promises to wreak havoc on property rights.

The ecosystem would become the repository of the balance of nature idea during the formative years of the environmental movement. "Ecosystems are capable of self-maintenance and self-regulation as are their component populations and organisms," wrote the Odums in *Fundamentals*.[45] "Homeostasis ... is the term generally applied to the tendency for biological systems to resist change and to remain in a state of equilibrium."[46] Note that in this pair of statements, the Odums imply that ecosystems are as homeostatic as organisms, and that, therefore, as "biological systems," ecosystems function very much as individual organisms do.

But none of these characteristics of the ecosystem, that it is a physical entity like a pond or a meadow, that it is self-balancing (homeostatic), or that it is a highly integrated, organism-like system, is true. According to S.T.A. Pickett and Mary L. Cadenasso, the ecosystem is not a physical entity that exists in nature but is, rather, a tool devised by ecologists.[47] It is defined as an assemblage of living things, the physical environment in which they live, and the interactions among all those living things and their physical environment. But beyond these very general characteristics, the term "ecosystem" denotes only what the ecologist wants it to denote, depending on the purposes of his study. The size of an ecosystem, for example, can be as small as a square yard of ground or as large as the whole earth.

> Because the conceptual definition of the ecosystem is neutral in scale and constraint, models are necessary to translate the definition into usable tools.[48]

The "model" that the ecologist constructs *is* the ecosystem. "Some of the features of ecosystem models are determined by the researcher, based on the questions guiding the research or application; whereas other features emerge from the nature of the material system under study."[49] Just as a climate researcher constructs a computer model to study climate dynamics, so the ecologist constructs a model, which he calls an ecosystem, to study the dynamics of living things in their natural environment. The ecologist might study a marsh by developing a marsh ecosystem, which is a model that emphasizes certain characteristics of the marsh and ignores others. But to refer to a specific marsh as itself an ecosystem, say Pickett and Cadenasso, is to speak metaphorically; it is to suggest that the marsh, a physical reality, is *like* an ecosystem, an abstract, man-made model.[50]

Consider a pond, the quintessential "ecosystem." Much of

what takes place in a pond—much of what determines its "functioning"—involves things that come from outside the "system" as delineated by the shoreline of the pond. Water flowing in from upstream or from precipitation run-off, for example. And wildlife; animals and insects come and go, as do fish via the streams that feed and empty the pond. On their migratory paths, birds can range many hundreds of miles from the pond. But all of this wildlife is part of the "unified whole" that Odum called an ecosystem.[51]

To understand all that takes place within the pond, an ecologist would need to define its ecosystem as the pond itself plus the watershed that feeds the pond, plus the ranges of all the wildlife that contribute to the pond's "functioning." Even defined in this way, though, the pond ecosystem would comprise a single unified whole only in an abstract sense, since the wildlife that come and go would also be a part of other "unified wholes" in the neighborhood of the pond.

As Robert O'Neill has said, "The simple fact is that the ecosystem is not an *a posteriori*, empirical observation about nature. The ecosystem concept is a paradigm, an *a priori* intellectual structure, a specific way of looking at nature."[52] All of a pond ecosystem's elements exist in nature, and all of their relationships with each other and with the pond exist in nature, but they only constitute a "unified whole" as they are mentally isolated from their surroundings and then treated as a single whole by the mind of man.

So what is gained by referring to a marsh as a "marsh ecosystem"? As Pickett and Cadenasso point out, the use of the term "ecosystem" in connection with a marsh can emphasize to the listener that "Everything is connected to everything else" in that marsh.[53] "In addition, the concept of the ecosystem can be used to stand for equilibrium, resistence or resilience, diversity, and adaptability," they write.[54] "Some of these connotations are only hypothetical from the perspective of science, whereas others *are clearly unsupported or highly problematical*. The point here is not to evaluate the veracity of such connotations, but only to point to *their ... power in the public discourse*." [55] However questionable has been the "veracity" of the ecosystem concept, it has more than proven its worth by virtue of its "power in the public discourse."

But the confusion associated with the ecosystem concept goes beyond the question of whether it designates a physical entity. How an ecosystem is defined by the ecologist can determine its characteristics. The size of wildlife populations tends to be more stable in larger ecosystems than in smaller ones, or in ones surrounded by similar ecosystems, as opposed to ones isolated by water or desert or human development. This is because smaller or isolated populations are more apt to disappear altogether if subject to disturbance. [56] An ecologist looking to find stability in wildlife populations can load the dice in his favor by delineating a larger rather than a smaller ecosystem, or one surrounded by similar ecosystems rather than one that is isolated.

According to O'Neill, an ecosystem can even be defined and studied entirely in terms of the "processes" that characterize it, such as energy flows, quite independently of whichever species that inhabit it. [57] Thus an ecosystem that saw radical changes in its species composition over time could nevertheless be characterized as highly "stable" if its energy flow patterns remained stable. One consequence of defining an ecosystem entirely in terms of its processes could be, according to O'Neill, that the boundary where one vegetationally-defined "ecosystem" gradually verges into another (called an ecotone), would tend to disappear altogether, since the "ecosystem processes" that occur on either side of an ecotone can to be similar to each other. [58] So much for the idea of the ecosystem as a patch of ground characterized by specific species of plants.

Eugene Odum had said, in effect, that if we were to go out

527

and investigate any random expanse of earth that contained a roughly homogeneous assortment of flora and fauna, we would discover that it "functioned" as a unified whole to maintain its own equilibrium, much as an organism does. It would do this rather in the way that a home heating system equipped with a thermostat functions to maintain a steady temperature within the home.

But this expression of the balance of nature idea has never been proven to be true. "Direct evidence that ecological systems are inherently systems in equilibrium, however, is still lacking. Indeed, individual organisms may be the only systems within which homeostatic mechanisms have been demonstrated to operate." [59] Or, as Kristen Shrader-Frechette has written, "there is no clear, confirmed sense in which natural ecosystems proceed toward homeostasis, stability, or balance." [60] Eugene Odum himself conceded in 1992 that "[a]n ecosystem is a thermodynamically open, far from equilibrium, system.... What is coming in and going out is as important as what is inside the tract. It is not a self-contained unit." He continued, "There are no thermostats, chemostats, or other set-point controls." [61] The idea of the homeostatic ecosystem that provided the "scientific" basis for the balance of nature idea that animated much of the environmental agitation, legislation, and regulation in the 1960s and 1970s was simply not valid.

The disagreement, or confusion, among ecologists about the ecosystem idea, including the question whether an ecosystem is an entity in nature or a man-made model—a state of affairs that continues to this day—suggests that the science of ecology has never yet reached a sufficiently advanced stage of development to entitle it to guide political decision-making. Again, Shrader-Frechette: "[E]cology does not appear to be "hard" enough or solid enough to be fully amenable [sic] to providing uncontroversial support for environmental ethics and policy." [62]

The ecosystem concept has been the repository of the balance of nature idea throughout the environmental era. It, along with the science of ecology, has provided "scientific" legitimacy for environmentalist efforts to alter fundamentally the economic and political landscape of the United States. In the end, it will not matter to environmentalists that ecologists are now abandoning the balance of nature idea altogether.[63] Environmentalism has never been about science. It is an ideology. It uses science to whatever extent science, whether valid or invalid, can be made to serve the purposes of nature preservation.

The Sixties

Finally, a word about the social and political context in which environmentalism arose. As I described in Chapter 3, the Progressive movement of the early twentieth century attacked the institution of private property across a broad front. It thereby helped pave the way for the conservation movement, which was based on the idea that natural resources ought, by right, to be public rather than private property. The conservation movement did not pursue this premise to its logical conclusion, which would have been the abolition of private ownership of land. Rather, the conservationists mostly limited themselves to carving out a major role for government in managing natural resources, a role that included permanent federal ownership of a vast domain of public lands. The early conservationists were content to leave the majority of the nation's lands in private hands. In the sixties, the left resumed the assault on private property where Progressivism and the conservation movement had left off. Sixties leftism performed the same function for environmentalism that Progressivism had performed for conservation.

The Civil Rights Act of 1964 prohibited *legally imposed* racial discrimination in genuinely public accommodations such as schools and city-owned buses.[64] This was a proper exercise of government power, as was the prohibiting of genuine discrimination in the administering of voter registration qualifications. But the prohibiting of such government-imposed discrimination should have been the sum total of the *legislative* responses to the Civil Rights movement. It was not. Title II of the Civil Rights Act of 1964 prohibited discrimination on the basis of race, color, religion, or national origin by *private* businesses such as hotels, theaters, and restaurants. Title VII also prohibited such discrimination in employment decisions by private businesses. And Title VIII of the Civil Rights Act of 1968 prohibited discrimination by the sellers and landlords of *private* housing.

However repulsive racism might be, people in a free society have every right to entertain racist ideas and to act accordingly, as long as they do not go about physically assaulting the persons or property of their fellow citizens. To classify restaurants as legally "public" accommodations that must admit all comers is to execute a partial taking of private property. In a free society, a person would have no more right to enter my restaurant against my wishes than he would to enter my home. To enact into law that a Black person—or any person—has such a right is to change the meaning of rights altogether.

In a free society, my only un-chosen obligation toward others is the negative one to refrain from physically assaulting them or their property. To posit a right that anyone may enter my restaurant against my wishes is to construe the idea of a right as an entitlement on the part of others to enter my restaurant, on the one hand, and an obligation on my part to admit all comers, on the other. While such a concept of rights might appear to increase the freedom of Blacks and the rest of the restaurant-going public, it diminishes the right of private property, without which all freedom is chimerical. What it does promote, though, is *equality of results*, which was a major theme of the sixties. If I may not keep Blacks out of my restaurant, then my rights are diminished, Blacks' "rights" are increased, and we are all a bit more equal in the end. The idea of equality of results stands diametrically opposed to the idea of individual rights.

Another product of the civil rights movement was school bussing to achieve racial integration. If students attending private schools had been forcibly bussed to other schools, it would have been a violation of the rights of those students and their parents, who were paying for their children's schooling. But since the schools involved were government-run schools, which the students attended for "free," it is more difficult to argue that their rights were violated by forced bussing. But bussing showed students and parents that the local public schools they supported with their tax dollars were not theirs to run as they chose. And it accustomed them to following orders in the name of some "greater good," in this case racial integration, even when it meant putting up with dramatically inferior schools than they had planned on when they bought homes in their neighborhoods. Busing also contributed to "white flight" from American cities, thereby contributing to the deterioration of those cities later on. And it demonstrated, in places like South Boston, how terribly damning—and politically effective—the charge of "racism" could be.

The civil rights movement also helped bring about a vast expansion of the welfare state. For the government to expropriate the wealth of one private citizen and give it to another private citizen amounts to institutionalized, legalized corruption. Because this power to redistribute wealth is inevitably used to buy votes for political parties, it becomes a cancer that metastasizes and progressively de-moralizes the political process. Welfare also sucks the life out of an economy by taking wealth out of productive hands and putting it into the hands of the non-productive, who constitute a dead weight on an economy.

Since the late sixties, welfare has laid waste to American cities, decimated the Black family, and helped to create a permanent underclass (which is not a normal part of a free economy). It has also helped to accustom Americans to surrendering their rights—to their wealth, in this case—for ever larger and more novel notions of the public good.[65] The erosion of the institution of private property in the interest of promoting the "greater good" of racial equality would make it easier, in the following decades, for environmentalists to win passage of a wide array of property rights-destroying measures in order to secure the "greater good" of a clean and healthy environmental.

The civil rights movement succeeded in having legally imposed racial discrimination abolished in the American south. By the 1960s, there were no analogous laws that legally discriminated against women. (If there had been, then women would have been entirely justified in resorting to political means to abolish those laws.) But in the feminist view, women were being unjustly excluded from the workplace in the 1960s, so the government needed to prohibit discrimination against women in hiring. This was accomplished by the Civil Rights Act of 1964.

But since women were going to be entering the workplace in unprecedented numbers, it also became necessary to legislate how men should behave toward women in the workplace. Thus was born the novel idea of sexual harassment law, to regulate what had formerly been a private matter of good manners. To legislate that a woman has a right to be treated in a certain way in the workplace is to turn a society of voluntary relationships, in which the government has a small and strictly limited role, into a society of politically regulated relationships in which the government has a ubiquitous and highly intrusive role in the most mundane of day-to-day affairs. This is precisely the kind of polity that environmentalism aspires to, and feminism helped to clear the way for it.

The anti-Vietnam War movement was another source of profound political and cultural change in the sixties. While there were many who opposed the war on the practical grounds that it was an unjustified waste of American lives and wealth, the most organized, sustained, and active opposition came from leftist student organizations. This element of the movement viewed America's goal of fighting the spread of communism, in the name of individual rights and capitalism, as insufficient justification for the war. They portrayed the United States, with considerable success, as unjust aggressors in the war, an argument that implied that communism was an equally morally legitimate form of government as capitalism. This kind of moral equivalence would have a profound influence on America's future. It underlies today's multiculturalism, which is a dagger aimed at the heart of America's English-speaking, western European culture. In the environmental era, this moral equivalence has helped to undermine Americans' confidence in the moral rightness of individual rights and capitalism and, thus, has helped make possible the environmentalist evisceration of property rights.

The Left has long wanted the individual to look directly to the state, rather than to other institutions like the family or the church, for all his important needs. The sixties spawned a whole range of phenomena that weakened the family. One was the sexual revolution. The winning of widespread acceptance for contraception, along with the invention of the birth control pill, the legalizing of abortion, and the diminution of religious influence over sexual matters all combined to make sexual relations available to persons outside of marriage, thus diminishing what until then had been an important motivation for people to get and to stay married. Combined with the liberalizing of divorce laws, the sexual revolution brought about less stable families, more families of step-parentage, and more single parents (including ones who choose not to marry to begin with).

Whereas the sexual revolution removed the need to get married for sexual reasons, government welfare programs diminished the need to get or to stay married for economic reasons. "Head Start" programs got kids out of their families (and into government schools) at an earlier age, and day care programs even earlier. Child safety laws (mandating child seats in cars and bicycle helmets) usurped an important parental function, the overseeing of children's safety. Sex education programs, free contraception, and even abortions without parental consent, have usurped another. And today Homer Simpson and a legion of other TV fathers make the case that fathers should not be entrusted with the responsibility of raising children. (The invention of electronic media—TV, radio, and the movies—have brought about a first in human history, the centralizing of the sources of popular culture. This, in turn, has made possible a pace of cultural, social, and political change that in earlier times could only have been achieved by violence.)

In 1960, 88 percent of children under 18 lived in two-parent households; in 2010, 66 percent did so.[66] In 1960, 2 percent of births in the United States were illegitimate; in 2008, 29 percent were illegitimate.[67] The radicals of the sixties could hardly have imagined how spectacularly they would succeed in transforming the family in just two generations. One consequence of the deterioration of the family has been a weakening of its all-important role as a primary transmitter of cultural values from one generation to the next, among them the values that support political institutions in a free society.

Another sixties phenomenon that weakened the family, specifically as a values-inculcating institution, was the generation gap. Rock and roll started it. Rock and roll is music created by teenagers for teenagers. It came on the scene at a time when teenagers began to have sufficient disposable wealth of their own to support an entire musical form devoted to their tastes. If art nourishes the soul, rock and roll is, at best, the musical equivalent of fast food. Most adults of the 1950s and 1960s recognized this, and many were offended by rock and roll. The Beatles' introduction of long hair to the clean-cut American culture of the time widened the generational separation, and when drugs and sex were added to the mix the generation gap widened further. The hippie phenomenon and the general slovenliness that it encouraged among the young, and then the anti-war movement, which was largely a phenomenon of the young, widened the generation gap into a yawning chasm.

In consequence of the generation gap, the young stopped listening to their elders, and the effectiveness of the family as an inculcator of values was diminished. The left encouraged this development; "Never trust anyone over thirty," they counseled. (Today, now that the left runs the culture and the media, one no longer hears the generation gap celebrated, although it seems to have become ingrained in the culture that the young have little to learn from their elders.) By diminishing the transmission of values from the older to the younger, the generation gap helped make it possible to introduce new values to the younger generation, among them those of environmentalism, which, in its early years, appealed mainly to the young.

The Concise Oxford Dictionary defines "bourgeois" as a member "of the shop-keeping middle class" or (in value-laden terms) a person "of hum-drum middle class values." What does a shop-keeper value? High on his list would be economic security. He would value law and order, and stability, in his own life and that of his family, and in the social, economic, and political life of his country. He would value reliability, responsibility, and sobriety, long-range planning, and a willingness to delay gratification. He would probably value restraint in sexual matters. And he would likely value his family. These are "middle class" values. They are not what a romantic adventurer would value, and they are the antithesis of what a sixties leftist would have valued.

Most societies in human history have had a small upper class, a small middle class, and a large lower class of physical laborers. A large middle class is a distinctive characteristic of a capitalist society.[68] The values usually attributed to the middle class, "bourgeois" values, are indispensable to running a capitalist society. One way to undermine a capitalist society would be to discredit and demoralize its middle class. In the sixties, the hippies were the shock troops in the left's assault on middle class morality. They represented the antithesis of just about every middle class value (with the exception that they did not tend to indulge in violent crime more than other segments of society). They were especially active, of course, in the sexual revolution and in the popularization of hallucinogenic drugs, both of which have had a lasting influence on American culture.

The sixties assault on middle class values manifested itself in the re-casting of the 1950s as a culturally sterile and "hum-drum" decade, an evaluation now firmly engraved in the popular imagination. In truth, the fifties were the last time that middle class values dominated the cultural landscape in the United States. It was a time of stable family life—as represented, for example, by the now-sneered at Cleavers of *Leave It to Beaver*, a time of relative social, economic, and political stability, and a time when schools were filled with clean-cut, well-dressed, well-behaved kids.

It was a time of sexual restraint, now the object of ridicule, as in the 1998 movie, *Pleasantville*, in which the 1950s characters appear on the screen in black and white until, one by one (or two by two), they discover the joys of sex and spontaneously burst into Technicolor. It was also a time when the vast majority of Americans agreed on the fundamental social, economic, and political principles that should govern life in the United States. Today, a half century later, that America has been swept away by a cultural revolution. The left's main indictment of that culture is that it was "conformist." Now, the conformity of the fifties has been replaced by the "diversity" of purple hair and nose rings and married homosexuals and unmarried mothers, but it is a diversity that is supported by a narrow, ruthlessly enforced intellectual conformity that would have embarrassed Cotton Mather.

Of a piece with the left's re-casting of the fifties as a bland and sterile decade is their long-running portrayal of America's suburbs as a social and cultural wasteland. The line stretches from Pete Seeger's 1963 hit, "Little Boxes (made of ticky-tacky)" to 2013's TV series not-so-subtly titled *Suburgatory*. Since the fifties, the suburbs have been the physical residence of middle class values in the U.S., all the more so since violent crime, race riots, and forced school busing drove the middle class out of the cities in the sixties and seventies. Life in the suburbs offered the chance to move one's family out of an urban apartment building and into a home of one's own with a back yard for the kids. It offered a quiet, low-crime environment with clean, fresh air and open spaces.

But the left has always hated the suburbs, for their alleged middle class conformity, and because "They paved over paradise and put up a parking lot," as Joni Mitchell put it. The left aims to stop the spread of these havens of middle class morality through such "smart growth" arrangements as Plan Bay Area, which will forcibly steer potential new middle class suburbanites away from what might have become new suburbs and back into existing urban areas (where their political power will be diluted).

The left also hate the suburbs for what they believe is the implicit racism of their ethnic whiteness. (The left argue that to be White *is* to be racist.) In the fall of 2013, the U.S. Department of Housing and Urban Development, to enforce the Fair Housing Act of 1968, began developing a plan that would spell the end of the suburbs as bastions of middle class values by forcing suburban communities to provide "affordable housing" specifically for non-middle class, ethnic minority urban dwellers. The State of

California already does something similar; each county must make a certain percentage of any new housing built there "affordable." But the HUD plan aims specifically to alter the *racial* makeup of the suburbs.

There is still another sense in which the left views the middle class as deficient; the middle class does not qualify as a group that has been victimized by injustice or misfortune. The Democratic Party has evolved into a coalition of victims' groups. The poor, Blacks and Hispanics, women, the disabled, and the sexually confused all have, in the left's telling, suffered injustice, and all merit special treatment by government and by their fellow citizens. (The middle class does not need government help, so they are of no use to the left.)

Accordingly, political power has flowed to such groups of alleged victims and away from non-victims, notable among them the middle class. Indeed, the middle class have been forced to surrender property rights and rights of free association in the name correcting "injustices;" they have seen their wealth expropriated and redistributed to the various groups of victims; and they have seen their children denied admission to colleges of their choice in the name of affirmative action. So marginalized have the middle class become that for the first time in U.S. history, an American president has been able to govern in open defiance of the middle class and their interests (by running up the national debt, vastly expanding the welfare state, and curtailing energy production, for example). The middle class are the *sine qua non* of a capitalist society. As they have become progressively more marginalized by the left, there have been fewer and fewer Americans who are able to defend property rights against the depredations of the environmentalists.

Public schools in the United States have, throughout our history, been called upon to perform three basic functions:

1. to teach basic mental skills (reading, writing, and 'rithmetic);
2. to teach basic civics, such as the history and fundamental principles of our republic;
3. to pass on our Western culture from one generation to the next.

(A fourth purpose, the inculcating of moral values, has usually been reserved to private schools, although public schools have until recently taught such basic moral principles as proper deportment and good manners.) As long as there was general agreement among Americans about the importance and the content of these basic functions, the public schools worked reasonably well. But the sixties explicitly challenged—and set out to subvert—America's fundamental values, including her basic civic principles and her grounding in the cultures of ancient Greece and Rome.

There are two problems with the idea of government-run schools. First, even when there is a general consensus about the nature and purpose of the public schools, any individual who disagrees with what is being taught in them is unjustly forced to pay to support the propagation of those ideas. This is the same reason that government-supported churches are unjust, and it is a decisive objection to government-run schools. Second, there is always the danger, which we are seeing realized today, that the schools will be turned to serving the purposes of the government and those who control it, rather than those of the families who send their children to them.

The sixties subversion of the schools, which continues unabated today, began with an assault on their prosecution of their most fundamental purpose, the imparting of basic knowledge and skills. It was often formulated as an attack on "rote" learning and usually advocated a greater emphasis on teaching students to think. It led to a great deal of often damaging experimentation in the schools, including such lunacies as the "whole language" method of teaching reading (as opposed to phonics), and the dispensing altogether with the teaching of basic English grammar. One result of all this tinkering with basic pedagogy has been steadily declining student performance since the sixties.

A second goal of the sixties assault on the schools has been to subvert the teaching of America's basic civic principles. By re-writing the history and literature curricula, the "reformers" succeeded in casting the United States as a nation of slave-owning, Indian-slaughtering racists, of sexist, homosexual-bashing bigots, and of rapacious, environment-destroying swindlers. America's old, outdated principles of individual rights and liberty, which, the reformers allege, had brought about intolerable inequality and oppression, would need to be changed in favor of radical equality enforced by a vastly expanded government.

Accordingly, the schools have been transformed into the means of engineering a new type of citizen. Sex education led the way, a means to accelerate and consolidate the sexual revolution (and usurp a crucially important function of parents). We now have diversity training and gay curricula in the schools, and nothing so vividly illustrates the reformers' intention to fundamentally remake our society and our culture as California's statute requiring that boys who want to be girls be allowed to use the girls' bath and locker rooms, and vice versa.

In 1969, Brown University adopted their "New Curriculum." Its chief architect was a student, Ira Magaziner, who would go on to become an advisor to President Bill Clinton. (In 1984 he promoted the Greenhouse Compact, which failed to win enactment, but which would have introduced comprehensive central planning into Rhode Island's economy.) Brown's New Curriculum did away with "distribution requirements," which had required students to take, in addition to courses studied in pursuit of their majors, a certain number of courses in the sciences, the humanities, and the social sciences. It also eliminated the requirement that students pass a final, comprehensive exam before graduation. The New Curriculum required simply that each student pass eight courses in his major field of study, and it left the rest of his course choices up to him.

The New Curriculum dispensed with the idea that there exists a body of knowledge that every well-educated person in Western society should master. Stanford University would follow suit in the late 1980s, to students' chants of "Hey, hey, ho, ho, Western Civ has got to go." The older idea that American universities have a sacred mission to pass on the best of Western culture from one generation to the next has since given way to the idea that they should promote "multi-culturalism." One way to overthrow a dominant culture is to portray it as no better and no worse than any other culture. Multiculturalism is moral relativism writ large. Once the academic gold standard was abolished, then anything and everything became worthy of inclusion in the college curriculum. So we have Black studies, women's studies, Chicano studies, and gay studies, and all are hostile to individual rights and to capitalism.

The history of public schools in the U.S. since the sixties has been one of ever more centralized power. The centralizing began before the sixties, as parents, for fiscal reasons, began to surrender control of their schools to large, unified school districts. Then in the seventies, courts manufactured out of thin air the pernicious and unjust idea that spending on schools needed to be equalized among the various public school districts throughout each state. In consequence, local tax receipts got funneled through the respective state governments, thus giving state authorities control over local schools. Then, under President Carter, the U.S. Department of Education came into being, and the federal role in education has grown ever since. Today, parents exercise almost no control over the schools they pay to send their children to.

Of all the various movements that sprung up the sixties, none, other than environmentalism itself, has so directly aided the propagation of the environmentalist ideology as the school reform effort has. It obliterated the schools' role as transmitters of the old culture of individualism and capitalism, on the one hand, and it introduced into the curriculum direct indoctrination in the ideology of environmentalism, on the other.

Conclusion

Life on earth presents man with certain needs, among them food, clothing, and shelter. The source of these things, and of all the material goods on which man's life depends, is the land. If the government owns all the land, then the government controls men's lives, and no freedom is possible. In 1689, John Locke identified the central role of property rights in a free society. A century later the Americans dedicated their new nation to the principles enunciated by Locke, and they framed a constitution of government specifically intended to secure individuals' rights of private property. Environmentalism aims to nullify this defining characteristic of the Constitution. Although it would allow legal title to privately owned land to remain in private hands, it would deprive owners of the all-important power to decide how to use their land. This process of expropriation-by-regulation has been going on since the start of the environmental era, and it is now well-advanced, especially in California and several other states.

The process of securing the food, clothing, shelter and other values upon which man's health and well-being depend requires that he manipulate nature and alter the land. In his famous dictum, Francis Bacon implied that it would be possible for man to command nature to a much greater extent than he was doing in 1620, and that it would be right for him to do so. Bacon's twin premises would underlie the rise of industrialism two centuries later. The constitution of government that the American founders framed created political conditions that are indispensable for the rise of an industrial economy. The unparalleled industrial might that would eventually characterize the United States was made possible by her founding principles.

In its opposition to property rights and to industrialism, environmentalism is antithetical to Americanism. It is not a good idea that is being carried to a harmful extreme, as it is so often characterized by its critics. It is a bad idea that is producing precisely the harmful consequences that it should be expected to produce. Unless it is stopped, it will vastly diminish Americans' health, wealth, comfort, safety, and security from foreign enemies, and it will ultimately deliver us to tyranny.

Notes

1. The estimate includes direct, indirect, and induced costs of regulation. Sanjay B. Varshney and Dennis H. Tootelian ,"Cost of State Regulations on California Small Businesses Study," September, 2009, accessed December 18, 2014, http://www.cmta.net/pdfs/20090922_smallbiz_regcost.pdf.
2. Ibid.
3. Ibid.
4. Jan Norman, "Report: 254 companies left California in 2011," The Orange County Register, March 2, 2012.
5. NERA Economic Consulting, "Macroeconomic Impacts of Federal Regulation of the Manufacturing Sector," Manufacturers Alliance for Productivity and Innovation, August 21, 2012, accessed November 10, 2014, http://www.mapi.net/research/publications/macroeconomic-impacts-federal-regulation-manufacturing-sector, p. 52. This study measured only the cost of what the authors termed "major"

federal regulations, and they stated that the cost of non-major regulations would increase the cost estimate considerably.

6. "All Employees: Manufacturing (MANEMP)," St. Louis Federal Reserve, March, 2013, accessed March, 2013, http://research.stlouisfed.org/fred2/graph/?s[1][id]=MANEMP #.

7. Ibid.

8. Ibid.

9. Robert D. Atkinson, Luke A. Stewart, and Stephen J. Ezell, "Worse than the Great Depression: What Experts Are Missing about the American Manufacturing Decline," The Information Technology & Innovation Foundation, March, 2012, accessed November 10, 2014,
http://www2.itif.org/2012-american-manufacturing-decline.pdf. The authors' target is not environmental regulation, *per se*. They argue, rather, that since 2000 the United States has been losing the international competition for manufacturing market share because of bad policies on the part of the Federal Government.

10. Ibid., 24-36.

11. Ibid., figure 45, p. 47. The authors cite "Fixed Assets Accounts" (table 3.7ES, Investment in Private Fixed Assets by Industry), Bureau of Economic Analysis, accessed by the authors January 23, 2012), ttp://www.bea.gov/iTable/index_FA.cfm and "National Income and Product Accounts" (table 1.15, Gross Domestic Product), Bureau of Economic Analysis, accessed by the authors January 23, 2012,
http://www.bea.gov/iTable/index_nipa.cfm.

12. Ibid., figure 52, p. 52. The authors cite "Fixed Assets Accounts" (table 3.2ES, Chain-type Quantity Indexes for Net Stock of Private Fixed Assets by Industry), Bureau of Economic Analysis, accessed by the authors January 23, 2012,
http://www.bea.gov/iTable/index_FA.cfm.

13. For steel, see No. HS-45. "Production Indicators for Agricultural, Fishery, Mineral, and Manufactured Products: 1900 to 2002," U.S. Census Bureau, Statistical Abstract of the United States: 2003, pp. 85-86, accessed November 10, 2014, http://www.census.gov/statab/hist/HS-45.pdf. For aluminum see Pat Plunkett, "Aluminum Statistical Compendium" USGS, January 11, 2013, accessed November 10, 2014, http://minerals.usgs.gov/minerals/pubs/commodity/aluminum /stat/, Table 1. (For aluminum statistics from 1991 onward, see "Aluminum Statistics and Information," USGS, October 22, 2014, accessed November 10, 2014, http://minerals.er.usgs.gov/minerals/pubs/commodity/alumin um/index.html.)

14. Atkinson, *et al*, "Worse Than," Figure 52.

15. "Table 7b—Production, Imports, Exports, and Consumption of Hardwood Timber Products, by Major Product, 1965–2005," Forest Products Laboratory, U.S. Forest Service. accessed November 10, 2014, http://www.fpl.fs.fed.us/products/products/datasets/fpl-rp-637/Table05-07.htm.

16. Atkinson, *et al*, "Worse Than," figure 52.

17. Ibid.

18. Ibid.

19. Atkinson, *et al*, "Worse Than," 54. The authors cite "Direct Investment and Multinational Companies (U.S. direct investment abroad, majority-owned nonbank foreign affiliates, capital expenditure; U.S. direct investment abroad, majority-owned foreign affiliates, capital expenditure)," Bureau of Economic Analysis, accessed by the authors February 14, 2012, http://www.bea.gov/iTable/index_MNC.cfm and "National Income and Product Accounts (table 1.7.5, relation of gross domestic product, gross national product, net national product, and personal income)," Bureau of Economic Analysis, accessed by

the authors February 14, 2012,
http://www.bea.gov/iTable/index_nipa.cfm.　　Analysis　　by
Atkinson, *et al.*

20. Atkinson, et al, figure 59, p. 58., "National Income and
Product Accounts (Table 6.17, Corporate Profits before Tax by
Industry)," Bureau of Economic Analysis, accessed by the authors
February 14, 2012,
http://www.bea.gov/iTable/index_nipa.cfm.
21. Atkinson, *et al*, pp. 46-47.
22. Michael Greenstone, John A. List, and Chad Syverson
estimate that environmental regulation raised the cost of doing
business by $21 billion per year between 1972 and 1993, which
amounts to 8.8 percent of corporate profits over that period. "The
Effects of Environmental Regulation on the Competitiveness of
U.S. Manufacturing," September, 2012,
home.uchicago.edu/syverson/enviroregsandproductivity.pdf,　p.
2.
23. They were also the most heavily unionized.
24. See *Rapanos v. United States.*
25.　*San Luis & Delta-Mendota Water Authority, et al. v. Natural
Resources Defense Council.*
26. In 1910, 28.3% of Americans lived in a city or in the suburbs.
Demographic Trends in the 20th Century, U.S. Census Bureau,
http://www.census.gov/prod/2002pubs/censr-4.pdf, p. 33. I
do not mean to suggest here that it would be possible to halt
economic growth without fatally deranging society.
27. "The Great Pandemic: The United States in 1918-1919," U.S.
Department of Health and Human Services," accessed November
10, 2014,
http://www.flu.gov/pandemic/history/1918/the_pandemic/in
dex.html.
28. *Demographic Trends,* 11.
29. Kristie M. Engemann and Michael T. Owyang, "Working

Hard or Hardly Working? The Evolution of Leisure in the United States" in "The Regional Economist," January, 2007, St. Louis Federal Reserve, accessed November 10, 2014,
http://www.stlouisfed.org/publications/re/articles/?id=43.
The authors cite "A Century of Work and Leisure" by Ramey, Valerie A., as well as Francis, Neville. Manuscript, University of California, San Diego, May 2006.
30. Joanna Short, "Economic History of Retirement in the United States," Economic History Association, accessed November 10, 2014,
http://eh.net/encyclopedia/economic-history-of-retirement-in-the-united-states/. Ms. Short cites Moen, Jon R., *Essays on the Labor Force and Labor Force Participation Rates: The United States from 1860 through 1950.* Ph.D. dissertation, University of Chicago, 1987 and Costa, Dora L. *The Evolution of Retirement: An American Economic History, 1880-1990,* (Chicago: University of Chicago Press, 1998), and the Bureau of Labor Statistics.
31. Frank N. Egerton, "Changing Concepts of the Balance of Nature" *Quarterly Review of Biology* 48 2 (June, 1973): 322-350.
32. Alston Chase has pointed out the kinship between classical Greek ideas of a teleological universe, that is, a universe subject to purposeful guidance, and more recent ideas of the balance of nature such as Linnaeus's. Alston
Chase, *In a Dark Wood*, (New Brunswick, NJ and London, England: Transaction Publishers, 2001).
33. Egerton, "Changing Concepts," 338. According to Egerton, "few naturalists" at the time seemed to have recognized the conflict between the idea of an unchanging nature and the fact of species extinction.
34. William W. Hay, *Experimenting on a Small Planet: A Scholarly Entertainment,* (Heidelberg, New York, Dordrecht, London: Springer, 2013), 76.
35. "Together, they did more than anyone else to define the

science in the postwar period." Donald Worster, *Nature's Economy: A History of Ecological Ideas*, (Cambridge: Cambridge University Press, 1994), 362.

36. By the third edition, published in 1971, Eugene Odum had become the sole author of *Fundamentals*.

37. Like teleology, ecological holism also derives from ancient Greek ideas, according to Chase. "Dark Wood," 97-98.

38. Eugene Odum, *Fundamentals of Ecology*, (Philadelphia, London, Toronto: W.B. Saunders Company, 1971), 5.

39. "The individual organism ... cannot survive for long without its population any more than the organ would be able to survive as a self-perpetuating unit without its organism." Ibid.

40. Donald Worster attributes the phrase "Declaration of Interdependence" in the ecological context to U.S. Secretary of Agriculture Henry Wallace. Worster, *Nature's Economy*, 320.

41. Worster, *Nature's Economy*, 324-325. For the Wheeler quotations, Worster cites "Ant Colony" and "Hopes in the Biological Sciences" in *Essays in Philosophical Biology.*"

42. Ecologist Warder Allee, ecologist Thomas Park, zoologist Karl Schmidt, biologist, entomologist, and zoology professor Alfred.E. Emerson, of the University of Chicago, and zoology professor Orlando Park of Northwestern University.

43. R. W. Gerard, "Higher Levels of Integration" in *Biological Symposia: A Series of Volumes Devoted to Current Symposia in the Field of Biology*, Jacques Cattell, ed., Vol. 8, Robert Redfield, ed, (Lancaster, PA: Jacques Cattell Press, 1942), 82.

44. See Ayn Rand, "Collectivized Ethics" in *The Virtue of Selfishness*, (New York, NY: New American Library, 1961).

45. Odum, *Fundamentals*, 33.

46. Ibid., 34.

47. S.T.A. Pickett and M.L. Cadenasso, "The Ecosystem As a Multidimensional Concept: Meaning, Model, and Metaphor" in *Ecosystems*, (2002) 5, accessed November 10, 2014,

http://link.springer.com/article/10.1007%2Fs10021-001-0051-y#page-1, p. 2.
48. Ibid.
49. Ibid.
50. Ibid., 5. "The ecosystem is often used as a metaphorical representation of some place on the earth's surface."
51. Robert V. O'Neill, Robert H. MacArthur Award Lecture to the Ecological Association of America, "Is It Time to Bury the Ecosystem Concept"? *Ecology* 82 12 (2001): 3277.
52. Ibid., 3276.
53. Pickett and Cadenasso, "The Ecosystem," 6.
54. Ibid.
55. Ibid. Emphasis added.
56. O'Neill, "Is It Time," 3277.
57. Ibid., p. 3278. See also William. S. Currie, "Units of Nature or Processes Across Scales? The Ecosystem Concept at Age 75" *New Phytologist* 190 1 (April, 2011): 21-34. "[A]n ecosystem is difficult to view rigorously as an entity in a locale, as common usage dictates, but instead should be viewed as a particular set of integrated processes in a re-conceived hierarchy of processes."
58. Ibid.
59. Jianguo Wu and Orie Loucks, "From Balance of Nature to Hierarchical Patch Dynamics: A Paradigm Shift in Ecology" *The Quarterly Review of Biology* 70 4 (December, 1995): 442.
60. Kristen Shrader-Frechette, "Ecology and Environmental Ethics" OpenMind, accessed November 10, 2014,
https://www.bbvaopenmind.com/en/article/ecology-and-environmental-ethics/?fullscreen=true.
61. Eugene Odum, "Great Ideas for Ecology in the 90s" *Bioscience* 42 7 (1992), 542-545.
But Odum was not prepared to let go altogether of the idea that ecosystems are, to one extent or another, homeostatic.
62. Shrader-Frechette. "Ecology and Environmental Ethics."

63. See, for example, Daniel B. Botkin, *The Moon in the Nautilus Shell* (Oxford: Oxford University Press, 2012).

64. Whether there ought to have been publicly owned accommodations such as schools or bus companies is another matter; had there been none, then there would have been no legally mandated racial discrimination.

65. The "General Welfare" clause of the U.S. Constitution ought to be read as prohibiting the welfare state; a government limited to serving the *general* welfare, such as by defending against foreign invasion, is not authorized to serve the welfare of *particular individuals* by redistributing tax receipts to them.

66. *Current Population Reports*, "America's Families and Living Arrangements – 2010," (Table C3), U.S. Census Bureau, accessed November 10, 2014, www.census.gov/population/www/socdemo/hh-fam/cps2010.html.

67. "The Decline of Marriage and the Rise of New Families, Pew Research Center," November 18, 2010, accessed November 10, 2014, http://pewsocialtrends.org/files/2010/11/pew-social-trends-2010-families.pdf.

68. In the United States, the "middle class" includes much of the laboring class, who share the same values as the white collar middle class, and who aspire to have their children join it.

Bibliography

Part I

Adams, John. *The Works of John Adams*. Edited by Charles Francis Adams. Boston, MA: Little, Brown, and Company, 1850-56.

Aristotle. *The Basic Works of Aristotle*. Edited by Richard McKeon. New York, NY: Random House, 1941.

Aristotle. *The Politics of Aristotle*. Translated by Ernest Barker. London: Clarendon Press, 1946.

Babbitt, Irving. *Rousseau and Romanticism*. New Brunswick, NJ: Transaction Publishers, 1991.

Bancroft, George. *History of the United States* New York, NY: D. Appleton and Company, 1885.

Beard, Charles A. *An Economic Interpretation of the Constitution of the United States*, New York, NY: MacMillan, 1913.

Becker, Carl L. *The Declaration of Independence: A Study in the History of Political Ideas*, New York, NY: Vintage Books, 1958.

Billias, George Athan. *Elbridge Gerry: Founding Father and Republican Statesman*. New York, NY: McGraw Hill, 1972.

Blackstone, William. *Blackstone's Commentaries*, Edited by Wayne Morrison. London: Cavendish Publishing, Ltd., 2001.

Bloom, Allan, ed. *Confronting the Constitution: The Challenge to Locke, Montesquieu, Jefferson, and the Federalists from Utilitarianism, Historicism, Marxism, Freudenianism, Pragmatism, Existentialism ....* Washington, D.C.: The AEI Press, 1990.

Cohen, Michael P. *The Pathless Way: John Muir and American Wilderness*. Madison, WI: University of Wisconsin Press, 1984.

Corwin, Edward. *The "Higher Law" Background of American Constitutional Law*. Ithaca, NY: Cornell University Press, 1928.

Corwin, Edward. *Liberty against Government*. Baton Rouge, LA: Louisiana State University Press, 1948.

Croly, Herbert. *The Promise of American Life*. Cambridge, MA: The Belknap Press, 1965.

Crosskey, William Winslow. *Politics and the Constitution*. Chicago, IL: The University of Chicago Press, 1953.

Dewey, John. *Logic: The Theory of Inquiry*. New York, NY: Henry Holt, 1938.

Dewey, Johns. *The Essential Dewey*. Edited by Larry A. Hickman and Thomas M. Alexander. Bloomington, IN: Indiana University Press, 1998.

Ely, Richard, T. *The Past and Present of Political Economy*. Baltimore, MD: N. Murray, Publications Agent, Johns Hopkins University, 1884.

Ely, Richard, T. *The Social Law of Service*. Cincinnati, OH: Curtis & Jennings, New York, NY: Eaton & Mains, 1896.

Emerson, Ralph Waldo. *Collected Works of Ralph Waldo Emerson*. Edited by Robert E. Spiller, and Alfred R. Ferguson. Cambridge, MA: Belknap Press of Harvard University, 1971.

Epstein, Richard. *Takings: Private Property and the Power of Eminent Domain*. Cambridge, MA: Harvard University Press, 1985.

Farrand, Max, ed. *Records of the Federal Convention of 1787*. New Haven, CT: Yale University Press, 1966).

Ford, Paul Leicester. *The Politics of John Dickenson*. Cambridge, MA: Da Capo Press, 1970.

Fraenkel, Osmond K. *The Curse of Bigness*. New York, NY: Viking Press, 1934), 68-69.

Freund, Ernst. The Police Power: Public Policy and Constitutional Rights. Chicago, IL: Callaghan & Co., 1904.

Goldman, Eric. *Rendezvous with Destiny*. New York: Alfred A. Knopf, 1952.

Graham, Frank, Jr. *The Audubon Ark*. New York, NY: Alfred A. Knopf, 1990.

Haines, Aubrey L. *The Yellowstone Story: A History of Our First*

National Park. Boulder: Yellowstone Library and Museum Association in cooperation with Colorado Associated University Press, 1977.

Haw, James. *John and Edward Rutledge of South Carolina*. Athens, GA: Univ. of Georgia Press, 1997.

Hays, Samuel P. *Conservation and the Gospel of Efficiency*. Cambridge, MA: Harvard University Press, 1959.

Henneberger, John. "State Park Beginnings." *George Wright Forum* 17, 3 (2000): 9-20.

Holmes, Oliver Wendell. *The Common Law*. Boston, MA: Little, Brown and Company, 1923.

Huyler, Jerome. *Locke in America*. Lawrence, KS: University Press of Kansas, 1995.

James, William. *The Works of William James*. Edited by Frederick H. Burkhardt and Fredson Bowers. Cambridge, MA: Harvard University Press, 1975.

Jefferson, Thomas. *The Writings of Thomas Jefferson*. Edited by Andrew A. Lipscomb & Albert Ellery Bergh. Washington, D.C.: The Thomas Jefferson Memorial Assocation, 1903.

Jefferson, Thomas. *Papers of Thomas Jefferson*. Edited by Julian Boyd. Princeton, NJ: Princeton University Press, 1950.

Jefferson, Thomas. *Thomas Jefferson: Writings*. Edited by Merrill D. Peterson. New York, NY: The Library of America, 1984.

Jones, W.T. *A History of Western Philosophy*. New York, NY; Chicago, IL: Harcourt Brace Jovanovich, Inc., 1969.

Kent, James. *Commentaries on American Law*. Boston, MA: Little, Brown, and Company, 1896.

Konefsky, Samuel J. *The Legacy of Holmes and Brandeis*. New York, NY: DaCapo Press, 1974.

Leopold, Aldo. *A Sand County Almanac*. New York, NY: Ballantine Books, 1970.

Locke, John. *Essay Concerning Human Understanding*. Edited by ed. Alexander Campbell Fraser. New York, NY: Dover Publications, 1959.

Locke, John. *Two Treatises of Government*. Edited by Peter Laslett. London: Cambridge University Press, 1960.

Madison, James. *The Papers of James Madison*. Edited by William T. Hutchinson and William M.E. Rachal. Chicago, IL: University of Chicago Press, 1975).

Marsh, George Perkins. *Man and Nature*. Edited by David Lowenthal. Cambridge, MA: The Belknap Press of Harvard University Press, 1965.

Marshall, John. *Life of Washington*, (Philadelphia, PA: Crissy & Markley, and Thomas, Cowperthwait and Company, 1850).

Mason, Alpheus Thomas. *Brandeis: A Free Man's Life*. New York, NY: The Viking Press, 1946.

Meine, Curt. *Aldo Leopold: His Life and Work*. Madison, WI: University of Wisconsin Press, 2010.

Mintz, Max M. *Gouverneur Morris and the American Revolution*. Norman, OK: University of Oklahoma Press, 1970.

Moore, Edward C. *American Pragmatism: Peirce, James, & Dewey*. New York, NY: Columbia University Press, 1961.

Morison, Samuel Eliot. *The Oxford History of the American People*. Oxford: Oxford University Press, 1965.

Morison, S.E., H.S. Commager, and Wm. E. Leuchtenburg. *The Growth of the American Republic*. New York, NY: Oxford University Press, 1969.

Muir, John. *Our National Parks*. Boston, MA: Houghton Mifflin, 1901.

Muir, John. *The Mountains of California*. New York, NY: The Century Company, 1907.

Muir, John. *The Yosemite*. New York, NY: The Century Company, 1912.

Muir, John. *A Thousand Mile Walk to the Gulf*. Edited by William Frederic Badé. Boston, MA: Houghton Mifflin Company,

1916.

Muir, John. *John of the Mountains: The Unpublished Journals of John Muir.* Edited by Linnie Marsh Wolfe. Boston, MA: Houghton Mifflin, 1979.

Muir, John. *Steep Trails.* San Francisco, CA: Sierra Club Books, 1994.

Nash, Roderick Frazier. *The Rights of Nature: A History of Environmental Ethics.* Madison, WI: The University of Wisconsin Press, 1989.

Nash, Roderick Frazier. American Environmentalism: Readings in Conservation History. New York, NY: McGraw Hill, 1990.

Nedelsky, Jennifer. Private Property and the Limits of American Constitutionalism: The Madisonian Framework and Its Legacy. Chicago, IL: The University of Chicago Press, 1990.

Newton, Julianne Lutz. *Aldo Leopold's Odyssey.* Washington, D.C.: Island Press/Shearwater Books 2006.

Paul, Ellen Frankel. *Property Rights and Eminent Domain.* New Brunswick, NJ: Transaction Books, 1987.

Paul, Ellen Frankel, and Howard Dickman, eds. *Liberty, Property and the Foundations of the American Constitution.* Albany, NY: State University of New York Press, 1989.

Pinchot, Gifford. *Breaking New Ground.* Washington, D.C.: Island Press, 1998.

Pinchot, Gifford. *The Fight for Conservation.* New York, NY: Doubleday, Page & Company, 1910.

Pollot, Mark L. *Grand Theft and Petty Larceny: Property Rights in America.* San Francisco, CA: Pacific Research Institute for Public Policy, 1993.

Pound, Roscoe. *The Spirit of the Common Law.* Francestown, NH: Marshall Jones Company, 1921.

Powell, John Wesley. "Institutions for the Arid Lands." *Century Magazine* (May, 1890): 111-116.

Thomas McCaffrey

Ratner, J., ed. *Intelligence in the Modern World*. New York, NY: Modern Library, 1939.

Rauschenbusch, Walter. *Christianizing the Social Order*. New York, NY: The MacMillan Company, 1912.

Reiger, John Franklin. *George Bird Grinnell and the Development of American Conservation, 1870-1901*. Unpublished doctoral thesis, Northwestern University, 1970.

Richardson, James, D., ed. *Messages and Papers of the Presidents*. New York, NY: Bureau of National Literature, Inc., 1897.

Richardson, Robert D. *Henry David Thoreau: A Life of the Mind*. Berkeley, CA: University of California Press, 1983.

Ross, Edward Alsworth. *Sin and Society*. Boston, MA: Houghton Mifflin Company, 1907.

Rossiter, Clinton, ed. *The Federalist Papers*. New York, NY: New American Library, 1961.

Rothbard, Murray. "Law, Property Rights, and Air Pollution." von Mises Institute. April 22, 2006. Accessed December 19, 2014. http://mises.org/library/law-property-rights-and-air-pollution.

Rousseau, Jean Jacques. *Emile*. Translated by Barbara Foxley. London: J.M. Dent and Sons, Ltd., 1911.

Rousseau, Jean Jacques. *Julie, or the New Héloïse*. Translated by Judith H. McDowell. University Park, PA: Pennsylvania State University Press, 1968.

Rousseau, Jean Jacques. "Discourse on the Arts and Sciences," in *The Essential Rousseau*. Edited by L. Bair. New York, NY: New American Library, 1975.

Runte, Alfred. *Yosemite: The Embattled Wilderness*. Lincoln, NB: University of Nebraska Press, 1990.

Ruskin, John. Sesame and Lilies, Unto This Last, and The Political Economy of Art. London: Cassell and Company, Ltd., 1907.

556

Russell, Carl P. *100 Years in Yosemite*. Berkeley, CA: University of California Press, 1931.

Schultz, David A. *Property, Power, and American Democracy*. New Brunswick, NJ: Transaction Publishers, 1992.

Schwartz, Bernard, ed. *The Roots of the Bill of Rights*. New York, NY: Chelsea House, 1980.

Siegan, Bernard H. *Economic Liberties and the Constitution*. Chicago, IL: University of Chicago Press, 1980.

Smith, Adam. *An Inquiry into the Nature and Causes of the Wealth of Nations*. Edited by Edwin Cannan. London: Methuen & Co., Ltd., 1904.

Smith, James Allen. *The Spirit of American Governmen*. Cambridge, MA: Belknap Press of Harvard University Press, 1965.

Steen, Harold K., ed. *The Origins of National Forests*. Durham, NC: Forest History Society, 1992.

Stegner, Wallace. *Beyond the Hundredth Meridian*. Boston, MA: Houghton Mifflin Company, 1953.

Stoebuck, William. "A General Theory of Eminent Domain." *Washington Law Review* 47, 4 (August, 1972): 554.

Strauss, Leo. *Natural Right and History*. Chicago, IL: University of Chicago Press, 1950.

Thompson, C. Bradley. *John Adams and the Spirit of Liberty*. Lawrence, KS: University of Kansas Press, 1998.

Thoreau, Henry David. *The Writings of Henry David Thoreau: Familiar Letters*. Edited by F.J. Sanborn. Boston, MA: Houghton Mifflin and Company, 1906.

Thoreau, Henry David. *Walden*. New York, NY: Holt, Rinehart, and Winston, 1961.

Thoreau, Henry David. *The Writings of Henry David Thoreau*. Edited by Robert Sattelmeyer. Princeton, NJ: Princeton University Press, 1984.

U.S. Council on Environmental Quality. First Annual

Report, August, 1970.

U.S. Department of the Interior. Annual Report of the Secretary of the Interior on the Operations of the Department for the Fiscal Year Ended June 30, 1877.

Wilkins, Thurman. *John Muir: Apostle of Nature.* Norman, OK: University of Oklahoma Press, 1995.

Williams, Dennis C. *John Muir, Christian Mysticism, and the Spiritual Value of Nature, 1866 to 1873.* Unpublished Masters Thesis, Texas Tech University, 1989.

Wood, Neal. *The Politics of Locke's Philosophy.* Berkeley, CA: University of California Press, 1983.

Wordsworth, William. *A Guide through the District of the Lakes in the North of England.* Kendal, England: Hudson and Nicholson; London, Longman and Co., Moxon and Whittaker and Co., 1835.

Wright, Benjamin Fletcher. *The Contracts Clause of the Constitution.* Cambridge, MA: Harvard University Press, 1938.

Veblen, Thorstein. *The Theory of the Leisure Class.* New York, NY: The Modern Library, 1934.

Part II

Avery, Dennis T. *Saving the Planet with Pesticides and Plastics.* Indianapolis, IN: Hudson Institute, 1995.

Bailey, Ronald, ed. *The True State of the Planet.* New York, NY: The Free Press, 1995.

Barringer, Richard. *A Maine Manifest.* Portland, ME: Tower Publishing Company, 1972.

Barrons, Keith C. *Are Pesticides Really Necessary?* Chicago, IL: Regnery Gateway, 1981.

Berry, Thomas. *The Dream of the Earth.* San Francisco, CA: Sierra Club Books, 1988.

Bookchin, Murray. *Toward an Ecological Society.* Montreal: Black Rose Books, 1980.

Borelli, Peter, ed. *Crossroads: Environmental Priorities for the Future*. Washington, D.C.: Island Press, 1988.

Bramwell, Anna. *Ecology in the 20ᵗʰ Century: A History*. New Haven, CT: Yale University Press, 1989.

Breemer, J. David. "What Property Rights: The California Coastal Commission's History of Abusing Land Rights and Some Thoughts on the Underlying Causes." *UCLA Journal of Environmental Law & Policy* 22, 2, (Winter, 2004): 289.

Botkin, Daniel B. *The Moon in the Nautilus Shell*. Oxford: Oxford University Press, 2012.

Bowler, Peter, J. The Earth Encompassed: A History of the Environmental Sciences. New York, NY: W.W. Norton and Company, 1992.

Brooks, Paul. *The House of Life: Rachel Carson at Work*. Boston, MA: Houghton Mifflin, 1972.

Cahn, Robert, ed. An Environmental Agenda for the Future. Washington, D.C.: Island Press, 1985.

Carson, Rachel. *Silent Spring*. Boston, MA: Houghton Mifflin, 1962.

Carson, Rachel. Always, Rachel: The Letters of Rachel Carson and Dorothy Freeman. Boston, MA: Beacon Press, 1994.

Carson, Rachel. *Lost Woods: The Discovered Writing of Rachel Carson*. Edited by Linda Lear. Boston, MA: Beacon Press, 1998.

Chase, Alston. *In a Dark Wood*. New Brunswick, NJ: Transaction Publishers.

Church, George, and Ed Regis. Regenesis: How Synthetic Biology Will Reinvent Nature and Ourselves. New York, NY: Basic Books, 2012.

Colinvaux, Paul. Why Big Fierce Animals Are Rare. Princeton, NJ: Princeton University Press, 1978.

Committee on Resources and Man of the Division of Earth Sciences, National Academy of Sciences—National Research Council. *Resources and Man: A Study and Recommendations*. San

Francisco, CA: W.H. Freeman and Company, 1969.

Commoner, Barry. *The Closing Circle*. Toronto: Bantam, 1972.

Comp, T. Allan. *Blueprint for the Environment: A Plan for Federal Action*. Salt Lake City, UT: Howe Brothers, 1989.

Currie, William S. "Units of Nature or Processes Across Scales? The Ecosystem Concept at Age 75." *New Phytologist* 190, 1 (April, 2011): 21-34.

Davis, Mark A. "Biotic Globalization: Does Competition from Introduced Species Threaten Biodiversity?" *Bioscience* 53, 5 (May 2003): 481.

Diamond, Jared. *Collapse: How Societies Choose to Fail or Succeed*. New York, NY: Penguin Books, 2005.

Efron, Edith. The Apocalyptics: How Environmental Science Controls What We Know About Cancer. New York, NY: Simon and Schuster, Inc., 1984.

Egerton, Frank N. "Changing Concepts of the Balance of Nature." *Quarterly Review of Biology* 48, 2 (June, 1973): 322-350.

Ehrlich, Paul. "An Ecologist's Perspective on Nuclear Power." *Federation of American Scientists Public Interest Report* 28, 5-6 (1975): 5.

El Dorado County General Plan Conservation and Open Space Element, Policy 7.1.2.7.

Energy Policy Staff of the White House Science Advisory Committee (with the Atomic Energy Commission, the Department of Health, Education, and Welfare, the Department of the Interior, the Federal Power Commission, the Rural Electrification Administration, the Tennessee Valley Authority, and the Council on Environmental Quality). Electric Power and the Environment, August, 1970.

Fitzsimmons, Allan K. *Defending Illusions: Federal Protection of Ecosystems*. Lanham, MD: Rowman and Littlefield Publishers, Inc., 1999.

Gardner, Martin. *Science: Good, Bad and Bogus.* Buffalo, NY: Prometheus Books, 1981.

Gerard, R.W. "Higher Levels of Integration." In *Biological Symposia: A Series of Volumes Devoted to Current Symposia in the Field of Biology.* Edited by Jacques Cattell. Vol. VIII. Edited by Robert Redfield. (Lancaster, PA: Jacques Cattell Press, 1942).

Goklany, Indur M. "Saving Habitat and Conserving Biodiversity on a Crowded Planet." *Bioscience* 48, 11 (November, 1998): 941.

Gore, Al. *Earth in the Balance: Ecology and the Human Spirit.* Boston, MA; New York, NY: Houghton Mifflin Company, 1992.

Haeckel, Ernst. *The Riddle of the Universe.* Translated by Joseph McCabe. Buffalo, NY: Prometheus Books, 1992.

Haig, S.M., *et al.* "Taxanomic Considerations in Listing Subspecies under the U.S. Endangered Species Act." *Conservation Biology* 20, 6 (December, 2006): 1585.

Hardin, Garrett. Living within Limits: Ecology, Economics, and Population Taboos. New York, NY: Oxford University Press, 1993.

Hay, William W. Experimenting on a Small Planet: A Scholarly Entertainment. Heidelberg: Springer, 2013.

Horwitz, Morton. *The Transformation of American Law, 1780-1860.* Cambridge, MA: Harvard University Press, 1977.

Katzenstein, Larry. "The Precautionary Principle: *Silent Spring's* Toxic Legacy." In *Silent Spring at 50: The False Crises of Rachel Carson.* Edited by Roger Meiners, Pierre Desrochers, and Andrew Morriss. Washington, D.C.: Cato Institute, 2012.

Krewitt, Wolfram, *et al,* "Health Risks of Energy Systems." *Risk Analysis* 18, 4 (August, 1998): 337.

Kricher, John. *The Balance of Nature: Ecology's Enduring Myth.* Princeton, NJ: Princeton University Press, 2009.

Lehr, Jay H, ed. *Rational Readings on Environmental Concerns.* New York, NY: Van Nostrand Reinhold, 1992.

Lieberman, Adam J., and Simona C. Kwon. *Facts Versus Fears: a Review of the* Greatest *Unfounded Health Scares of Recent Times.* New York, NY: American Council on Science and Health, 2004.

McCormack, Meredith C., *et al.* "In-home Particle Concentrations and Childhood Asthma Morbidity." *Environmental Health Perspectives* 117, 2 (February, 2009): 296.

Odum, Eugene. *Fundamentals of Ecology.* Philadelphia, PA: W.B. Saunders Company, 1971.

Odum, Eugene. "20 Great Ideas for Ecology in the 90s" *Bioscience* 42 (1992): 542-545.

O'Neill, Robert V. "Is It Time to Bury the Ecosystem Concept?" *Ecology* 82, 12 (2001): 3277.

Paavolo, Juoni. "Water Quality as Property: Industrial Water Pollution and Common Law in Nineteenth Century United States." *Environment and History* 8, 3 (August, 2002): 304.

Paehlke, Robert C. *Environmentalism and the Future of Progressive Politics.* New Haven, CT: Yale University Press, 1989.

Phalen, Robert F. "The Particulate Air Pollution Controversey." *Dose Response* (October, 2004): 266.

Phalen, Robert F., and Robert N. Phalen. *Introduction to Air Pollution Science: A Public Health Perspective.* Burlington, MA: Jones and Bartlett Learning, 2013.

President's Science Advisory Committee. Restoring the Quality of Our Environment: Report of the Pollution Panel, November, 1965.

Rand, Ayn. "Collectivized Ethics" in *The Virtue of Selfishness.* New York, NY: New American Library, 1961.

Senate Report No. 1196, 91st Cong., 2d Sess., at 2 (1970).

Shabecoff, Philip. *A Fierce Green Fire: The American Environmental Movement.* New York, NY: Hill and Wang, 1993.

Shrader-Frechette, K.S., and E.D. McCoy. *Method in Ecology: Strategies for Conservation.* Cambridge: Cambridge University Press, 1993.

Simon, Julian L., and Herman Kahn. The Resourceful Earth: A Response to Global 2000. Oxford, England: Basil Blackwell Publisher Limited, 1984.

Singer, S. Fred. *Global Climate Change: Human and Natural Influences*. New York, NY: Paragon House, 1989.

Simon, Julian L. *Population Matters: People, Resources, Environment and Immigration*. New Brunswick, NJ: Transaction Publishers, 1990.

Stirling, M. David. *Green Gone Wild: Elevating Nature above Human Rights*. Bellevue, WA: Merrill Press, 2008.

Thomas, William L., ed. *Man's Role in Changing the Face of the Earth*. Chicago, IL: The University of Chicago Press, 1956.

Tren, Richard, and Roger Bate. *When Politics Kills: Malaria & the DDT Story*. New Delhi, India: Liberty Institute, 2000.

van Loon, A.J. "The Dubious Role of Man in a Questionable Mass Extinction." Earth-Science Reviews 62 (2003): 180.

White House Conference on Natural Beauty. *Beauty for America*. Washington, D.C.: U.S. Government Printing Office, 1965.

Wildavsky, Aaron. *Searching for Safety*. New Brunswick, NJ: Transaction Publishers, 1988.

Worster, Donald. *Nature's Economy: A History of Ecological Ideas*. Cambridge: Cambridge University Press, 1977.

Wu, Jianguo, and Orie Loucks, "From Balance of Nature to Hierarchical Patch Dynamics: A Paradigm Shift in Ecology." *The Quarterly Review of Biology* 70, 4 (December, 1995): 442.

Zubrin, Robert. Merchants of Despair: Radical Environmentalists, Criminal Pseudo-Scientists, and the Fatal Cult of Antihumanism. New York, New Atlantis Books, 2012.

Index

www.ingramcontent.com/pod-product-compliance
Lightning Source LLC
Chambersburg PA
CBHW020129290326
R18043300002B/R180433PG41927CBX00010B/5